Evolution & Environment
in Tropical America

Evolution & Environment in Tropical America

Edited by
Jeremy B. C. Jackson
Ann F. Budd
Anthony G. Coates

THE UNIVERSITY OF CHICAGO PRESS

CHICAGO & LONDON

Jeremy B. C. Jackson is director of the Center for Tropical Paleoecology and Archeology of the Smithsonian Tropical Research Institute. Ann F. Budd is professor of geology at the University of Iowa. Anthony G. Coates is deputy director of the Smithsonian Tropical Research Institute.

The University of Chicago Press, Chicago 60637
The University of Chicago Press, Ltd., London
© 1996 by The University of Chicago
All rights reserved. Published 1996
Printed in the United States of America
05 04 03 02 01 00 99 98 97 96   1 2 3 4 5
ISBN (cloth): 0-226-38942-1
ISBN (paper): 0-226-38944-8

Library of Congress Cataloging-in-Publication Data

Evolution & environment in tropical America / edited by Jeremy B.C. Jackson, Ann F. Budd, and Anthony G. Coates.
    p.   cm.
   Includes bibliographical references and indexes.
   ISBN 0-226-38942-1 (cloth). — ISBN 0-226-38944-8 (pbk.)
   1. Paleoecology—Latin America. 2. Paleoecology—Tropics.
3. Paleoecology—quarternary. I. Jackson, Jeremy B. C., 1942– .
II. Budd, Ann F. III. Coates, Anthony G. (Anthony George)
QE720.E93 1996
560'.45'098—dc20   96-21641
                            CIP

Contents

Preface

In his *Principles of Geology,* Charles Lyell argued that understanding environmental and biological processes in the present is essential for reconstructing the past. The success of his perspective is manifest in the remarkable achievements in historical geology and paleontology of the past 150 years. But it is also increasingly apparent that the present is a very strange time by the standards of the past few million years—an aberrantly warm, interglacial interlude in an otherwise glacially cold but madly fluctuating world. Most modern species passed the majority of their evolutionary careers in different places, under different conditions, and in association with different suites of species than at present. Understanding their ecology and adaptations is possible only in the context of the conditions in which they first evolved.

A historical framework is also essential for trying to understand the ecological consequences of human activities and the prospects of further global change. This is because the same abiotic event can have myriad different biological effects contingent upon the unique set of environmental conditions and biota present at the time and place that the event occurs. Global climate models are also extraordinarily sensitive to their initial starting conditions. Thus, the only way to evaluate empirically the likelihood of alternative scenarios is to reconstruct the ecological consequences of earlier environmental changes and look for parallel situations.

Integration of geological and biological perspectives for understanding patterns of biodiversity is at least as old as Darwin's studies of coral reefs and *The Origin of Species.* However, biologists long ago abandoned large-scale investigations of evolution and environment in favor of small-scale, controlled, manipulative experiments that excel in revealing ecological and evolutionary mechanisms but do not always scale up to explain the diversity and structure of natural communities. Growing environmental concerns demand answers to questions on larger scales than are amenable to experiment, and long-term perspectives to separate signal from noise that stretches beyond our lifetimes and patience. So it is with a certain ironic satisfaction that paleontologists now enjoy the rediscovery of history as essential to understanding the origin, dynamics, and conservation of biological diversity and communities.

Three major events during the past ten million years profoundly shaped the tropical American biota. The first was the gradual formation of the Central American Isthmus, which led to different oceanographic conditions and marine biotas on opposite sides of the developing barrier and ultimately joined the terrestrial biotas of North and South America, with dramatic consequences. The developing isthmus also diverted global oceanic circulation, strengthening the Gulf Stream and eventually triggering Northern Hemisphere glaciation. Subsequent intensification of glaciation probably caused the second great event—the burst of extinction and speciation that fundamentally altered the tropical American biota about two million years ago. Third, the invasion of people at the end of the last glaciation almost certainly caused the extinction of most species of large tropical vertebrates, which in turn may have altered vegetation as much as climate change. All of these events affected living species or their closest relatives, as well as extinct forms.

The interchange of animals and plants between North and South America has been intensively studied ever since Darwin's *Voyage of the Beagle*. However, most of the fieldwork was done in the temperate zones of the two continents, where conditions were more familiar and the fossils easier to find, and the great tropics between north and south were *terra incognita*. David Webb and Paul Colinvaux are among the handful of scientists who have set out to fill this great void. Their results reveal a dynamic and ever-changing tropical landscape, without apparent refuges from climatic fluctuations, and seem to demand a careful rethinking of explanations of the origin and maintenance of Amazonian diversity.

Events in the shallow oceans were ignored even more. The generally accepted but untested chronology of events grew out of a few intensively described molluscan faunas and almost no environmental data. The three of us, along with Laurel Collins and Peter Jung, founded the Panama Paleontology Project (PPP) to correct this imbalance by constructing a well-constrained stratigraphic framework and extensive collections of fossils for the entire region. First in the Dominican Republic, and more recently along the Central American Isthmus, we are compiling data on a massive scale to reconstruct the environmental and biological history of tropical American seas. Most of the chapters in this volume are a first attempt by the PPP to assemble the new environmental and paleontological data that have already profoundly altered our understanding of the magnitude and chronology of events. We hope these efforts will lead to a new, more rigorous synthesis and understanding of the subtle and complex interdependence of evolution and environment.

This book grew out of a symposium held as part of the Fifth North American Paleontological Convention in Chicago in 1992. We thank Scott Lidgard, Peter Crane, and the other organizers of the convention for their encouragement and support, and the Field Museum of Natural History for the splendid facilities and hospitality for the symposium. We are also grateful to the numerous reviewers of all the papers for their thoughtful criticisms and suggestions and to our editors at the University of Chicago Press for their patience, encouragement, and believing in this book.

1

Evolution and Environment: Introduction and Overview

Jeremy B. C. Jackson and Ann F. Budd

Environments are always changing, and the rates of change are themselves extremely variable. New discoveries of the magnitude and speed of changes in climate and ocean circulation are challenging Lyellian and Darwinian dogma that slow and gradual changes in earth processes drive similar changes in species and communities (Crowley and North 1988; Kennett and Stott 1991). Abrupt shifts in the composition of fossil assemblages that were formerly dismissed as artifacts of a bad fossil record (Darwin 1859; Levinton 1988) are now attributed to rapid environmental change (but see MacLeod and Keller 1991), and there is increasing understanding of the importance of nonlinearities, thresholds, and the historical sequence of environmental events for the nature and magnitude of their biological effects (Jackson 1994). All of this is as true of the Paleogene (Prothero and Berggren 1992; Zachos et al. 1993) as the Quaternary (Shackelton et al. 1984; GRIP 1993; Valentine and Jablonski 1993), and on the land as in the sea (Webb and Bartlein 1992; Vrba 1993).

It is equally apparent, however, from the papers in this volume and elsewhere, that inadequate sampling, taxonomic ignorance, and bad stratigraphy have greatly biased our understanding of the relationship between large-scale biotic and climatic change and have often led to wrong conclusions (Vermeij 1993). Indeed, the much ballyhooed imperfections of the fossil record are commonly less problematic than biases introduced by the inherent assumptions of our research, and the failure to recognize those biases. These problems are particularly great in traditional, opportunistic studies of fossils in museum collections obtained from studies widely varying in quality of sampling and dating (Koch 1987; Jackson et al. 1993; Allmon et al. 1993). In contrast, studies based on the Deep Sea Drilling Project are distinguished by generally adequate sampling, precise dating, consistent taxonomy, and a multidisciplinary approach, which permit rigorous, quantitative investigation of the ecological and evolutionary processes that mold the history of life in the sea (Kennett and Srinivasan 1983; Berggren et al. 1985; Stanley et al. 1988; Norris et al. 1993; Wei 1994a,b), as well as of the motions of the oceans and the sea floor.

Extensive new surveys and collections of Neogene and Quaternary nearshore deposits in the Dominican Republic (Saunders et al. 1986) and by the Panama Paleontology Project (PPP) along the Central American Isthmus (Coates et al. 1992) are attempts to apply the same rigorous standards to traditional marine paleontological studies, and comparable efforts are being made to document the Neogene and Quaternary history of terrestrial tropical climates and biotas (Stehli and Webb 1985; Bush et al. 1992; Colinvaux 1993, this volume, chap. 13; Leyden et al. 1993; Marshall and Sempere 1993; Webb and Rancy, this volume, chap. 12). Most of the insights reported here are based on material from these surveys. But new problems, such as the variable timing of faunal turnover in different parts of the tropical western Atlantic at the end of the Pliocene (Allmon et al., this volume, chap. 10; Jackson et al., this volume, chap. 9), the extreme variability in oxygen isotopic ratios and other indicators of paleoclimate in approximately contemporary samples from adjacent basins in Central America (Borne et al. 1994; Collins et al. 1995; Teranes et al., this volume, chap. 5), and complex geographic variation in terrestrial aridity and temperature during glacial maxima (Colinvaux 1993, this volume, chap. 13; Webb and Rancy, this volume, chap. 12), underline the need for even more extensive sampling, replication, and stratigraphic precision to establish environmental causes and biological effects. Just as for Recent marine ecology (Paine and Levin 1981; Jackson 1991; Levin 1992; Angel 1993; NRC 1995), small-scale paleontological studies of individual taxa cannot hope to elucidate the ecological and evolutionary consequences of paleoenvironmental change.

WHAT DO WE KNOW ABOUT ENVIRONMENTAL CHANGE?

Two phenomena dominate the Late Neogene and Quaternary environmental history of tropical America: (1) isolation and oceanographic divergence of the eastern Pacific and Caribbean due to the emergence of the Central American Isthmus, and (2) fluctuations in sea level, oceanographic conditions, and terrestrial climate associated with the abrupt intensification of Northern Hemisphere glaciation about 2.4 million years ago (Ma). The timing of these events in the tropics is still controversial, and we can only guess at the relative magnitude of changes in parameters like temperature and primary productivity in the oceans, or temperature and moisture on the land.

Formation of the isthmus involved a series of transitions over roughly 15 million years (my) from a deep ocean seaway, to growth of an island arc and accumulation of accreting terrains, to progressively more isolated sedimentary basins along the developing barrier (Coates and Obando, this volume, chap. 2). Thus, the isthmus not only separated the Caribbean from the eastern Pacific, but also greatly increased the amount of shallow, nearshore, geographically fragmented habitat for benthic organisms. Deep circulation between the eastern Pacific and Caribbean persisted until the Late Miocene, and shallow seaways remained open until about 3.5 Ma (Keigwin 1982; Hasson and Fischer 1986; Whittaker 1988), although the isthmus may have been temporarily breached at the end of the Pliocene owing to fluctuating sea levels (Cronin and Dowsett, this volume, chap. 4).

Isolation of the two oceans resulted in small but significant increases in Caribbean oceanic salinity, carbon-13 enrichment, and carbonate deposition compared to the eastern Pacific, and probably in a large increase in eastern Pacific seasonality and productivity compared to the southwestern Caribbean (Keigwin 1982; L. Collins, this volume, chap. 6; Cronin and Dowsett, this volume, chap. 4; Teranes et al., this volume, chap. 5). There was also a marked decline in productivity along the Gulf of Mexico coast of Florida during the Late Pliocene (Allmon et al. 1995; Allmon et al. this volume, chap. 10), and possibly along the Caribbean coast of Costa Rica and Panama, although the evidence from there is conflicting (Borne et al. 1994; Collins et al. 1995). But none of these patterns are well replicated geographically, and all are almost certainly more regionally complex, locally variable, and diachronous than originally described.

The long debate about the environmental significance of oxygen isotopic fluctuations in response to Northern Hemisphere glaciation is swinging back in favor of Emiliani's (1955) original estimates of a 5–6°C drop in tropical sea surface temperatures (SSTs) in addition to fluctuations in sea level (CLIMAP 1976; Shackelton et al. 1984; Emiliani 1992; Beck et al. 1992; Guilderson et al. 1994). Moreover, Late Pliocene tropical SSTs appear to have been generally as warm as they are today (Dowsett and Poore 1991; Cronin and Dowsett, this volume, chap. 4). Thus, the onset at the end of the Pliocene of intense, glacially induced fluctuations of perhaps 3–5°C in tropical SSTs (Jackson et al., this volume, chap. 9) was a major event relative to the preceding several million years. The only exceptions would have been areas of upwelling where seasonal temperature fluctuations already exceeded those due to glaciation (Jackson 1994; Jackson et al., this volume, chap. 9). Other oceanographic changes that probably occurred in tropical America along with fluctuations in sea level and temperature are little studied. Nevertheless, fluctuations in CO_2 concentrations in ice cores and model calculations suggest that primary productivity increased greatly during glacial maxima (Mix 1989).

The confusion about oceanic cooling has also complicated reconstruction of terrestrial environments of the Amazon Basin, and throughout the rest of tropical America (Bush et al. 1992; Leyden et al. 1993; Colinvaux, this volume, chap. 13; Webb and Rancy, this volume, chap. 12). Ice Age vegetation was profoundly disturbed by climate change associated with glacial cycles. Floral associations based on extensive pollen and phytolith records demonstrate that temperatures were at least 5°C cooler than present levels during the last glacial maximum. However, sedimentologic and other evidence, including the widespread drying up of lakes and associated changes in clay minerals and diatoms, also points toward extreme aridity around the Amazon Basin (Bush et al. 1992; Leyden et al. 1993; Colinvaux, this volume, chap. 13), and perhaps within it as well (Absy et al. 1991). Dominance of many Middle to Late Pleistocene Amazonian vertebrate faunas by large, pastoral grazers also suggests the occurrence of more arid, savanna-like conditions over large areas of the Amazon Basin (Webb and Rancy, this volume, chap. 12). However, Colinvaux disagrees with this aridity hypothesis based on palynological data. Cooling definitely occurred, but whether or not cooling was accompanied by significant aridity hinges on the comparative value of vertebrate teeth versus pollen floras from large catchment areas to resolve terrestrial climates and vegetational mosaics.

The terrestrial story is even further complicated, however, by the extinction roughly 11 thousand years ago (Ka) of virtually all large, tropical American herbivores. This was almost certainly due to hunting by the first humans in the Americas (Janzen and Martin 1982; Martin and Klein 1984; Azzaroli 1992). Archaeological arguments to the contrary ignore the overwhelming evidence that the same large vertebrates had survived extreme climatic fluctuations comparable to the last glacial maximum many times before with no obvious evolutionary effects. Before this extinction, parts of tropical American forests may have more closely resembled open forest and grassland, like that maintained today by elephants in Africa or by cattle throughout Latin America, than the dense tropical forest we imagine as more natural (Janzen and Wilson 1983). Thus, vegetation may have changed owing to sudden elimination of herbivores, *independently* of any of the rapid changes in climate that were also going on at the same time (Webb and Rancy, this volume, chap. 12).

The only way to sort out these different factors is to reconstruct climatic and vegetational changes during earlier glacial-interglacial transitions, when the herbivores were definitely present and people were absent. If hunting was indeed responsible for extinction, then the modern flora and fauna (what's left of it) must have evolved in more open habitats than the dense "green mansions" of Hudson's (1904) and professional ecologists' imaginations. Conservation strategies for protection of rapidly diminishing tropical biodiversity probably require adjustment accordingly.

WHAT DO WE KNOW ABOUT BIOLOGICAL CHANGE?

Benthic foraminifers, reef corals, mollusks, and some cheilostome bryozoans have been sampled sufficiently in the tropical western Atlantic to have a rough idea of the major events of the last 10 my, despite our continuing inability to estimate their total diversity at any one time or place (Allmon et al. 1993, this volume, chap. 10; Jackson et al. 1993; Jackson et al. this volume, chap. 9; Budd et al. 1994; Budd et al. this volume, chap. 7; Cheetham and Jackson, this volume, chap. 8; L. Collins, this volume, chap. 6). However, the precision of regional stratigraphic correlation is still only about 1 my, so that we do not know whether major environmental and biological events were synchronous or diachronous across the Caribbean (Allmon et al., this volume, chap. 10; Dowsett and Cotton, this volume, chap. 3). Moreover, the stratigraphic record is much less complete along the Pacific coast of Central America owing to the subduction of active margin deposits and lack of well-developed sedimentary basins (Coates et al. 1992; Coates and Obando, this volume, chap. 2). That is why the only transisthmian study in this volume (chap. 9) is based on a single clade of snails that had been previously collected intensively from California to Peru (Jung 1989). Comparable eastern Pacific collections for total macrofauna do not exist.

Despite these problems, there are two important general patterns that are unlikely to change qualitatively with further study:

Most Biological Change Occurs in Pulses

The marine biota of tropical and subtropical America was dramatically transformed roughly 2 Ma (Jackson 1994), although the tempo, and perhaps the timing, of turnover varied within and among taxa. For example, eleven genera of Caribbean reef corals became extinct during this brief interval, as many as during the entire preceding 20 my, and only one new coral genus first appeared later than 3–4 Ma versus twenty-one new genera before. Thus, extinctions of coral genera during the Neogene were confined to the Late Pliocene, whereas first appearances of genera were more evenly spread out over Miocene time. In contrast, Neogene originations of Caribbean coral species were exceptionally high during the Late Pliocene and during the Early and Late Miocene; but as for genera, species extinctions were concentrated in a pulse at the end of the Pliocene (Budd et al., this volume, chap. 7).

Molluscan faunas of the Caribbean and southeastern North America also experienced an intense pulse of origination and extinction about 2 Ma (Stanley and Campbell 1981; Stanley 1986; Allmon et al. 1993; Allmon et al., this volume, chap. 10; Jackson et al. 1993). In lower Central America, the modern molluscan fauna appears in the Lomas del Mar section of the Moin Formation, only a few hundred thousand years later than the classic Pliocene fauna (Coates et al. 1992; Jackson et al. 1993; Robinson 1993), and the same is probably true for the Mare Formation in Venezuela (Gibson-Smith and Gibson-Smith 1979). More molluscan subgenera and prosobranch gastropod species appear to have originated and become extinct within this brief interval than at all other times during the Pliocene. Extinction of Caribbean strombinid gastropods and origination of eastern Pacific strombinids also peaked during the same interval (Jackson et al. 1993; Jackson et al., this volume, chap. 9).

Whatever the cause, the apparent restriction of much of the speciation in tropical American seas during the last 10 my to a brief interval near the Pliocene-Pleistocene boundary has profound implications for evolutionary biology. For example, if most species alive today originated 2 Ma or more, then we should expect to see little significant evolutionary change at present; and whatever is going on now may not be very helpful for understanding what actually happens during speciation in the sea (Jackson 1995). Similarly, if most surviving species in a clade originated at the same time, then the inability of systematists to distinguish ancestry and descent based only on living species (unresolved "polytomies" in cladistic jargon) may reflect real biological ambiguities, not just a lack of resolution due to inadequate characters or methodologies.

The Magnitude and Timing of Biological Change Varies Greatly among Taxa and Environments

Rates of evolution vary greatly among different taxa for reasons that are only poorly understood. Among living species, molecular divergence between closely related species across the Central American Isthmus varies several-fold among fishes, crustaceans, and echinoderms (Cunningham and Collins 1994; T. Collins, this volume, chap. 11); and the Caribbean fossil record shows even greater differences among higher taxa. For example, common species of benthic foraminifers hardly changed throughout the entire Neogene (L. Collins, this volume, chap. 6) whereas common subgenera of mollusks experienced a major evolutionary pulse at the end of the Pliocene (Jackson et al. 1993; Jackson 1994). Likewise for cheilostome Bryozoa: none of the ten Late Miocene species of *Stylopoma* that survived into the Pliocene became extinct afterward, whereas all nine species of *Metrarabdotos* that survived into the Pliocene became extinct by the Early Pleistocene (Jackson and Cheetham 1994; Cheetham and Jackson 1995, this volume, chap. 8). This difference was part of a more general pattern of mass extinction of bryozoans with arborescent colony form versus the continued diversification of encrusting bryozoans; and also of the greater extinction of taxa associated with subtidal seagrass beds versus those from other nearshore habitats (Johnson et al. 1995; Budd et al., this volume, chap. 7; Cheetham and Jackson, this volume, chap. 8).

There were even greater differences in the fate of different higher taxa between the eastern Pacific and Caribbean, which are obvious despite the discrepancy in the quality of the fossil records and sampling effort in the two oceans. Reef corals suffered nearly complete extinction in the eastern Pacific during the Pliocene, and the present depauparate fauna includes species of Indo-Pacific as well as Caribbean affinities (Budd 1989). In contrast, "paciphile" mollusks, which amounted to roughly 10% of the Neogene Caribbean fauna, became extinct in the Caribbean at the end of the Pliocene but radiated greatly in the eastern Pacific at the same time (Jackson et al. 1993; Jackson et al. this volume, chap. 9).

The timing of the turnover of western Atlantic reef corals may have begun 1 my before that of mollusks, and turnover of both groups in Florida may have preceded that in the southwestern Caribbean by the same amount (Allmon et al. 1993, this volume, chap. 10; Jackson et al. 1993; Jackson et al., this volume, chap. 9; Budd et al. 1994; Budd et al., this volume, chap. 7). However, all of these differences may be methodological artifacts or due to varying stratigraphic precision among sites. For example, the turnover of mollusks in Panama and Costa Rica, first dated at 1.8–1.6 Ma, was based just on fauna from two relatively well dated basins in that region (Coates et al. 1992), all studied directly by the authors. In contrast, the turnover of reef corals dated at 4–2 Ma was based on almost the entire western Atlantic fauna from old and new studies, and on well-dated and poorly dated sites combined. In the latter case, Budd and others employed a variety of statistical techniques to spread out first and last appearances through the entire range of stratigraphic uncertainty for poorly dated faunas, with the inevitable result of blurring any peak in evolutionary rates. Under the circumstances, the sharpness of the turnover of coral species is remarkable. A major event in coral evolution and ecology obviously took place, but the timing of the event is inevitably more suspect than for the mollusks, at least for Panama and Costa Rica.

EVOLUTION AND ENVIRONMENT

The study of ancient environments and the history of life are necessarily descriptive sciences. Thus, inference of cause and effect depends on correlations between independent measures of environmental conditions and evidence of biotic change. Our confidence in such correlations naturally depends on the stratigraphic precision and replicability of the observations, understanding of possible mechanisms, and ability to predict new patterns based on what we have already learned. The study of evolution and environments in tropical America is still very much in the early stages of defining basic patterns, searching for correlations, and cautious inference. Nevertheless, several general, if tentative, conclusions are possible.

The Effects of the Closure of the Central American Isthmus on Marine Evolution and Biodiversity Were Gradual and Unremarkable

Fragmentation of the tropical American seascape by the gradual formation of the Central American Isthmus should have greatly increased rates of allopatric speciation and diversity (Mayr 1954, 1963; Rubinoff 1968; Vermeij 1978; Lessios 1981), but the supporting data are surprisingly equivocal. The most frequently cited evidence is the occurrence of "geminate" species pairs from opposite sides of the Central American Isthmus, which are supposed to be each other's closest relatives (Jordan 1908; Vermeij 1978). However, very few supposed geminates have been studied genetically or paleontologically, and many are highly suspect (T. Collins, this volume, chap. 11). Indeed, phylogenetic analysis of bivalve mollusks in the family Cardiidae demonstrates that only one of the five pairs cited by Vermeij (1978) are geminates, and the rest are only distantly related offshoots of lineages that either diverged millions of years before the final closure of the isthmus or may even represent independent southward migrations from the Arctic in the two oceans (Schneider 1995, unpublished data)! Caribbean and eastern Pacific strombinid gastropod subgenera were also substantially different by the Early Pliocene (Jackson et al. 1993; Jackson et al., this volume, chap. 9).

Independent genetic, ecological, and behavioral measures of divergence in seven carefully chosen geminate pairs of the snapping shrimp *Alpheus* strongly suggest that transisthmian evolution was staggered over millions of years, with species restricted to deeper or more offshore environments exhibiting the greatest divergence (Knowlton et al. 1993). Moreover, several members of these pairs also have sympatric relatives that show comparable amounts of genetic divergence. Fossil cardiid bivalves show similar patterns of morphological divergence drawn out over millions of years (Schneider 1995), but there are as yet no molecular data on living cardiids for comparison. All of these patterns suggest that diversification was strongly influenced by increasing habitat complexity associated with the formation of the isthmus, but not just by the simple division of the oceans as has been commonly assumed. Further understanding will require careful comparison of paleontological versus molecular phylogenies of highly diverse clades, such as the cardiids, that have an excellent fossil record (T. Collins, this volume, chap. 11).

Additional support for the evolutionary importance of the developing isthmus comes from regional increases in total diversity of major taxa over time. Numbers of species of Caribbean reef corals approximately doubled 10 Ma but increased only slowly thereafter until the major turnover at the end of the Pliocene (Budd et al. 1994; Budd et al., this volume, chap. 7). Similarly, the combined number of strombinid gastropod species from both oceans tripled during the Late Miocene but was steady thereafter (Jackson et al. 1993; Jackson et al., this volume, chap. 9); and the number of Caribbean molluscan subgenera increased during the Late Miocene and Early Pliocene, as measured by the slopes of cumulative diversity curves plotted against sampling effort (Jackson et al. 1993, unpublished data). Nevertheless, it is difficult to evaluate how much these changes reflect genuine increases in diversity instead of relatively inadequate sampling of older faunas.

All of these marine results are vastly more subtle and drawn out than the dramatic consequences of the joining of North and South America on the terrestrial biota (Stehli and Webb 1985; Webb 1991; Marshall and Sempere 1993; Vrba 1993). Breakdown of biogeographic barriers has more immediate and greater biological impact than their formation (Vermeij 1991).

Biological Consequences of Changes Correlated with the Intensification of Northern Hemisphere Glaciation Were Sudden and Dramatic

The coincidence of both marine and terrestrial faunal turnover with the onset of intense Northern Hemisphere glaciation is so obvious that, at first glance, it is hard to understand all the fuss. Surely, refrigeration must have been important (Stanley and Campbell 1981; Stanley 1986; Vrba 1993)! Differential evolutionary events in the eastern Pacific and Caribbean, and the survival of Caribbean paciphiles in areas of upwelling, further support the cooling hypothesis. The problem, however, is that it is not at all obvious how a drop in tropical SSTs of only a few degrees, no matter how sudden, could have caused a massive turnover of the biota (Clarke 1993; Jackson et al., this volume, chap. 9); and many other factors besides temperature changed during glacial cycles (Mix 1989; Bush et al. 1992; Webb and Bartlein 1992; Leyden et al. 1993; Colinvaux, this volume, chap. 13). Thus, we have a strong correlation of environmental and biological events, but no real understanding of mechanisms.

We need much better sampling and stratigraphy to exploit local variance in the timing, sequence, magnitude, and correlation of environmental and biological events. If the onset of cooling and faunal turnover both began substantially earlier in the north than in the south, whereas changes in upwelling and productivity did not, then this would strongly support the cooling hypothesis. But establishing such patterns, whatever they really are, will require much more work to distinguish signal from noise (e.g., compare the conflicting results and interpretations in Borne et al. 1994; Collins et al. 1995; Cronin and Dowsett, this volume, chap. 4; and Teranes et al., this volume, chap. 5). Tens to hundreds of samples will be required from every stratigraphic horizon in every basin of interest in order to separate local basinal variation in space and time from events in other basins and stratigraphic horizons. This is especially true because changes in salinity, temperature, and possibly other factors all potentially influence isotopic ratios in shells, and other paleoenvironmental indicators as well, and replication is necessary to try to tease apart these different factors.

Likewise, a minimum of 50,000 specimens will be necessary to reliably estimate molluscan subgeneric diversity during any interval of time in a single tropical American basin (fig. 1 in Jackson et al. 1993, unpublished data), and this number would easily rise to at least 100,000 specimens to estimate species diversity. But as Robinson (1993) has rightly noted, biogeographic comparisons are rarely based on more than a few thousand specimens, and taxonomy is generally inadequate besides. Lumping taxa unwittingly leads to ecological and evolutionary noise (Knowlton 1993; Knowlton and Jackson 1994). We urgently need well-curated taxonomic reference collections and on-line, electronic database systems for all taxa and regions of interest, along with explicit protocols for consistently recognizing and identifying species and higher taxa.

Last, there has been virtually no study of preservational bias in important tropical American deposits, but understanding taphonomic processes is critical to ecological and evolutionary interpretations of any paleontological pattern (Kidwell 1985, 1989; Koch 1987; Kidwell and Behrensmeyer 1988; Behrensmeyer and Hook 1993). For example, the famous Gatun Formation in Panama contains well over 300 genera and subgenera of mollusks (Woodring 1982), but shells are commonly leached, and most of the thin-shelled fauna may be lost (Jonathan Todd, personal communication).

History Constrains Ecological and Evolutionary Responses to Climate Change

Historical contingency fundamentally distorts any simple cause-and-effect relationship between the magnitude of shifts in environmental conditions and the subsequent biological response (Jackson 1994). This is because long-standing regional environmental processes may determine how organisms respond to change more than the amount of change per se (Ricklefs 1987; Ricklefs and Schluter 1993; Brey et al. 1994). The most important example in our context concerns whatever factors associated with intensifying Northern Hemisphere glaciation precipitated the massive marine faunal turnover at the end of the Pliocene. Virtually all living tropical species have survived much stronger environmental fluctuations during subsequent glacial cycles (Shackelton et al. 1984; Jansen and Sjoholm 1991; Coope 1995), with no apparent evolutionary effect. Thus, once past the initial evolutionary filter of rapid environmental change, all that were left were relatively eurytopic species (as defined by whatever factors caused the turnover in the first place). This reasoning suggests that faunal stasis over the past two million years reflects adaptations, or more probably exaptations (Gould and Vrba 1982), to rapid changes in temperature, sea level, and other environmental disturbances associated with glaciation, rather than the disruption of processes leading to speciation as proposed by Potts (1983, 1984). This conclusion is further supported by faunal stasis *before* the turnover and the onset of extreme environmental fluctuations. Plants and animals in tropical forests presumably have a similar history, but there are few data (Vrba 1993; Webb and Rancy, this volume, chap. 12).

Many modern species may actually benefit from glacially induced environmental fluctuations. One possible example is the extraordinary increase in diversity and dominance of shallow coral reefs worldwide by the genus *Acropora* sometime during the last two million years (Rosen 1993). Shallow Caribbean forereef deposits dominated by tall staghorn and elkhorn acroporids are less than 2 Ma, whereas reefs in the same environment dominated by short pocilloporid finger corals are 2 Ma at least (Jackson et al. 1996). Pocilloporid dominance lasted between 5 and 6 my, then turned over completely in less than 1 my. Afterward, acroporids were always more abundant, although they may not have reached their present extreme dominance until the Late Pleistocene (Budd and Kievman 1994), and *Pocillopora* did not finally become extinct in the Caribbean until 60 Ka (Geister 1977; Budd et al. 1994). Thus, the shift in branching coral dominance was an ecological rather than an evolutionary event.

This ecological interpretation is further supported by the unique characteristics of acroporids compared with all other reef corals (Rosen 1993; Jackson et al. 1996). Staghorn and elkhorn corals, for example, grow up to ten times faster than all other Caribbean species and several times faster than living relatives of the branching pocilloporids that they apparently displaced about 2 Ma, and the same is true for *Acropora* in general (Buddemeier and Kinzie 1976; Tunnicliffe 1983; Stimson 1985; Jackson 1991). This extraordinary growth potential must have greatly increased the success of *Acropora* relative to other branching corals when glacial cycles and sea level fluctuations intensified again about 1 Ma (Shackelton et al. 1984). In the latter event, the success of these recently dominant corals is an accidental side effect of characters selected for other reasons, rather than adaptations for their present circumstances.

One important moral of this *Acropora* story, whether right or wrong, is that paleoecologists need to pay more attention to the relative abundance of species if we are ever going to understand what is going on ecologically. Virtually all of the important questions about the biological consequences of climate change depend on this perspective. Pocilloporids were abundant until they were replaced by acroporids, arborescent bryozoans were abundant until they were replaced by encrusting bryozoans in the Late Pliocene (Cheetham and Jackson, this volume, chap. 8), and Caribbean strombinids were abundant until they suddenly became extinct at the end of the Pliocene (Jackson et al., this volume, chap. 9). Explanations for all these events must be fundamentally different than if these taxa had gradually faded away. Acroporids and pocillo-porids are not just "ships that pass[ed] in the night" (Gould and Cal-loway 1980). Palynologists have long and successfully relied on the relative abundance of pollen grains to infer past climates and plant asso-ciations (Colinvaux 1993, this volume, chap. 13). However, marine pa-leontologists have often shied away from examining long-term changes in relative abundance with the argument that sampling bias makes such counts not only useless but misleading.

CONCLUSIONS

We have a long way to go in describing basic patterns (what happened, where, and when) and ascribing cause and effect (why a change oc-curred). The Panama Paleontology Project was organized to assemble the kinds of large empirical databases required to answer these simple but elusive questions for marine environments in a consistent and coor-dinated fashion. This is possible only when the different kinds of data are collected from the same sites, at more or less the same time, so that they can be compared directly. It is not particularly useful, for example, to have diversity data for one set of samples and paleoenvironmental data for another. In the brief comments below, we outline the most pressing problems for describing and interpreting patterns that have emerged from the chapters which follow.

What happened? To determine what happened we need to be able to estimate at some acceptable level of confidence the frequency of oc-currence or abundance of a species, the diversity of a higher taxon, the value of a geochemical indicator of paleoclimate or any other character of interest. This requires an adequate number of replicates to calculate a mean and a variance, or to determine the shape of a cumulative diver-sity curve. In the latter case, as for the mollusks, the diversity may be so staggeringly great that modern reference collections are essential for every region of interest. We also need to assess possible taphonomic or diagenetic bias for every sample. There has been no rigorous evaluation of the extent of taphonomic bias for any important tropical American fauna.

Where did it happen? Spatial control is essential because of the fundamentally discontinuous and heterogeneous nature of marine coastal environments, and the consequently patchy distribution of or-ganisms in space and time. Much less attention has been paid to spatial control than to dating, but they are different sides of the same coin. Sedimentary basins are a natural, large-scale sampling unit for marine deposits, and we need to be much more careful about treating them as spatial replicates. For example, our confidence in the generality of the molluscan turnover at the end of the Pliocene increased dramatically with our recent, unpublished discovery that the Early Pleistocene fauna at Swan Cay, in the Bocas del Toro Basin, is as modern as that from the Lomas del Mar beds at Limón. The argument about Amazonian aridity is another example of the need for much more rigorous spatial control.

When did it happen? Nearshore deposits are notoriously hard to date well enough to compare the timing of events between adjacent sedimentary basins, such as Limón and Bocas del Toro, much less between Florida, Venezuela, and the isthmus. In addition, most stratigraphic sections are short and disconnected so that use of simple superposition is rarely practical. However, geographic variation in the onset of faunal turnover, or of speciation compared with extinction during turnover, is almost certainly much less than 1 my. Thus, the only way to attempt such comparisons is to use all possible forms of dating together (particularly nannofossils, planktonic foraminifers, and paleomagnetism) for eventual graphic correlation.

Why did it happen? All the previously discussed precision is essential for inferring cause and effect because we are totally dependent on correlations as explanations. This is the only way to exploit local and temporal differences in variation to reject or accept alternative hypotheses, especially when everything seems to happen at once, as during the great turnover at the end of the Pliocene.

But whatever the limitations of the data, paleontology continues to contribute to basic ecological and evolutionary understanding, be it through the discovery of punctuated evolution of species (Gould and Eldredge 1993), the synchronous turnover of entire biotas (Jackson 1994), or challenging the authority of the molecular clock (Aitkin et al. 1993; T. Collins, this volume, chap. 11). Paleontology still provides the only empirical test of the long-term biological effects of environmental change.

ACKNOWLEDGMENTS

We thank Tim Collins and two anonymous reviewers for their very helpful reviews of an earlier draft of the manuscript.

REFERENCES

Absey, M. L., A. Cleef, M. Fournier, L. Martin, M. Servant, A. Sifeddine, M. F. da Silva, F. Soubies, K. Suguio, B. Turcq, and T. Van der Hammen. 1991.

Mise en évidence de quatre phases d'ouverture de la forêt dense dans le sud-est de l'Amazonie au cours des 60,000 dernières années: Première comparaison avec d'autres regions tropicales. *C.R. Acad. Sci. Paris,* ser. 2, 312: 673–78.

Aitkin, M. J., C. B. Stringer, and P. A. Mellars. 1993. *The origin of modern humans and the impact of chronometric dating.* Princeton: Princeton University Press.

Allmon, W. D., S. D. Emslie, D. S. Jones, and G. S. Morgan. 1995. Occurrence and decline of Plio-Pleistocene upwelling on the west coast of Florida: Paleontological and isotopic evidence and oceanographic mechanisms. *J. Geol.* In press.

Allmon, W. D., G. Rosenberg, R. W. Portell, and K. S. Schindler. 1993. Diversity of Atlantic coastal plain mollusks since the Pliocene. *Science* 260:1626–29.

Angel, M. V. 1993. Spatial distribution of marine organisms: Patterns and processes. In *Large-scale ecology and conservation biology,* ed. P. J. Edwards, R. M. May, and N. R. Webb. London: Blackwell Scientific Publications.

Azzaroli, A. 1992. Ascent and decline of monodactyl equids: A case for prehistoric overkill. *Ann. Zool. Fennici* 28:151–63.

Beck, J. W., R. L. Edwards, E. Ito, F. W. Taylor, J. Recy, F. Rougerie, P. Joannot, and C. Henin. 1992. Sea-surface temperature from coral skeletal strontium/calcium ratios. *Science* 257:644–47.

Behrensmeyer, A. K., and R. W. Hook. 1993. Paleoenvironmental contexts and taphonomic modes. In *Terrestrial ecosystems through time,* ed. A. K. Behrensmeyer, J. D. Damuth, W. A. DiMichele, R. Potts, H.-D. Sues, and S. L. Wing, 15–136. Chicago: University of Chicago Press.

Berggren, W. A., D. V. Kent, J. J. Flynn, and J. A. Van Couvering. 1985. Cenozoic chronology. *Geol. Soc. Am. Bull.* 96:1407–18.

Borne, P. F., T. M. Cronin, and H. J. Dowsett. 1994. Microfaunal evidence of Late Pliocene–Early Pleistocene coastal oceanic upwelling in the Moin Formation, Costa Rica. *Geol. Soc. Am. Ann. Meeting, Abstr. Progr.* 26:A-170.

Brey, T., M. Klages, C. Dahm, M. Gorny, J. Gutt, S. Hain, M. Stiller, W. E. Arntz, J.-W. Wagele, and A. Zimmermann. 1994. Antarctic benthic diversity. *Nature* 368:297.

Budd, A. F. 1989. Biogeography of Neogene Caribbean reef corals and its implications for the ancestry of eastern Pacific reef corals. *Mem. Assoc. Australas. Palaeontols.* 8:219–30.

Budd, A. F., and C. M. Kievman. 1994. Coral assemblages and reef environments in the Bahamas Drilling Project cores. Part 3 of the *Final draft report of the Bahamas Drilling Project.* Coral Gables: Rosentiel School of Marine and Atmospheric Sciences, University of Miami.

Budd, A. F., T. A. Stemann, and K. G. Johnson. 1994. Stratigraphic distributions of genera and species of Neogene to Recent Caribbean reef corals. *J. Paleontol.* 68:951–77.

Buddemeier, R. W., and R. A. Kinzie III. 1976. Coral growth. *Oceanogr. Mar. Biol. Ann. Rev.* 14:183–225.

Bush, M. B., D. R. Piperno, P. A. Colinvaux, P. E. De Oliveira, L. A. Krissek, M. C. Miller, and W. E. Rowe. 1992. A 14,300-yr paleoecological profile of a lowland tropical lake in Panama. *Ecol. Monogr.* 62:251–75.

Cheetham, A. H., and J. B. C. Jackson. 1995. Process from pattern: Tests for selection versus random change in punctuated bryozoan speciation. In *New approaches to speciation in the fossil record*, ed. D. H. Erwin and R. L. Anstey, 184–207. New York: Columbia University Press.

Clarke, A. 1993. Temperature and extinction in the sea: A physiologist's view. *Paleobiology* 19:499–518.

CLIMAP Project Members. 1976. The surface of the ice-age earth. *Science* 191:1131–37.

Coates, A. G., J. B. C. Jackson, L. S. Collins, T. M. Cronin, H. J. Dowsett, L. M. Bybell, P. Jung, and J. A. Obando. 1992. Closure of the Isthmus of Panama: The near-shore marine record of Costa Rica and western Panama. *Geol. Soc. Am. Bull.* 104:814–28.

Colinvaux, P. A. 1993. Pleistocene biogeography and diversity in tropical forests of South America. In *Biological relationships between Africa and South America*, ed. P. Goldblatt, 473–99. New Haven: Yale University Press.

Collins, L. S., D. H. Geary, and K. C. Lohmann. 1995. A test of the prediction of decreased Caribbean coastal upwelling caused by the emergence of the Isthmus of Panama, using stable isotopes of neritic Foraminifera. *Geol. Soc. Am. Ann. Meeting, Abstr. Progr.* 27:A-156.

Coope, G. R. 1995. Insect faunas in ice age environments: Why so little extinction? In *Extinction rates*, ed. J. H. Lawton and R. M. May. Oxford: Oxford University Press.

Crowley, T. J., and G. R. North. 1988. Abrupt climate change and extinction events in earth history. *Science* 240:996–1002.

Cunningham, C. W., and T. M. Collins. 1994. Developing model systems for molecular biogeography: Vicariance and interchange in marine invertebrates. In *Molecular ecology and evolution: Approaches and applications*, ed. B. Schierwater, B. Streit, and R. DeSalle, 405–33. Basel, Switz.: Birkhauser.

Darwin, C. 1859. *On the origin of species by natural selection*. London: John Murray.

Dowsett, H. J., and R. Z. Poore. 1991. Pliocene sea surface temperatures of the North Atlantic ocean at 3.0 Ma. *Quatern. Sci. Revs.* 10:189–204.

Emiliani, C. 1955. Pleistocene temperatures. *J. Geol.* 63:538–78.

———. 1992. Pleistocene paleotemperatures. *Science* 257:1462.

Geister, J. 1977. Occurrence of *Pocillopora* in the Late Pleistocene coral reefs. *Mems. Bur. Recherch. Géol. Min.* 89:378–88.

Gibson-Smith, J., and W. Gibson-Smith. 1979. The genus *Arcinella* (Mollusca: Bivalvia) in Venezuela and some associated faunas. *Geos* 24:11–32.

Gould, S. J., and C. B. Calloway. 1980. Clams and brachiopods—ships that pass in the night. *Paleobiology* 6:383–96.

Gould, S. J., and N. Eldredge. 1993. Punctuated equilibrium comes of age. *Nature* 366:223–27.

Gould, S. J., and E. S. Vrba. 1982. Exaptation—a missing term in the science of form. *Paleobiology* 8:4–15.

GRIP. 1993. Climate instability during the last interglacial period recorded in the GRIP ice core. *Nature* 364:203–7.

Guilderson, T. P., R. G. Fairbanks, and J. L. Rubenstone. 1994. Tropical temperature variations since 20,000 years ago: Modulating interhemispheric climate change. *Science* 263:663–65.

Hasson, P. F., and A. G. Fischer. 1986. Observations on the Neogene of northwestern Ecuador. *Micropaleontology* 32:32–42.

Hudson, W. H. 1904. *Green Mansions*. N.p.

Jackson, J. B. C. 1991. Adaptation and diversity of reef corals. *BioScience* 41: 475–82.

———. 1994. Constancy and change of life in the sea. *Phil. Trans. Roy. Soc. Lond.* B-343:55–60.

———. 1995. The fossil record of speciation in the sea. *J. Cell. Biochem. Suppl.* 19B:336

Jackson, J. B. C., A. F. Budd, and J. M. Pandolfi. 1996. The shifting balance of natural communities? In *Evolutionary paleobiology: Essays in honor of James W. Valentine*, ed. D. Jablonski, D. H. Erwin, and J. Lipps. Chicago: University of Chicago Press.

Jackson, J. B. C., and A. H. Cheetham. 1994. Phylogeny reconstruction and the tempo of speciation in cheilostome Bryozoa. *Paleobiology* 20:407–23.

Jackson, J. B. C., P. Jung, A. G. Coates, and L. S. Collins. 1993. Diversity and extinction of tropical American mollusks and emergence of the Isthmus of Panama. *Science* 260:1624–26.

Jansen, E., and J. Sjøholm. 1991. Reconstruction of glaciation over the past six Myr from ice-borne deposits in the Norwegian Sea. *Nature* 349:600–603.

Janzen, D. H., and P. S. Martin. 1982. Neotropical anachronisms: The fruits the gomphotheres ate. *Science* 215:19–27.

Janzen, D. H., and D. E. Wilson. 1983. Mammals: Introduction. In *Costa Rican natural history*, ed. D. H. Janzen, 426–42. Chicago: University of Chicago Press.

Johnson, K. G., A. F. Budd, and T. A. Stemann. 1995. Extinction selectivity and ecology of Neogene Caribbean reef corals. *Paleobiology* 21:52–73.

Jordan, D. S. 1908. The law of the geminate species. *Am. Nat.* 42:73–80.

Jung, P. 1989. Revision of the *Strombina*-group (Gastropoda: Columbellidae), fossil and living. *Schweiz. Paläontol. Abh.* 111:1–298.

Keigwin, L. D., Jr. 1982. Isotopic paleoceanography of the Caribbean and east Pacific: Role of Panama uplift in Late Neogene time. *Science* 217:350–53.

Kennett, J. P., and L. D. Stott. 1991. Abrupt deep-sea warming, paleoceanographic changes, and benthic extinctions at the end of the Palaeocene. *Nature* 353:225–29.

Kennett, J. P., and M. S. Srinivasan. 1983. *Neogene planktonic Foraminifera: A phylogenetic atlas*. Stroudsburg, Pa.: Hutchinson Ross.

Kidwell, S. M. 1985. Palaeobiological and sedimentological implications of fossil concentrations. *Nature* 318:457–60.

———. 1989. Stratigraphic condensation of marine transgressive records: Origin of major shell deposits in the Miocene of Maryland. *J. Geol.* 97:1–24.

Kidwell, S. M., and A. K. Behrensmeyer. 1988. Overview: Ecological and evolutionary implications of taphonomic processes. *Palaeogeogr., Palaeoclimatol., Palaeoecol.* 63:1–13.

Knowlton, N. 1993. Sibling species in the sea. *Ann. Rev. Ecol. Syst.* 24:189–216.

Knowlton, N., and J. B. C. Jackson. 1994. New taxonomy and niche partitioning on coral reefs: Jack of all trades or master of some? *Trends Ecol. Evol.* 9:7–9.

Knowlton, N., L. A. Weigt, L. A. Solorzano, D. K. Mills, and E. Bermingham. 1993. Divergence in proteins, mitochondrial DNA, and reproductive compatibility across the Isthmus of Panama. *Science* 260:1629–32.

Koch, C. F. 1987. Prediction of sample size effects on the measured temporal and geographic distribution patterns of species. *Paleobiology* 13:100–107.

Lessios, H. A. 1981. Divergence in allopatry: Molecular and morphological differentiation between sea urchins separated by the Isthmus of Panama. *Evolution* 35:618–34.

Levin, S. A. 1992. The problem of pattern and scale in ecology. *Ecology* 73: 1943–67.

Levinton, J. 1988. *Genetics, paleontology, and macroevolution.* Cambridge: Cambridge University Press.

Leyden, B. W., M. Brenner, D. A. Hodell, and J. H. Curtis. 1993. Late Pleistocene climate in the Central American lowlands. In *Climate change in continental isotopic records,* 165–78. American Geophysical Union Geophysical Monograph, no. 78.

MacLeod, N., and G. Keller. 1991. How complete are Cretaceous/Tertiary boundary sections? A chronostratigraphic estimate based on graphic correlation. *Geol. Soc. Am. Bull.* 103:1439–57.

Marshall, L. G., and T. Sempere. 1993. Evolution of the neotropical Cenozoic land mammal fauna in its geochronologic, stratigraphic, and tectonic context. In *Biological relationships between Africa and South America,* ed. P. Goldblatt, 329–92. New Haven: Yale University Press.

Martin, P. S., and R. G. Klein, eds. 1984. *Quaternary extinctions: A prehistoric revolution.* Tucson: University of Arizona Press.

Mayr, E. 1954. Geographic speciation in tropical echinoids. *Evolution* 8:1–18.

———. 1963. *Animal species and evolution.* Cambridge: Belknap Press of Harvard University Press.

Mix, A. C. 1989. Influence of productivity variations on long-term atmospheric CO_2. *Nature* 337:541–44.

Norris, R. D., R. M. Cornfield, and J. E. Cartlidge. 1993. Evolution of depth ecology in the planktic Foraminifera lineage *Globorotalia (Fohsella). Geology* 21:975–78.

NRC (National Research Council). 1995. *Understanding marine biodiversity.* Washington, D.C.: National Academy Press.

Paine, R. T., and S. A. Levin. 1981. Intertidal landscapes: Disturbance and the dynamics of pattern. *Ecol. Monogr.* 51:145–78.

Potts, D. C. 1983. Evolutionary disequilibrium among Indo-Pacific corals. *Bull. Mar. Sci.* 33:619–32.

———. 1984. Generation times and the Quaternary evolution of reef-building corals. *Paleobiology* 10:48–58.

Prothero, D. R., and W. A. Berggren. 1992. *Eocene-Oligocene climatic and biotic evolution.* Princeton: Princeton University Press.

Ricklefs, R. E. 1987. Community diversity: Relative roles of local and regional processes. *Science* 235:167–71.

Ricklefs, R. E., and D. Schluter, eds. 1993. *Species diversity in ecological communities: Historical and geographical perspectives.* Chicago: University of Chicago Press.

Robinson, D. G. 1993. The zoogeographic implications of the prosobranch gastropods of the Moin Formation of Costa Rica. *Am. Malacol. Bull.* 10:251–55.

Rosen, B. R. 1993. Change in coral reef communities: The Late Cainozoic emergence of *Acropora* as an ecologically dominant coral. In *Abstracts of papers: First European Regional Meeting of the International Society of Reef Studies.* Vienna: International Society of Reef Studies.

Rubinoff, I. 1968. Central American sea-level canal: Possible biological effects. *Science* 161:857–61.

Saunders, J. B., P. Jung, and B. Biju-Duval. 1986. Neogene paleontology in the northern Dominican Republic. Part 1, Field surveys, lithology, environment, and age. *Bulls. Am. Paleontol.* 89:1–79.

Schneider, J. A. 1995. Phylogenetic relationships of transisthmian Cardiidae (Bivalvia) and the usage of fossils in reinterpreting the geminate species concept. *Geol. Soc. Am. Ann. Meeting, Abstr. Progr.* 27:A-52.

Shackleton, N. J., J. Backman, H. Zimmerman, D. V. Kent, M. A. Hall, D. G. Roberts, D. Schnitker, J. G. Baldauf, A. Desprairies, R. Homrighausen, P. Huddlestun, J. B. Keene, A. J. Kaltenback, K. A. O. Krumsiek, A. C. Morton, J. W. Murray, and J. Westberg-Smith. 1984. Oxygen isotope calibration of the onset of ice-rafting and history of glaciation in the North Atlantic region. *Nature* 307:620–23.

Stanley, S. M. 1986. Anatomy of a regional mass extinction: Plio-Pleistocene decimation of the western Atlantic bivalve fauna. *Palaios* 1:17–36.

Stanley, S. M., and L. D. Campbell. 1981. Neogene mass extinction of western Atlantic molluscs. *Nature* 293:457–59.

Stanley, S. M., K. L. Wetmore, and J. P. Kennett. 1988. Macroevolutionary differences between the two major clades of Neogene planktonic Foraminifera. *Paleobiology* 14:235–49.

Stehli, F. G., and S. D. Webb, eds. 1985. *The Great American Biotic Interchange.* New York: Plenum Press.

Stimson, J. 1985. The effect of shading by the table coral *Acropora hyacinthus* on understory corals. *Ecology* 66:40–53.

Tunnicliffe, V. 1983. Caribbean staghorn populations: Pre–Hurricane Allen conditions in Discovery Bay, Jamaica. *Bull. Mar. Sci.* 33:132–51.

Valentine, J. W., and D. Jablonski. 1993. Fossil communities: Compositional variation at many time scales. In *Species diversity in ecological communities,* ed. R. E. Ricklefs and D. Schluter, 341–49. Chicago: University of Chicago Press.

Vermeij, G. J. 1978. *Biogeography and adaptation.* Cambridge: Harvard University Press.

———. 1991. When biotas meet: Understanding biotic interchange. *Science* 253:1099–1104.

————. 1993. The biological history of a seaway. *Science* 260:1603–4.

Vrba, E. S. 1993. Mammal evolution in the African Neogene and a new look at the Great American Interchange. In *Biological relationships between Africa and South America*, ed. P. Goldblatt, 393–434. New Haven: Yale University Press.

Webb, S. D. 1991. Ecogeography of the Great American Interchange. *Paleobiology* 17:266–80.

Webb, T., III, and P. J. Bartlein. 1992. Global changes during the last three million years: Climatic controls and biotic responses. *Ann. Rev. Ecol. Syst.* 23: 141–73.

Wei, K.-Y. 1994a. Allometric heterochrony in the Pliocene-Pleistocene planktic foraminiferal clade *Globoconella*. *Paleobiology* 20:66–84.

————. 1994b. Stratophenetic tracing of phylogeny using SIMCA pattern recognition technique: A case study of the Late Neogene planktic Foraminifera *Globoconella* clade. *Paleobiology* 20:52–65.

Whittaker, J. E. 1988. *Benthic Cenozoic Foraminifera from Ecuador.* London: British Museum (Nat. Hist.).

Woodring, W. P. 1982. Geology and paleontology of Canal Zone and adjoining parts of Panama: Description of Tertiary mollusks (Pelycypods: Propeamussiidae to Cuspidariidae; additions to families covered in P 306-E; additions to gastropods; cephalopods). *U.S. Geol. Soc. Prof. Pap.* 306-F:541–759.

Zachos, J. C., K. C. Lohmann, J. C. G. Walker, and S. W. Wise. 1993. Abrupt climate change and transient climates during the Paleogene: A marine perspective. *J. Geol.* 101:191–213.

2

The Geologic Evolution of the Central American Isthmus

Anthony G. Coates and Jorge A. Obando

INTRODUCTION

The formation of the Isthmus of Panama has been a complex and extended process stretching over the last 15 million years, with fundamental consequences to ocean circulation, global climatic patterns, and the biogeography, ecology, and evolution of both terrestrial and marine organisms (Jones and Hasson 1985; Stehli and Webb 1985). The closure of the isthmus was the culmination of an extended geologic history involving the growth and migration of the Central American volcanic arc, at the junction of the Pacific and Caribbean Plates, and its collision with South America. The arc was also profoundly modified during this process by accretion of exotic terranes and the subduction of aseismic oceanic ridges. This led to the formation of numerous sedimentary basins, initially formed as fore- and back-arc basins interconnected along an extended archipelago, later reduced to a subcontinuous isthmus with two or three marine corridors, and finally raised to a complete marine barrier about 3.1–2.8 Ma. As will be described elsewhere in this volume, temporary breaching of this barrier may also have occurred in the Late Pliocene as a result of eustatic sea-level changes.

Thus, the closure was not a single event and its biological effects on marine organisms are likely to have been spread over several million years as oceanic circulation, nutrient distribution, temperature, salinity, and habitats were successively modified. These effects were probably different for different taxa according to their ecological and life-history characteristics. In order to understand the nature and timing of these profound oceanographic and biological changes, we will, in this chapter, synthesize the geologic history of the relevant sedimentary basins. The data for this synthesis come from our own extensive fieldwork in Panama, Costa Rica, and Nicaragua (Coates et al. 1992; Collins et al. 1995) and from the literature. From this analysis, we attempt a regional paleogeographic reconstruction of the emergence of the isthmus during the

Late Neogene. This physical model can then be related to the patterns of evolutionary and ecological change within the various taxa described in other chapters of this volume.

The principal geologic units and neotectonic features of the region studied are shown in figure 2.1. The location and names of the sedimentary basins are given in figure 2.2, which also displays the structure of the volcanic arc and the location of the exotic terranes (Nicoya, Osa, Burica, and Azuero) and older ophiolitic and ultramafic subduction complexes (Mahe, Baudo, San Blas, and Dabeiba). The last connections between the Pacific Ocean and the Caribbean must have lain between the northwesternmost continental margin of the South American Plate in the south, and the southern margin of the continental Chortis Block, located close to the present Costa Rica–Nicaragua border, in the north (fig. 2.1). This region forms part of the trailing margin of the Caribbean Plate and consists of the Chorotega and Choco Blocks (Dengo 1985; Duque-Caro 1990a; Escalante 1990; Coates et al. 1992), separated by the Gatun Fault Zone (or Canal Fault Zone, de Boer et al. 1988). The Chorotega Block (fig. 2.1) has been the site of a volcanic arc since the Late Cretaceous. The Choco Block (fig. 2.1) became defined later, as the Central American arc collided in the Middle Miocene with South America. During the Pliocene, the Choco Block and part of the Chorotega Block became detached from the Caribbean Plate to form what is here called the Costa Rica–Panama (CRP) microplate. The principal structural features of the CRP microplate are illustrated in figure 2.1. It is bounded in the south by the Central American Trench (CAT), which to the east of the Panama Fracture Zone (PFZ) passes into the South Panama Deformed Belt (SPDB) (Silver et al. 1990), which in turn flexes sharply southward to form the Colombia Trench (CT). On the eastern margin, the CAT is defined by the Uramita Fault Zone (UFZ) (Duque-Caro 1990a,b), which is the site of the collision with the South American Plate. The northern margin of the microplate is formed by the overthrust of the North Panama Deformed Belt (NPDB) (Bowland 1984; Bowland and Rosencrantz 1988), and this appears to pass westward into a sinistral strike-slip fault complex, the Costa Rica Transcurrent Fault System (CRTFS) (Astorga et al. 1991), which impinges on the CAT to the south of the Nicoya Peninsula.

The remaining autochthonous portion of the Chorotega Block forms a relatively narrow zone of oceanic arc rocks on the Caribbean Plate, between the detached CRP microplate and the southern margin of the Chortis Block (fig. 2.1). From west to east this zone (fig. 2.2), which forms the southern termination of the Nicaragua Depression (Mc-

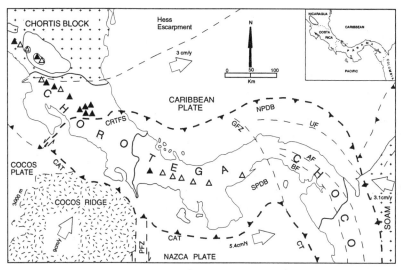

Fig. 2.1 Principal structural features of the Chorotega and Choco oceanic blocks. The thicker dashed line marks the limits of the Costa Rica–Panama microplate. The Chortis Block, in the northwest, and the South American Plate (SOAM), in the east, are defined by a pattern of crosses. They form the continental crustal limits of the oceanic isthmian crust and thus constrain the area of the last marine connections between the Pacific and Caribbean. The Cocos Ridge, outlined by random dashes, is an aseismic 15-km-thick oceanic ridge rising 2–2.5 km above the surrounding ocean floor. AF = Atrato Fault; BF = Baudo Fault; CRTFS = Costa Rica Transcurrent Fault System; GFZ = Gatun Fault Zone, a poorly defined zone passing through the area of the sedimentary Panama Canal Basin and dividing the Chorotega and Choco Blocks; CAT = Central American Trench; NPDB = the North Panama Deformed Belt, the northward-thrusting compressive margin of the Costa Rica–Panama microplate; PFZ = Panama Fracture Zone, which forms the transform boundary between the Cocos and Nazca Plates; SPDB = South Panama Deformed Belt, which forms the relatively aseismic margin of the Nazca and Caribbean Plates; UF = Uramita Fault. Closed triangles are active volcanoes; open triangles are volcanoes that have become inactive.

Birney and Williams 1965; Mann and Burke 1984; Mann et al. 1990), consists of an exotic Nicoya terrane (Nicoya High); a fore-arc basin (Tempisque); a rift basin (San Carlos), and a normal-faulted back-arc basin (Northern Limón). The extensional tectonics of this zone may be related to movement along a strike-slip fault system which forms the extension of the Hess Escarpment into the Santa Elena Peninsula (Astorga 1992).

On the Chorotega Block, a series of marine connections joined the fore- and back-arc basins across the isthmus until the Pliocene. These marine corridors comprise the Tempisque, San Carlos, Northern Li-

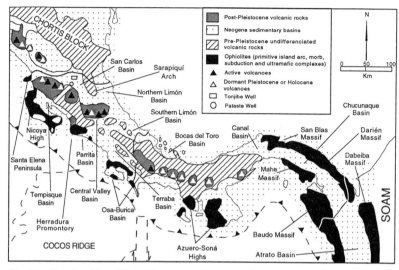

Fig. 2.2 Relationship of Late Neogene sedimentary basins to the volcanic arc of the Chorotega Block and the subduction complexes of the Choco Block.

món Basins; the Central Valley Basin; and the Burica, Terraba, Southern Limón Basins. On the Choco Block, marine connections existed through the Chucunaque and Atrato Basins, which run parallel to the axis of the isthmus flanked by subduction complexes. A marine connection also existed into the Pliocene at the junction of the two blocks, which is the location of the Canal Basin (fig. 2.2).

We will first describe the principal tectonic mechanisms which have affected the geologic evolution of the sedimentary basins of the Central American Isthmus, then describe the stratigraphic sequences for each basin, and last, propose paleogeographic reconstructions of the isthmus for the Middle Miocene, the Late Miocene, and the Early Pliocene.

LATE NEOGENE TECTONIC EVOLUTION OF THE CHOROTEGA AND CHOCO BLOCKS

There are three principal tectonic mechanisms that appear to have controlled the development of the Chorotega and Choco Blocks, the subsequent detachment of the CRP microplate, and the location of the Neogene sedimentary basins. These are the formation of the volcanic arc, the subduction of the Cocos Ridge, and the collision of the arc with South America.

Development of the volcanic arc began during the Late Cretaceous and continued throughout much of the Cenozoic, as a result of subduction of the Farrallon Plate, and later the Cocos and Nazca Plates, beneath the Caribbean Plate. Immediately south of the continental Chortis Block in Nicaragua and Costa Rica, strong Cenozoic subduction by the northward-moving Cocos Plate sustained an active volcanic arc and development of a well-defined set of fore- and back-arc basins along the western part of the Chorotega Block (Dengo 1962; Lonsdale and Klitgord 1978; Case and Holcombe 1980; Escalante 1990; Astorga et al. 1991).

However, in the eastern Chorotega Block, in western Panama, which lies to the east of the Panama Fracture Zone, volcanism peaked in the Late Miocene (about 12 Ma) and was largely absent for about 10 my until the beginning of the Pleistocene (de Boer et al. 1988). In this region, the east-northeastward-moving Nazca Plate is subducted very obliquely under the similarly eastward-moving Caribbean Plate (fig. 2.1). The resulting very small differentials in vectors and rate of plate movement make the plate boundary weakly defined, relatively aseismic, and without active volcanism (Lonsdale and Klitgord 1978; McCann and Pennington 1990; Silver et al. 1990; Kolarsky and Mann 1995b).

The Neogene history of the Choco Block is quite different. To the east of the Gatun Fault Zone, a bathyal to abyssal oceanic crust of unknown width existed until the Middle Miocene (Donnelly 1992), after which foreland basins formed. In this area (fig. 2.2), the Central American Trench appears to have been much less well defined, volcanism has been largely absent, and fore- and back-arc basins were not developed (Duque-Caro 1990a,b).

The second primary tectonic feature controlling basin development has been the shallow subduction of the Cocos Ridge (Corrigan et al. 1990; Kolarsky et al. 1995; Collins et al. 1995). The subduction of this anomalously thick (15 km) aseismic oceanic ridge, approximately 200–300 km wide and rising some 2–2.5 km above the surrounding ocean floor, has rapidly uplifted and deformed the entire volcanic arc from about the Azuero Peninsula in western Panama to the approximate line of the Costa Rica Transcurrent Fault System along the Central Valley of Costa Rica (fig. 2.3). The extension of the CRTFS to the Central American Trench, which occurs south of the Nicoya Peninsula, has been called the East Nicoya Fault Zone (Crowe and Buffler 1984; Corrigan et al. 1990; Kolarsky et al. 1995).

Kolarsky and others (1995) have documented six major tectonic effects of the shallow subduction of the Cocos Ridge under the Chorotega Block:

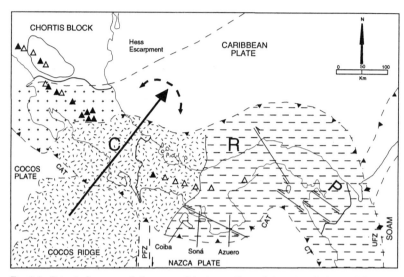

Fig. 2.3   Principal structural elements of the lower Central American Isthmus. That part of the CRP microplate strongly influenced by buoyant shallow subduction of the Cocos Ridge is indicated by random dashes; that principally affected by collision with SOAM by horizontal dashes. The pattern of small crosses defines the area characterized by the extensional tectonics of the southern portion of the Nicaragua Depression. Symbols as for figure 2.1. The arrowed faults are left-lateral, strike-slip faults mapped by Kolarsky and Mann (1995a).

1. There is a 175 km gap in the chain of stratovolcanoes that runs through southern central Central America (fig. 2.3). These volcanoes are otherwise very closely spaced, averaging 26 km between peaks. This gap corresponds with an elevated "cold" magmatic arc, the Cordillera of Talamanca, which rises more than 1800 m above the rest of the southern Central American Isthmus, steadily diminishing in altitude (fig. 2.4) as far as Arenal, 115 km to the west in Costa Rica, and to El Valle, 300 km to the east in Panama (de Boer et al. 1988). The cumulative amounts of young volcanic rocks and the K_2O content of the volcanics also decrease from Arenal and El Valle toward the Talamanca Cordillera.

2. The Cocos Ridge shows a marked shallowing in subduction angle and an abrupt decrease in seismicity below 75 km (Mann et al. 1990; Burbach et al. 1984; Guendel 1986; Kolarsky et al. 1995), compared to other segments of the Cocos Plate.

3. The fore-arc has been indented (fig. 2.3) and, along with the back-arc, uplifted in its submarine and onshore shelf sections from

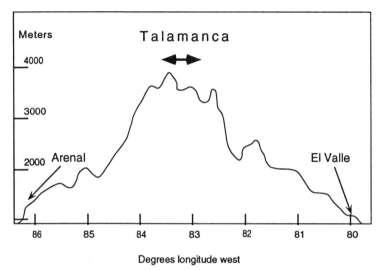

Fig. 2.4 Profile of the Chorotega volcanic arc from northwest (Arenal Volcano, Costa Rica) to southeast (El Valle, Panama). The volcanic arc-gap (arrows) of the Talamanca Cordillera is aligned with the long axis of the Cocos Ridge and has the maximum elevation. Volcanic peak heights diminish in both directions away from this axis. Modified from de Boer et al. 1988.

the CRTFS, south of the Nicoya Peninsula, to the east of the Burica Peninsula (Corrigan et al. 1990).

4. Both the fore-arc and back-arc basins have been structurally "inverted." In this region, structural inversion is attributed to original basin-controlling divergent faults reversing their movement because of the convergent tectonics caused by the collision of the Cocos Ridge (Kolarsky et al. 1995). The sedimentary basins of Terraba on the fore-arc and southern Limón on the back-arc have thus been converted to topographic highs.

5. Structural highs in the basement of the western Colombian Basin and the region of Osa-Burica are aligned with the Cocos Ridge on both the Pacific and Caribbean side of the isthmus (fig. 2.3).

6. The nature and orientation of the regional stress field strongly suggest compressional effects aligned with the direction of Cocos Ridge movement.

It seems evident that Cocos Ridge subduction has had pervasive and extensive effects over a distance of some 400 km along the isthmus. Earlier workers have assumed that these effects of Cocos Ridge subduction were initiated at about 1 Ma (Lonsdale and Klitgord 1978; Corrigan et al. 1990; Kolarsky et al. 1995). However, nannofossil and planktonic foraminiferal dates of formations involved in the rapid uplift caused by Cocos Ridge subduction indicate that these processes started as early as 3.0 Ma on the Pacific side although they did not affect the Caribbean side until about 1.9–1.8 Ma (Collins et al. 1995).

The third major tectonic element affecting the development of the Chorotega and Choco Blocks was the collision of the eastern end of the Central American arc with the South American Plate (fig. 2.3). Several authors have proposed models in which an approximately east-west oriented arc began to collide with the continental South American Plate during the Middle Miocene (Lonsdale and Klitgord 1978; Pindell and Dewey 1982; Wadge and Burke 1983; Mann and Burke 1984; Silver et al. 1990; Kolarsky and Mann 1995a). The three characteristic features of this compressive phase are (1) the S-shaped oroclinal bending and uplift of the CRP microplate, (2) the development of the North Panama Deformed Belt with northward-verging thrusts, and (3) a series of northwest-striking folds and thrusts, many of which have a sinistral strike-slip component. The latter are particularly well developed in the easternmost part of Panama (Kolarsky and Mann 1995a,b), and they also occur in Coiba Island and the Soná and Azuero Peninsulas. We suggest that the region characterized by these structures defines the eastern part of the CRP microplate and the zone directly affected by the collision with South America. The western part of the CRP microplate has been strongly influenced by the tectonic effects of the subduction of the Cocos Ridge (fig. 2.3). Thus, the northern boundary of the CRP microplate passes westward from a compressional northward-verging NPDB into the extensional strike-slip CRTFS complex on the western side of the Cocos Ridge.

NEOGENE SEDIMENTARY BASINS OF THE CHOROTEGA AND CHOCO BLOCKS

The stratigraphic sequences of the Neogene sedimentary basins on the Chorotega Block have been recently analyzed by Astorga and others (1991). They suggest that the basins can be divided into those whose structural control was primarily related to the compressive forces of Cocos-Caribbean subduction (these basins are primary fore- and back-arc basins, developed in the Late Cretaceous and throughout the Cenozoic) and those which were subsequently developed or modified as a result of later, more regional transisthmian tectonic effects. The latter, we suggest, are primarily the effects of subduction of the Cocos Ridge and the extensional tectonics associated with the extension of the Hess Escarpment along the southern margin of the Chortis Block.

By grouping those basins that align across the isthmus, we can define corridors that represent marine connections between the Pacific and the Caribbean across the Chorotega volcanic arc. This is more difficult for the Choco Block, which represents an uplifted oceanic terrane in which subduction complexes have isolated borderland basins for the most part parallel to the axis of the isthmus. The basins are associated as follows (fig. 2.2);

Chorotega Block
    1. Tempisque, San Carlos, Northern Limón Basins, Costa Rica
    2. The Central Valley Basin, Costa Rica
    3. The Burica, Terraba, Southern Limón Basins, Costa Rica
Choco Block
    4. Canal Basin, Panama
    5. The Chucunaque and Atrato Basins, Colombia and Panama

The Tempisque, San Carlos, Northern Limón Basins
The Neogene of the Tempisque Basin (fig. 2.5) consists of a sequence of regressive shelf deposits in which a gradual transition occurs from shale- to sandstone-dominated facies through the Late Miocene. During the Pliocene, a diachronous transition to fan-deltaic marginal marine and estuarine deposits occurred. The Middle to Late Miocene Punta Carballo Formation is a tuffaceous shale and calcareous sandstone, frequently fossiliferous and about 200 m thick near its type locality in the south of the basin. The overlying Pliocene El Real Formation is increasingly sandier and volcaniclastic. Volcanic activity with extrusion of lavas and accompanying laharic and pyroclastic flows is extensively developed in the Pleistocene. The Tempisque Basin in the north is extensively covered by Quaternary volcanic deposits in the region of Junquilla and Salinas Bay. At this point, a traverse of the isthmus does not rise above 45 m in altitude, but marine deposits, confirming a pre-Pleistocene connection of the Tempisque Basin, via the southern limit of Lake Nicaragua, to the San Carlos Basin in north-central Costa Rica, are not exposed. There is, however, a dextral offset of the volcanic arc at the point of the presumed connection (fig. 2.2).

Seismic sections across the San Carlos Basin (Astorga et al. 1991) indicate normal faults and grabens typical of a rift basin. A shallow marine to brackish Neogene sequence is known extensively throughout the San Carlos Basin from a series of boreholes where deposits ranging from Middle Miocene to Pleistocene are lumped together in the Venado Formation. The Venado Formation has a very restricted outcrop located in the extreme west of the basin, close against the volcanic arc, only 40–50 km from the Pacific Ocean. The section comprises 600–650 m of marginal marine to brackish calcarenitic limestone, shale, siltstone, and sandstone (Sen Gupta et al. 1986). Benthic foraminifers in these sediments suggest a shoreline near Venado during the Late Miocene, and there is a clear transition from dominantly shelf mudstone to marginal marine and brackish sand (Sen Gupta et al. 1986; L. Obando 1986).

Analysis of the surface geology of the San Carlos Basin indicates that the Venado Formation deepened from marginal marine to outer neritic depths from Early to Middle Miocene, then sharply regressed in the Late Miocene to a marginal marine environment (Sen Gupta et al. 1986; L. Obando 1986). However, subsurface data in the Tonjibe and Pataste Wells (fig. 2.2) also showed a similar cycle recurred in the Early Pliocene, and again in the latest Pliocene (Aguilar and Fernandez 1989). Thus, marine sediments were still being deposited in the western part of the San Carlos Basin during the early part of the Pleistocene (fig. 2.5).

The San Carlos embayment may not have been completely open to the Caribbean on its eastern margin, where it appears to have been partly separated from the Northern Limón Basin by the Sarapiquí Arch (fig. 2.2), which may be a relict earlier volcanic arc (Astorga et al. 1991). Moreover, the marine connection to the Caribbean via the Northern Limón Basin may have been restricted to a narrow strait at the southern ends of the San Carlos and Northern Limón Basins as shown in fig. 2.2. To the east, the Northern Limón Basin forms a back-arc basin with northeast-stepping normal listric faults typical of the extensional tectonics of a passive margin basin (Astorga et al. 1991). In the Late Miocene and Pliocene, the basin was the site of a large sedimentary clastic wedge formed by the Rio San Juan proto-delta and the source of a large Late Neogene submarine fan deposit identified as the Costa Rica Fan in the western Colombian Basin (Bowland 1993).

The San Carlos and Northern Limón Basins lie at the southernmost end of the 600-km-long Nicaragua Depression (McBirney and Williams 1965). This is a half-graben delimited on the western side by a high-angle normal fault of considerable displacement and containing the line of Quaternary volcanoes which runs to the southwestern tip of Nicaragua, and Managua and Nicaragua Lakes. In northern Costa Rica,

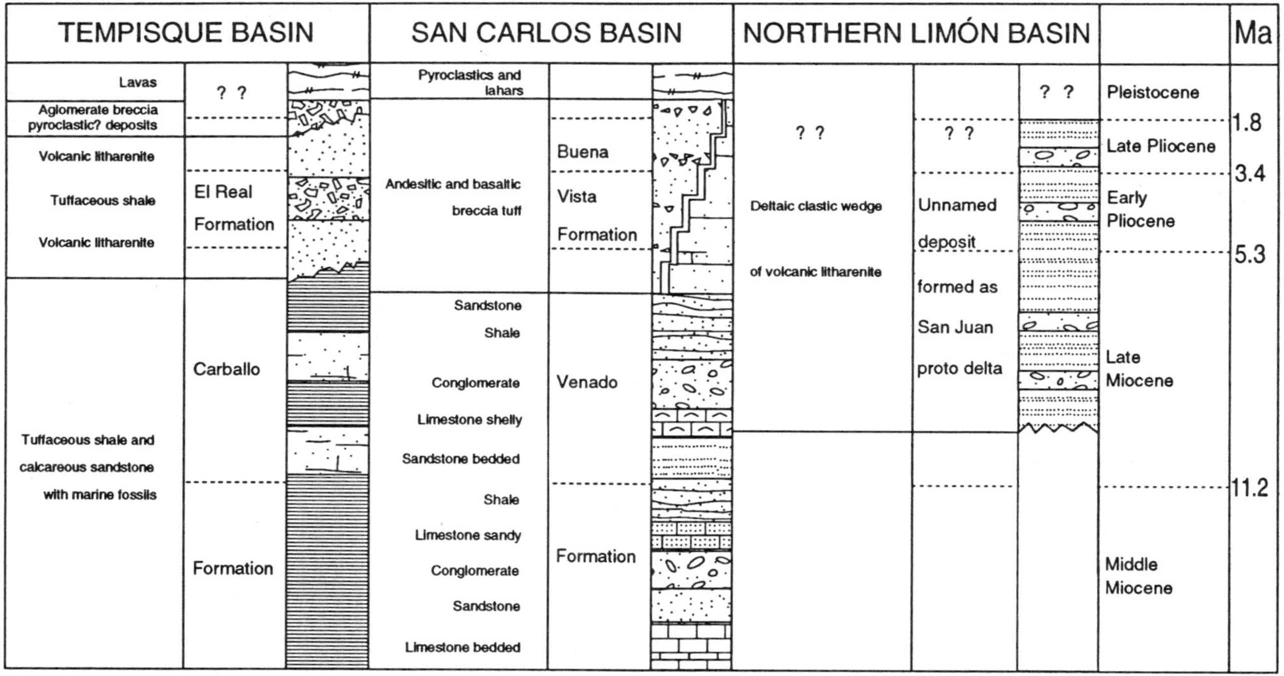

TEMPISQUE BASIN | SAN CARLOS BASIN | NORTHERN LIMÓN BASIN | Ma

TEMPISQUE BASIN:
- Lavas
- Agglomerate breccia pyroclastic? deposits
- Volcanic litharenite — El Real Formation
- Tuffaceous shale
- Volcanic litharenite
- Carballo Formation
- Tuffaceous shale and calcareous sandstone with marine fossils

SAN CARLOS BASIN:
- Pyroclastics and lahars
- Andesitic and basaltic breccia tuff — Buena Vista Formation
- Sandstone
- Shale
- Conglomerate
- Limestone shelly
- Sandstone bedded — Venado
- Shale
- Limestone sandy
- Conglomerate
- Sandstone
- Limestone bedded — Formation

NORTHERN LIMÓN BASIN:
- ? ?
- Deltaic clastic wedge of volcanic litharenite
- Unnamed deposit formed as San Juan proto delta

Ma / Age:
- Pleistocene — 1.8
- Late Pliocene — 3.4
- Early Pliocene — 5.3
- Late Miocene — 11.2
- Middle Miocene

Fig. 2.5 Correlation of transisthmian sedimentary basins in northwestern and northern Costa Rica. This region forms the southern limit of the Nicaragua Depression.

the Nicaragua Depression crosses the isthmus to the Caribbean side via the San Carlos and Northern Limón Basins. This area is marked by the lowest topography of the Central American Isthmus, extensional tectonics in the San Carlos rift, and the normal-faulted Northern Limón Basin. It is also outside the zone of uplift tectonics due to Cocos Ridge subduction (fig. 2.3). These features, combined with the subsurface evidence of extensive marine Pliocene and Pleistocene in the San Carlos Basin, suggests that this region may have remained a shallow marine connection between the Pacific Ocean and the Caribbean through most of Pliocene time.

The Central Valley Basin

The Central Valley Basin has been interpreted to be a second-order pull-apart basin created by the Costa Rica Transcurrent Fault System (Astorga et al. 1991). However, the present sinistral strike-slip motion of the CRTFS would appear to be in the wrong sense for a pull-apart basin to develop in the Central Valley. The present sense of the CRTFS may, however, reflect extensional forces developed along the western margin of the northeastward-moving Cocos Ridge and hence date only from the Early Pliocene, since which time the Central Valley Basin has been uplifted and closed.

Stratigraphy within the Central Valley Basin is difficult to unravel because of scattered outcrops and rapid facies changes. Three formations have been described from the later Neogene of the Central Valley. The Early Miocene San Miguel Formation consists of a variety of bioclastic and nodular limestone, calcareous sandstone, and shale, indicating a shallow nearshore marine environment with offshore bars (Carballo and Fischer 1978). The Middle to Late Miocene Turrucares Formation consists of fossiliferous limestone, conglomerate, and volcaniclastic sandstone (Woodring and Malavassi 1961; Castillo 1969; Montero 1974; Fischer 1981a,b). The Coris Formation, of similar age, is characterized by quartz arenite, tuffaceous sandstone, and minor shale (Franco 1978; Fischer and Franco 1979; Obando et al. 1991). These three formations indicate that between the southern Gulf of Nicoya, near the Herradura Promontory and the Southern Limón Basin, a shallow marine connection existed until Late Miocene time (fig. 2.2). A series of measured sections in the Coris Formation immediately to the south of San José, expose mixed (shelf) facies and a strand-plain lower, middle, and upper shoreface facies (Obando et al. 1991). The stratigraphic arrangement of these facies in several sections across the Central Valley suggests a shoreline prograding southeastward across the Central Valley at the end of the Miocene. Thus, uplift of the volcanic arc to the northwest and the severing of the marine connection with the Caribbean must have occurred by the earliest Pliocene.

The Osa-Burica, Terraba, Southern Limón Basins

This suite of basins (fig. 2.6) lies across the isthmus in alignment with the Cocos Ridge (fig. 2.2) and appears to have been strongly influenced by its subduction since the Early Pliocene (Corrigan et al. 1990; Coates et al. 1992; Collins et al. 1995; Kolarsky et al. 1995). The Osa-Burica Basin forms part of the outer Central American fore-arc (J. Obando 1986; Obando and Baumgartner 1986; Obando et al. 1990; Corrigan et al. 1990) and consists of a sequence of shoreface to deep-trench turbidites, up to 4000 m thick, that have accumulated since the Early Pliocene. These volcaniclastic sediments are best exposed on the Burica and Osa Peninsulas but are mostly covered or submerged in the Golfo Dulce and the alluvial plains north of the Osa and Burica Peninsulas. The Neogene sediments overstep the basement complex which consists in Osa of a Cenozoic subduction complex and in Burica of a non-subducted seamount complex (Baumgartner et al. 1989; Mora et al. 1989).

The Neogene sediments of the Osa-Burica Basin consist of three formations (Coates et al. 1992). The oldest is the Peñita Formation, which consists of inner-shelf to shoreface conglomerate, massive to thick-bedded shelly sand and silty sand, and is more than 3.5 Ma in age (Early Pliocene). The Burica Formation conformably overlies the Peñita Formation, is about 2800 m thick, and consists of mostly fine-grained, volcaniclastic turbidites with local megabreccias formed by large-scale intraformational slumps. Benthic foraminifers analyzed throughout the sequence (Corrigan et al. 1990; Collins et al. 1995) indicate deposition in the lower part of the formation was in water depths of about 2000 m. The overlying Armuelles Formation strongly resembles the Peñita Formation in lithofacies, indicating rapid shallowing through the upper portion of the Burica Formation (Collins et al. 1995). The rapid shallowing evident in the Burica Formation and the nearshore to estuarine conditions represented by the Armuelles Formation, whose base is dated at 1.7–1.9 Ma, suggest that the uplift resulted from the insertion of the Cocos Ridge in the Central American Trench as early as 3 Ma (Collins et al. 1995).

The Terraba Basin forms part of the inner fore-arc basin (fig. 2.2) and contains a thick sequence of Eocene to Pliocene sedimentary and volcanic rocks deposited between the Talamanca Cordillera and the

Ballena-Celmira Fault Zone, which separates the basin from the alloch-thonous Golfito and Burica terranes and their sedimentary cover (J. Obando 1986). The stratigraphy of the region has been extensively studied (Mora 1979; Lowery 1982; Phillips 1983; Yuan 1984; Kolarsky et al. 1995). There are 3500–4000 m of marine, fluvial, and alluvial fan facies with minor sills and dikes, remnants of which also occur on the Talamanca Cordillera, thereby constraining the timing of uplift of this magmatic arc.

There are four Neogene formations, whose stratigraphic relations are shown in figure 2.6. The oldest is the Terraba Formation, which ranges from Oligocene to Middle Miocene and is more than 1000 m thick. It consists of black marine shale in the lower part and grades up-ward into volcaniclastic sandstone-dominated turbidites (Phillips 1983; Yuan 1984). The Middle to Late Miocene Curre Formation conforma-bly overlies the Terraba Formation and is about 800 m at maximum thickness. It consists mostly of medium-grained, quartz-poor, lithare-nite, conglomerate, breccia, and mudstone, with shallow marine mol-lusks and plant fragments. The succession of facies records shoaling from bathyal to inner neritic conditions with local deep-sea fan con-glomerate and breccia (Lowery 1982). Local development of shallow marine, shoreface, and fluvial facies (Mora 1979) indicates infill of the basin by the Late Miocene, but with oceanic connections to the South-ern Limón Basin possible until this time.

The Pliocene Paso Real Formation, estimated to be approximately 900 m thick, overlies the Curre Formation with strong angular uncon-formity and is formed principally of massive volcanic breccia with a tuffaceous matrix, conglomerate, and volcanic flows (Dengo 1962; Lowery 1982). Unconformably overlying all the other Neogene forma-tions is the Late Pliocene to Pleistocene Brujo Formation, at least 600 m thick. It consists of poorly stratified coarse gravels and debris-flow deposits of alluvial origin.

The pre–Paso Real formations' deposits show compressive folds and reverse faults indicating maximum compression aligned to N35°E (Ko-larsky et al. 1995), which coincides with the vector of the axis of the Cocos Ridge and the compression in the outer arc basins (Corrigan et al. 1990). This strongly implies that uplift is due to the subduction of the Cocos Ridge, as suggested by Corrigan and others (1990). However, the collision and resulting uplift occurred much earlier than 1 Ma and probably during the Late Pliocene (Lonsdale and Klitgord 1978; Corri-gan et al. 1990; Collins et al. 1995; Kolarsky et al. 1995). Scattered out-liers of the Curre Formation on the Talamanca Cordillera also suggest

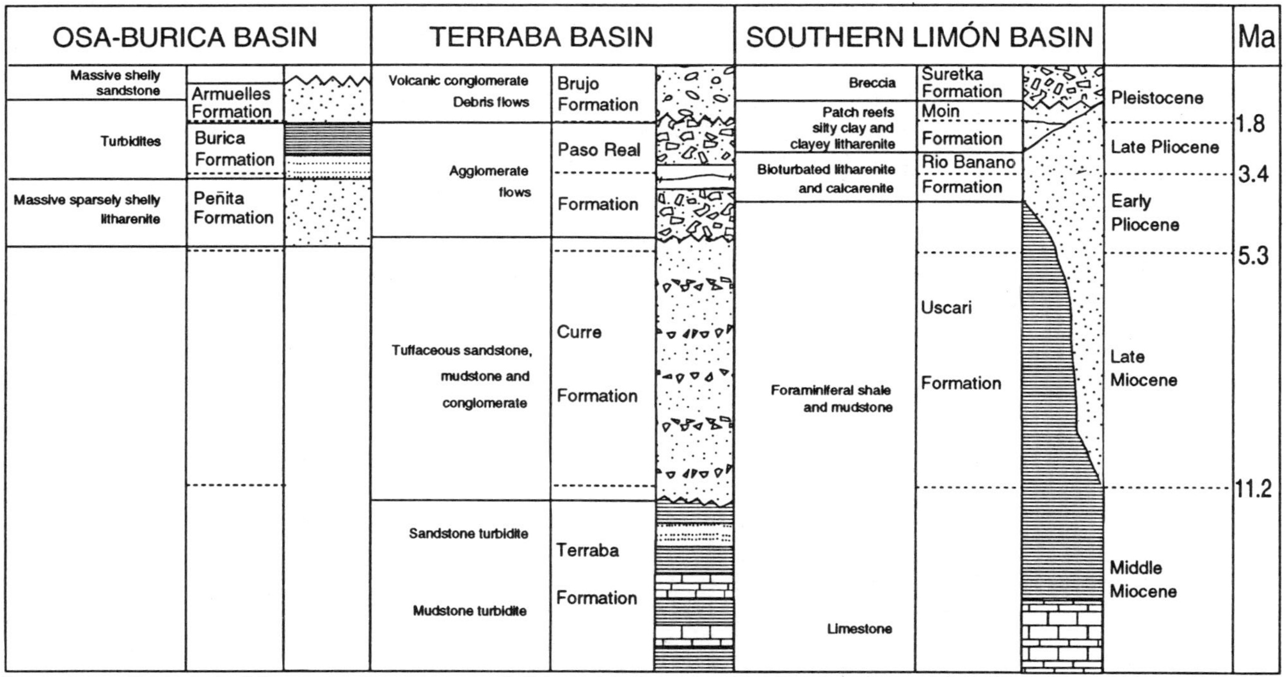

Fig. 2.6 Correlation of transisthmian sedimentary basins in Costa Rica aligned with the Cocos Ridge.

that marine connections from the Terraba Basin to the Southern Limón Basin existed until late Miocene time and that the uplift of the Talamanca magmatic arc, like the structural inversion of the Terraba Basin, is a result of the Pliocene subduction of the Cocos Ridge and hence is the same or slightly younger in age.

The Southern Limón Basin forms the northern foothills and the coastal plain north of the Talamanca Ridge, and displays compressional tectonics with northeasterly verging thrusts. Like the Terraba Basin, it was the site of marine deposition throughout the Miocene. However, the subsequent history of regression, uplift, and structural inversion differs in timing from that of the Terraba Basin.

The Neogene stratigraphy of the basin has been described by many authors (Escalante 1990) and was recently revised by Coates and others (1992) on the basis of extensive new sections and new planktonic foraminiferal and nannofossil data. The Neogene series consists of a largely marine Limón Group, which contains the Uscari, Rio Banano, and Moin Formations, and a continental Suretka Formation, which unconformably and diachronously overlies the other formations (fig. 2.6). The oldest and most widespread unit is the Early to Late Miocene Uscari Formation. It is more than 550 m thick and consists mostly of soft gray foraminiferal shale representing bathyal to abyssal depths (Laurel Collins, personal communication 1993). It has been assumed (Cassell and Sen Gupta 1989a,b) that a major hiatus exists between the top of the Uscari Formation and the overlying Rio Banano Formation occupying most of the Late Miocene, a regional break that occurs widely in South America (Duque-Caro 1990b). Recent subsurface work shows that the top of the Uscari Formation is strongly diachronous, ranging from early Late Miocene to earliest Pliocene (German Gonzalez, personal communication 1993).

The overlying Rio Banano Formation varies considerably in thickness, up to 1500 m, and represents a regressive sequence of shallow inner-shelf to delta-top environments, composed of litho- and calcarenite that is massive, bioturbated, and often rich in mollusks, echinoids, and benthic foraminifers (Coates et al. 1992). The base ranges from early Late Miocene to Early Pliocene (3.5–3.4 Ma), reflecting the different times of onset of deltaic deposition. The youngest beds dated are 2.4 Ma (Late Pliocene), although some of the deltaic facies may be as young as earliest Pleistocene. The deltaic deposits are thickest and most fully developed immediately to the west of Puerto Limón and thin rapidly to the north and south. The source was presumably the rising Talamanca Cordillera.

Locally developed in the Puerto Limón area, the partly time-equivalent and overlying Moin Formation (Coates et al. 1992) consists of a series of silty clay and shelly volcaniclastic litharenite, interbedded with several patch-reef members containing eighty-six species of corals (Ann Budd, personal communication 1995). The corals indicate shallowing during the deposition of the Moin Formation from outer-shelf to less than 30 m water depth during the Late Pliocene to earliest Pleistocene (3.5–1.7 Ma). The lower part of the Moin Formation may be laterally equivalent to the deltaic upper portion of the Rio Banano Formation. The Suretka Formation, a Plio-Pleistocene terrestrial volcanic breccia, unconformably overlies both the Rio Banano and Moin Formations in the northern part of the Southern Limón Basin.

Age and depth relations of sections in the Limón Group (Moin and Rio Banano Formations) and their time-equivalent units in northwestern Panama (Bocas del Toro Group) demonstrate a negligible rate of Pliocene emergence in the Limón and Bocas del Toro Basins. In contrast, uplift was extremely rapid (1100 m/my) in the Burica Basin on the Pacific side of the uplift (Collins et al. 1995). Uplift of the Southern Limón Basin (72 m/my) began in the Early Pleistocene owing to the later arrival of compressive and uplift effects of the subducting Cocos Ridge on the Caribbean margin of the isthmus. This 1.9 my (3.5–1.6 Ma) offset in time of arrival across the 150 km from Burica to Limón perfectly matches the present rate of motion of the Cocos Ridge (8 cm/yr × 1.9 my = 152 km).

The regressive sequences outlined above indicate that the last marine connections from the Burica and Terraba Basins to the Southern Limón Basin across the isthmus were cut off by the rise of the Talamanca Cordillera in latest Miocene to earliest Pliocene time. This isolation is manifested by the onset of terrigenous deposition in the Terraba Basin (Paso Real Formation) and the development of a large delta complex (Rio Banano and Suretka Formations) in the Southern Limón Basin.

Canal Basin, Panama

The sequence of Neogene rocks forming the Canal Basin (fig. 2.7) was extensively exposed during the construction of the Panama Canal (Woodring 1957, 1970, 1982). The construction of the canal revealed an extensive transisthmian marine sedimentary basin, which appears to have been controlled by a major fault system, the Gatun Fault Zone, separating the Chorotega and Choco Blocks (fig. 2.1; Mann and Corrigan 1990). The GFZ is apparently manifested by continued seismic activity (Eduardo Camacho, personal communication 1993).

Marine and volcanic rocks are complexly interbedded in the Canal Basin, particularly the Oligocene and Early Miocene deposits. During the Early Miocene in the central part of the isthmus (Gaillard Cut and Madden Basin), a variable sequence of areally restricted units, including the Culebra, Cucaracha, and La Boca Formations, show fossiliferous, dark shale, mudstone, calcareous sandstone, and diverse coral limestone lithofacies, indicating marine conditions were widespread (Stewart and Stewart 1980). However, agglomerate, tuffaceous debris, and basalt flows indicate contemporary nearby volcanism, presumably along the Chorotega arc in western Panama (de Boer et al. 1988). Overlying these deposits, and representing perhaps the latest Early Miocene (Stuart and Stuart 1980) is the Alhajuela Formation, which crops out extensively in the Madden Basin and consists of tuffaceous sandstone, calcareous sandstone, and limestone. None of these formations extends to the Caribbean coast, but the marine La Boca Formation extends from the Madden Basin across the axis of the isthmus to the Pacific mouth of the Panama Canal, indicating a marine connection across the isthmus in the Early Miocene.

Neogene deposits on the Caribbean side begin with the Late Miocene Gatun Formation, characterized by inner- to middle-shelf, fine-grained, calcareous, richly fossiliferous siltstone and sandstone, which rests on either the Cretaceous volcanic basement or the Late Oligocene marine and volcanic Caimito Formation. The Gatun Formation is about 500 m thick, contains a cross-bedded basal comglomerate 2–3 m thick, and is strongly unconformable on the underlying units. Woodring (1957) indicates the Gatun is a transgressive deposit, and there appear to have been regional tectonic movements in the Middle Miocene preceding its deposition. He also notes the absence of tuffs in the Gatun, which corroborates the Miocene (10 Ma) to Pleistocene (1.6 Ma) phase of quiescence in western Panamanian volcanism proposed by de Boer and others (1988). These Middle Miocene movements coincided with the beginning of collisional tectonics to the east (Duque-Caro 1990b) and the formation of the San Blas–Darién massif as part of the Dabeiba Arch, and thus the beginning of constriction of Pacific to Caribbean marine connections in the Canal and Chucunaque-Atrato Basin areas.

Conformably overlying the Gatun Formation is the Chagres Sandstone. The base consists of a coarse, cross-bedded coquina of dominantly echinoid and barnacle plates (the Toro Point Member), which is overlain by homogeneous massive silty sandstone with scattered mollusks, apparently of Pliocene age (Woodring 1957). Benthic foramini-

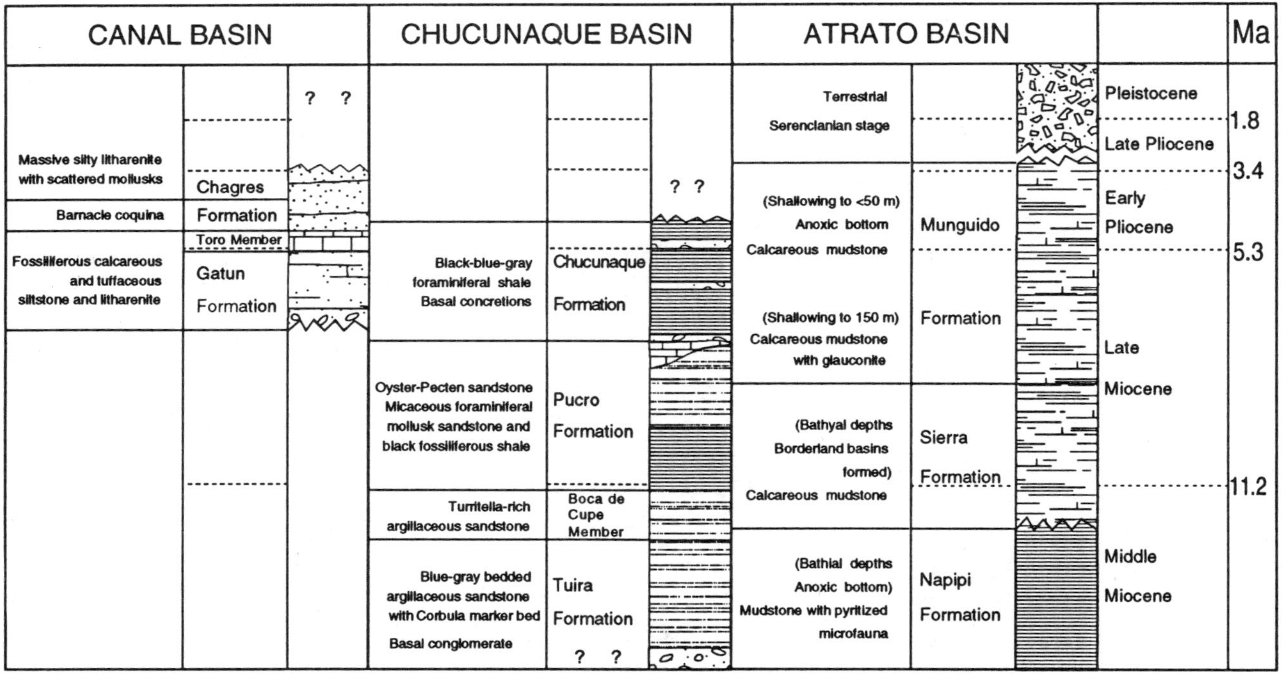

Fig. 2.7 Correlation of transisthmian sedimentary basins in central and eastern Panama and northwestern Colombia.

fers in the Gatun Formation indicate water depths of 20–40 m. Surprisingly, in view of the lithology of the Toro Point Member, the benthic foraminifers of the Chagres Formation indicate depths of 175–200 m, and the majority are clearly of Pacific affinity (Collins and Coates 1993). In addition, Peter Jung (personal communication 1993) has identified a strombinid, *Cotonopsis* sp., in the Chagres Formation, that is known dominantly from the Pacific. Furthermore, coeval deposits elsewhere in Panama and Costa Rica have a benthic foraminiferal fauna of clearly Caribbean affinity. The Chagres Formation may represent a major eustatic rise in sea level at about 5 Ma (Haq et al. 1987) that allowed Pacific outer-shelf benthic faunas to be introduced locally through the Canal Basin into the Caribbean (Collins and Coates 1993).

Chucunaque and Atrato Basins, Eastern Panama and Northwestern Colombia

The Atrato and Chucunaque Basins (fig. 2.7) form the central portion of the Choco Block, whose tectonic evolution and relationship to northwestern South America has been studied by several authors (Lonsdale and Klitgord 1978; Pennington 1981; Dengo 1985; Case et al. 1984; Howell et al. 1985; Duque-Caro 1990a). Although there are differences of opinion as to precise boundaries of the tectonostratigraphic units, there is general agreement that the Choco Block is an exotic terrane accreted to South America in the Neogene. It consists of three tectonostratigraphic units:

1. The Dabeiba Arch is a sequence of basalt, tuff, and granitoid plutons, associated with pelagic, turbidite, and conglomerate lithologies, that form elongate mountain ranges from San Blas to the northwesternmost Western Cordillera. In the Atrato region, these deposits involve extensive mélange zones (Duque-Caro 1990a) and form the northeasterly and easterly limits of the Choco Block. The Uramita Fault Zone on the eastern limit of the Dabeiba Arch in Colombia is the suture of the Choco exotic terrane to the South America Plate.

2. The Baudo Arch comprises the western margin of the Choco Block, where it forms the poorly known massifs of Mahe and Baudo, in eastern Panama and Colombia respectively, and appears to be very similar in origin and composition to the Dabeiba Arch.

3. The Chucunaque and Atrato Basins lie between the Dabeiba and Baudo Arches (fig. 2.2) and are distinguished by up to 10,000 m of sediment. From Oligocene to Middle Miocene, the deposits

are mostly pelagic and hemipelagic and from Middle Miocene to Pliocene they are hemipelagic to terrigenous. The upper 4000 m crops out at the surface; the rest is known only from subsurface exploration. Prior to the collision of the Choco Block with north-westernmost South America, no basins were defined in this region, and the sedimentary sequence consists of a deep-water cycle of pelagic sedimentation (Bandy 1970; Duque-Caro 1990b).

The Chucunaque Basin

The Chucunaque Basin (fig. 2.7) is less well known than the Atrato Basin but has been described in part (Office of Interoceanic Canal Studies 1966; Bandy 1970; Esso Exploration and Production 1971; Bandy and Casey 1973). The deep-water pelagic phase of sedimentation is represented by a series of massive, blue-black calcareous shale, rich in radiolarians and foraminifers, which are exposed in the easternmost part of Panama in both the Sambu Valley and the Chucunaque and Tuira Valleys (Terry 1956; Bandy 1970). The shale has been separated into two formations. The lower sequence, which may be Oligocene or Early Miocene, contains small thin-shelled bivalves and carbonaceous fragments and is known as the Arusa Formation. Although no type section has been defined, it presumably lies along the Arusa River, a southerly tributary of the Tuira River.

The Aquaqua Formation overlies the Arusa conformably and contains ridge-forming limestones. The unit is bituminous with asphaltic oil seeps, is distinguished by fairly frequent bentonite horizons, and contains an abundant and diverse benthic foraminiferal assemblage (Terry 1956). Northwestward, in the extreme west of the Chucunaque Basin in the region of Lake Bayano the deposits consist of sandstone containing sharks teeth and fish vertebrae (Stewart 1966). The Aquaqua Formation is apparently Early Miocene (Terry 1956; Bandy 1970). In general lithology and faunal assemblage, including the presence of radiolarians, as well as abyssal depth range, the Arusa and Aquaqua Formations are paleoecologically equivalent to the Uva Formation in the Atrato Valley (Duque-Caro 1990b) and are similar in thickness (about 1400 m). They indicate that open oceanic conditions at near abyssal depths prevailed throughout eastern Panama and western Colombia during the Early Miocene.

In the Chucunaque Basin, the overlying sediments comprise the Lara Group (Esso Exploration and Production 1971), consisting of four formations. The lower units are the Tapaliza Formation and the overlying Tuira Formation. Together these represent the "Lower Gatun" unit described by earlier authors (Terry 1956; Shelton 1952; Bandy 1970). However, they are different in both age and lithology from the Gatun Formation of the Canal Basin and the term "Lower Gatun" should not be used for these units.

The type locality of the Tapaliza Formation lies along the Tapaliza River, about 40 km southeast of Yaviza, where the section is about 1100 m thick. There is a distinctive basal conglomerate about 100 m thick in easternmost Panama, which thickens to 800 m in the region of the lower reaches of the Chucunaque River but thins again northwestward toward the western margin of the Chucunaque Basin in the area of Lake Bayano. Overlying the conglomerate, the formation consists of medium to coarse sandstone with small limestone lenses and concretionary shale (Terry 1956; Shelton 1952; Office of Interoceanic Canal Studies 1966), containing thick-shelled mollusks and plant remains in its type area and the lower Chucunaque Valley. Northwestward, in the region of Lake Bayano, the equivalent of the Tapaliza consists of a thick sequence of siltstone (Stewart 1966) of abyssal to lower bathyal origin (Bandy 1970). The Lake Bayano region thus corresponds paleoecologically to the Napipi Formation in the Atrato Basin (Duque-Caro 1990b) and represents the continuation of abyssal to lower bathyal conditions in the Middle Miocene. However, these two areas are separated by more shallow water facies in the lower Chucunaque Valley and easternmost Panama, where conglomeratic, thick-shelled mollusk-bearing deposits characterize the Tapaliza Formation.

The overlying Tuira Formation extends from its type locality in the Tuira River valley south of Yaviza through the lower Chucunaque Valley. It consists of a lower *Corbula*-bearing blue-gray foraminiferal concretionary sandy and partly conglomeratic shale, which is overlain conformably by distinctive concretionary shaly sandstone, containing numerous turritellids and plant remains, that has been separated as the Boca de Cupe member (Shelton 1952). The upper part of the formation is a blue-green cross-bedded sandstone rich in mollusks. The Tuira Formation correlates with the lower part of the middle bathyal Sierra Formation of the Atrato Basin (Duque-Caro 1990a) but appears to be much shallower in origin. As is the case for the Tapaliza Formation, the equivalent of the Tuira Formation in the western Chucunaque Basin in the region of Lake Bayano was also bathyal (Bandy 1970). The deeper-water deposits in the Lake Bayano form the lower portion of the Sabana beds (Bandy 1970).

The Tuira Formation is overlain by the Pucuro Formation, which is about 700–1000 m thick in the area of its type locality west of Pucuro near the junction of the Tuira and Tapaliza Rivers (Esso Exploration and Production 1971). In this region the formation consists of a lower sequence of black shale and sandstone, rich in mollusks, and an upper distinctive strongly ledge-forming sandy limestone and sandstone, rich in oysters and pectinids. To the northwest, the limestone unit rapidly thins and disappears. The Pucuro Formation is poorly constrained biostratigraphically but is probably latest Middle Miocene to Late Miocene.

The uppermost formation of the Lara Group is the Chucunaque Formation, whose type locality is vaguely defined as the central Chucunaque Valley north of the Chico River. The formation consists of about 5–700 m of indurated gray calcareous shale, rich in foraminifers, with interbedded sandstone and sandy limestone. It ranges from Late Miocene to Early Pliocene. The Pucuro and Chucunaque Formations together correlate with the upper portion of the Sabana beds (Bandy 1970). The Pucuro Formation again represents a more shallow-water facies than the Sabana beds but by the Pliocene the Sabana beds indicate shoaling to outer-shelf depths (Bandy 1970), so that the depth differential between the eastern and western Chucunaque Basin deposits has disappeared.

The Atrato Basin
The Atrato Basin (fig. 2.7) in Colombia, which runs into the Chucunaque Basin of eastern Panama (fig. 2.2), has been studied in detail by Duque-Caro (1990a,b). The pelagic cycle of deep-water sedimentation is represented by two formations (fig. 2.7). The oldest is the Oligocene to earliest Middle Miocene Uva Formation (Haffer 1967; Duque-Caro 1990b), which consists of white to gray radiolarian and foraminiferal limestone and mudstone, with inclusions of volcanic glass, and is approximately 1600 m thick. The fauna suggests open oceanic conditions in water depths greater than 2000 m. The Napipi Formation conformably overlies the Uva Formation, is about 550 m thick, and ranges in age from earliest Middle Miocene (zone N9) to Middle Miocene (zone 11). Consisting mostly of hemipelagic mudstone with minor lenticular limestone interbeds, the Napipi Formation indicates similar depth conditions as the Uva Formation but with a lower ratio of planktonic to benthic foraminifers, pyritization of planktonic foraminifers, and the common occurence of *Uvigerina* spp., all of which suggest strong influence of nutrient-rich waters during the deposition of the Napipi Formation (Duque-Caro 1990b).

After deposition of the Napipi Formation, an extensive marine unconformity occurs in the coastal sediments of both the Pacific and Caribbean regions of northwest South America (Weiss 1955; Blow 1969; Sigal 1969; Duque-Caro 1972, 1975, 1979, 1990b; Keller and Barron 1983; Keller 1986). Benthic foraminifers suggest that bottom-water circulation and sedimentation changed, and major tectonic disturbances coincide with this change in Colombia (Haffer 1970; Van Houten 1976; Duque-Caro 1984, 1990a).

These events suggest the beginning of collision of the Choco Block with South America, the emergence of the Dabeiba Arch, and the formation of inner borderland basins similar to those of southern California. Initial deposition in these borderland basins is represented by the Sierra Formation (Duque-Caro 1990b), which ranges in age from Middle Miocene (zone N13) to early Late Miocene (zone N17). The lithology is massive dark gray siltstone and silty mudstone, and the formation is about 800 m thick. The major changes in environment are a shallowing from middle to upper bathyal depths in an inner borderland, the development of conditions for the formation of glauconite, and the segregation of the benthic taxa into exclusively Pacific forms.

The conformably overlying Munguido Formation is about 1400 m thick and consists of calcareous mudstone, medium sandstone, and minor conglomerate with occasional mollusk and fish deposits that are pyritized and carbonaceous (Duque-Caro 1990b). There is a marked decrease in size of benthic foraminifers and in the diversity and abundance of planktonic foraminifers. The Munguido Formation ranges from early Late Miocene to Pliocene (3.4 Ma). The benthic foraminiferal assemblage at the bottom of the Munguido Formation indicates intermingling of the Caribbean and Pacific faunas and shallowing of the basin to less than 150 m, restricted circulation, and a very rich input of carbonaceous debris. By the top of the Munguido Formation, during the Early Pliocene, deposition was in water less than 50 m, with anoxic bottom. Overlying the Munguido with a marked unconformity are terrestrial deposits correlated with the Late Pliocene to Pleistocene Sincelejian stage in northwest Colombia (Duque-Caro 1984).

In summary, there was marked similarity in sedimentation patterns between the Atrato Basin and the western Chucunaque Basin, in that pelagic abyssal to bathyal sediments continued from the Late Oligocene until near the end of the Miocene. However, from the Middle Miocene, the Atrato Basin shows evidence of restriction into inner borderland basins with anoxic bottoms (Duque-Caro 1990), whereas the western Chucunaque Basin suggests continuously open oceanic conditions that are steadily shallowing. Both regions shallow rapidly in the latest Miocene–Early Pliocene. In contrast, from the Middle Miocene onward, the deposits of the eastern Chucunaque Basin, in the area of the lower Chucunaque and Tuira river valleys, indicate shallowing to perhaps inner- to middle-shelf depths, with possible deltaic deposits locally. This strongly suggests there was an active tectonic axis separating the Atrato and Chucunaque Basins that perhaps delimited the northern extent of the inner borderland basins.

NEOGENE PALEOGEOGRAPHY

From the Late Cretaceous until the Middle Miocene, volcanism, accretion, compression, and uplift along the Central American arc were the primary factors controlling sedimentary basins within the Chorotega Block. Broad marine corridors ran across the arc connecting the primary fore-arc basins, such as Tempisque, Nicoya, and Terraba with their back-arc complements, Northern and Southern Limón (Astorga et al. 1991). During the Late Cretaceous and Eocene, a volcanic arc may have extended as far as Colombia. Maury and others (1955) record typical ocean-arc volcanic rocks from several localities along the San Blas massif, which K-Ar radiometric dates indicate are all Eocene. They suggest this timemarked the separation of the Caribbean Plate from the Pacific Farallon Plate. However, since that time, active volcanism does not appear to have extended further east than western Panama (de Boer et al. 1988), and until the Middle Miocene there was an open ocean of abyssal depth, connecting what is now the Canal Basin in central Panama, and the Western Cordillera of northwestern South America (fig. 2.8). This extensive bathyal oceanic platform is characterized by foraminiferal mudstone deposited in water deeper than 2000 m.

At this time, the northwestern boundary of South America was an active convergent margin following the present San Jacinto Belt, an exotic terrane accreted to South America in the Eocene (Duque-Caro 1979, 1984). In the Middle Miocene, the Panama Arc began to be compressed by the South American Plate, the Choco Block was formed, and subduction then followed the line of the Panama Arc. Given the unusually thick nature of the Caribbean crust (Case et al. 1990), it may have been relatively hard to subduct, resulting in repeated backstepping of the suture. Thus, the suture moved to the Dabeiba Arch, beginning the current rise of the San Blas–Darién massif and the constriction of circu-

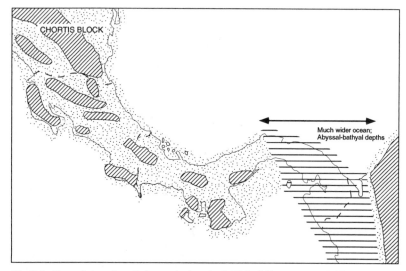

Fig. 2.8 Central American Isthmus during the Middle Miocene, 16–15 Ma. Emergent land is represented by oblique parallel lines, shelf sediments by dots, and abyssal oceanic sediments by horizontal parallel lines.

lation between the Caribbean and the Pacific. Subsequently, the suture may have jumped again to the line of the Mahe-Baudo massifs and then to the present Colombian Trench (fig. 2.2).

These processes formed elongate borderland basins where the arc abutted the continental South American Plate, namely, in the Atrato Basin, but did not do so where the arc was entirely oceanic, as in the Chucunaque Basin, northwest of the Gulf of Uraba (fig. 2.2). In both regions, water depth shallowed steadily from the Middle Miocene and the basins were filled and uplifted by the Middle Pliocene. Between these basins, in easternmost Panama, shallow-water sediments were developed from early in the Miocene. Thus, the accretion, growth, and uplift of the Choco Block is due to the compressional tectonics of collision between the Panama Arc and South America since the Middle Miocene. The process is apparently continuing today as evidenced by the fact that the area is out of isostatic equilibrium owing to an excess of mass (Case et al. 1990). The development of northwest-striking sinistral strike-slip faults as far as the Soná and Azuero Peninsulas (Kolarsky et al. 1995) suggests that the zone affected by collisional tectonics and uplift since the Middle Miocene extends to the eastern part of the Chorotega Block (fig. 2.3).

During the Late Pliocene, subduction of the Cocos Ridge strongly affected the Chorotega Block, except in its northwest portion between the Costa Rica Transcurrent Fault System and the Chortis Block. Here, there is a slight gap and offset of the volcanic chain, and the Chorotega Block has been subject to mainly extensional tectonics as evidenced by the seismic profiles of the San Carlos rift basin and the normal-faulted, passive-margin Northern Limón Basin (Astorga et al. 1991). These movements have been explained by renewed strike-slip motion along the extension of the Hess Escarpment (fig. 2.1) into the Santa Elena Peninsula (Astorga 1992). The topography of this zone today does not rise above 34 m, and is mostly constructed of Quaternary volcanic flows and lahars that form part of the chain stretching through Nicaragua and El Salvador. Extensive subsurface marine Pliocene and Pleistocene deposits in the northern San Carlos Basin reach to within 25–30 km of the Pacific. The combination of the Late Neogene stratigraphy, extensional tectonics, and present topography strongly suggests that the northern Tempisque, San Carlos, and Northern Limón Basins provided an elongate marine connection from the Pacific Ocean to the Caribbean well into the Pliocene (figs. 2.8, 2.9, 2.10).

From the CRTFS southeastward, however, the Chorotega Block is strongly influenced after the Miocene by the subduction of the Cocos Ridge. The zone structurally affected by the Cocos Ridge is outlined in figure 2.3. Marine corridors via the Central Valley and the Terraba Basin were in connection with the Southern Limón Basin until the Late Miocene. However, the volcanic, terrestrial nature of the Pliocene in the Terraba Basin (Paso Real Formation) and the rapid growth of the Rio Banano delta at the end of the Early Pliocene suggest that uplift of the Talamanca Cordillera had severed marine connections by the beginning of the Pliocene at the latest.

The general pattern that emerges is that up to the Middle Miocene frequent Pacific to Caribbean marine connections existed through the Central American arc (Chorotega Block) as well as a wide abyssal oceanic connection between the end of the arc and the Western Cordillera of northwestern South America (fig. 2.8). By the Late Miocene, an extended archipelago with restricted interconnections between multiple basins stretched from the Chortis Block to the Atrato Valley (fig. 2.9). At the beginning of the Pliocene, marine corridors disappeared in that portion of the Chorotega Block affected by Cocos Ridge subduction. No marine sediments are known after the beginning of the Pliocene in this region. Thus, marine connections across the Central American

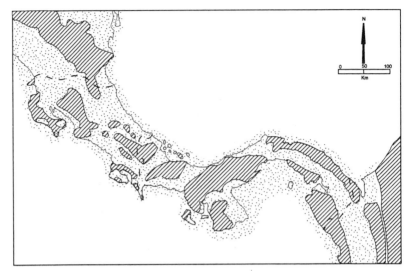

Fig. 2.9 Central American Isthmus during the Late Miocene, pre–Cocos Ridge configuration, 7–6 Ma. Symbols as for figure 2.8.

Isthmus were then restricted to the extreme western (Tempisque and San Carlos Basins) and the eastern (Canal, Chucunaque, Atrato Basins) margins of the Costa Rica–Panama microplate (fig. 2.10).

Timing of Final Closure

The age of the top of the youngest marine unit of the Atrato Basin, the Munguido Formation, is 3.4 Ma on the basis of the last appearance datum (LAD) of *Globorotalia margaritae margarita* (Duque-Caro 1990a). Until this time, however, 23 species of benthic foraminifers in the Munguido Formation (Duque-Caro 1990b, fig. 6) are common to the Borbón and Manabí Basins of Ecuador, the Gatun Formation of the Canal Basin, Panama, and the Tubara and Cuesta Formations of the Caribbean Colombian coast. These broad distributions indicate that there was a marine connection between the Caribbean and the Pacific as late as 3.4 Ma. According to Duque-Caro (1990b), this marine connection was severed at 3.1 Ma.

Collins and Coates (1993) have noted that 50% of the benthic foraminifers in the probably Early Pliocene (5.3–3.4 Ma) Chagres Formation on the Caribbean Panama coast are only known otherwise from the Pacific. They do not occur in the coeval Bocas del Toro or Limón Group sediments to the west. One otherwise dominantly Pacific strombinid genus, *Cotonopsis* sp. (Peter Jung, personal communication 1993),

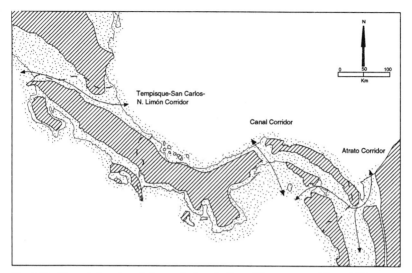

Fig. 2.10 Central American Isthmus during the Late Pliocene, post–Cocos Ridge configuration, about 3 Ma. Symbols as for figure 2.8.

also occurs in the Chagres Formation. This local distribution of Pacific outer-shelf benthic foraminifers coincides with the similarly localized clastic wedge of massive sand and *Balanus* coquina that defines the Chagres Formation lithologically, and may represent a deepening and flushing of the Canal Basin from the Pacific to the Caribbean during the 5 Ma high stand recorded by Haq and others (1987) and estimated to be 115 m.

On the other hand, analysis of Pacific coast molluscan faunas originally described by Olsson (1922, 1942) and redefined biostratigraphically (Coates et al. 1992) shows that none of 137 later Pliocene species recognized from the Peñita and Burica Formations (more than 3.5 to less than 1.9–1.7 Ma) of the Burica Peninsula occur in any Caribbean formation. At present there are no adequately dated Late Miocene to Early Pliocene molluscan faunas on the Pacific coast to compare to the Gatun fauna. The above faunal patterns strongly confirm the conclusions of Keigwin (1978, 1982) and Duque-Caro (1990b) that the first complete closure of the Isthmus of Panama occurred in the Middle Pliocene at around 3.5–3.1 Ma.

However, it is also clear that the rising isthmus became an ecological barrier much earlier. Deeper-water organisms should have been affected first, a pattern apparently confirmed by alpheid shrimp whose deeper-water cognate pairs are genetically, behaviorally, and morphologically more different than the more shallow-water forms (Knowlton et al. 1993). Thus, in the Middle Miocene, schematically reconstructed in figure 2.8, broad marine connections existed across much of the Chorotega Block and a wide bathyal open-ocean passage existed between the volcanic arc and South America. During the Late Miocene, the growth of the arc and the beginning of collision with South America formed a large and elongated archipelago of increasingly isolated and sedimentologically different basins. Figures 2.9 and 2.10 show that during the Late Miocene and Pliocene there was markedly decreased connection between the Pacific and Caribbean and increased geographic isolation of the sedimentary basins. Increasing physical heterogeneity along as well as across the isthmus suggests that major opportunities for increased speciation would have existed during the Late Miocene and earliest Pliocene. Increasing uplift in the Late Pliocene, due to Cocos Ridge subduction, steadily reduced transisthmian marine connections, and inverted sedimentary basins in the central portion of the CRP microplate left only narrow connections, possibly via the Tempisque, San Carlos, Northern Limón Basins, in the west, and the Canal and Atrato Basins in the east (fig. 2.10). Marine environmental heterogeneity should have decreased with the conversion of the archipelago into a subcontinuous isthmus. The differences between the Pacific and Caribbean marine systems would have become more clearly defined and stabilized by the end of the Pliocene, especially after the isthmus evolved tectonically enough to become a complete barrier at high stands of the eustatic cycles.

REFERENCES

Aguilar, A., and J. A. Fernández. 1989. Convenio de Asistencia Técnica RECOPE-PCIAC. *Rev. Geol. Am. Central* 10:95–99.

Astorga, A. 1992. Descubrimiento de Corteza Oceánica Mesozoica en el norte de Costa Rica y el sur de Nicaragua. *Rev. Geol. Am. Central* 14:109–12.

Astorga, A., J. A. Fernández, G. Barboza, L. Campos, J. A. Obando, A. Aguilar, and L. G. Obando. 1991. Cuencas sedimentarias de Costa Rica: Evolución geodinámica y potencial de hidrocarburos. *Rev. Geol. Am. Central* 13:25–59.

Bandy, O. L. 1970. Upper Cretaceous Cenozoic paleobathymetric cycles, eastern Panama and northern Colombia. *Trans. Gulf Coast Assoc. Geol. Soc.* 20: 181–93.

Bandy, O. L., and R. E. Casey. 1973. Reflector horizons and paleobathymetric history, eastern Panama. *Geol. Soc. Am. Bull.* 84:3081–86.

Baumgartner, P. O., J. A. Obando, C. Mora, and A. Steck. 1989. Paleogene accretion and suspect terranes in southern Costa Rica (Osa, Burica, Central America). *Transactions of the Twelfth Caribbean Geological Conference* St. Croix. ed. D. K. Larue and G. Draper, 529.

Blow, W. A. 1969. Late Middle Eocene to Recent planktonic foraminiferal biostratigraphy. In *Proceedings of the First International Conference on Planktonic Microfossils, Geneva*, ed. P. Bronnimann and H. H. Renz, 1:199–421. Leiden: E. J. Brill.

Bowland, C. L. 1984. Seismic stratigraphy and structure of the western Colombian Basin, Caribbean Sea. M.S. thesis, University of Texas, Austin, Texas.

———. 1993. Depositional history of the western Colombian Basin, Caribbean Sea, revealed by seismic stratigraphy. *Geol. Soc. Am. Bull.* 105:1321–45.

Bowland, C. L., and E. Rosencrantz. 1988. Upper crustal structure of the western Colombian Basin, Caribbean Sea. *Geol. Soc. Am. Bull.* 100:534–46.

Burbach, G. V., C. Frolich, W. D. Pennington, and T. Matsumoto. 1984. Seismicity and tectonics of the subducted Cocos Plate. *J. Geophys. Res.* 89: 7719–35.

Carballo, M., and R. Fischer. 1978. La Formación San Miguel. *Inst. Geo. Nac. Info. Semestr.* (January–June):48–144.

Case, J. E., and T. L. Holcombe. 1980. *Geologic-tectonic map of the Caribbean Region.* U.S. Geological Survey Miscellaneous Investigations Series, map I-1100, scale 1:250,000. Reston, Va.

Case, J. E., T. L. Holcombe, and R. G. Martin. 1984. Map of geologic provinces in the Caribbean Region. In *The Caribbean–South America Plate boundary and regional tectonics*, ed. W. E. Bonini, R. B. Hargraves, and R. Shagam, 1–30. Geological Society of America Memoir 162. Boulder, Colo.

Case, J. E., W. D. MacDonald, and P. J. Fox. 1990. Caribbean crustal provinces: Seismic and gravity evidence. In The Caribbean Region, Vol. H of *The Geology of North America*, ed. G. Deugo and J. E. Case. Boulder, Colo.: Geological Society of America.

Cassell, D. T., and B. Sen Gupta. 1989a. Foraminiferal biostratigraphy and paleoenvironments of the Tertiary Uscari Formation, Limón Basin, Costa Rica, *J. Foram. Res.* 19(1):2–71.

———. 1989b. Pliocene Foraminifera and environments, Limón Basin of Costa Rica. *J. Paleontol.* 63:146–57.

Castillo, R. 1969. *Geología de Costa Rica: Una sinopsis.* San José: Editorial de la Universidad de Costa Rica.

Coates, A. G., J. B. C. Jackson, L. S. Collins, T. M. Cronin, H. J. Dowsett, L. M. Bybell, P. Jung, and J. A. Obando. 1992. Closure of the Isthmus of Panama: The near-shore marine record of Costa Rica and Panama. *Geol. Soc. Am. Bull.* 104:814–28.

Collins, L. C., and A. G. Coates. 1993. Marine paleobiogeography of Caribbean Panama: Last Pacific influences before closure of the Tropical American Seaway. *Geol. Soc. Am. Ann. Meeting, Abstr. Progr.* 25:A-428.

Collins, L. C., A. G. Coates, and J. A. Obando. 1995. Timing and rates of emergence of the Limón and Bocas del Toro Basins. Caribbean effects of Cocos subduction? In *Geologic and tectonic development of the Caribbean Plate boundary in southern Central America*, ed. P. Mann., 263–89. Geological Society of America Special Paper 295. Boulder, Colo.

Corrigan, J. D., P. Mann, and J. C. Ingle, Jr. 1990. Forearc response to subduction of the Cocos Ridge, Panama–Costa Rica. *Geol. Soc. Am. Bull.* 102:628–52.

Crowe, J. C., and R. T. Buffler. 1984. *Regional seismic reflection profiles across the Middle America Trench and convergent margin of Costa Rica.* University of Texas Institute of Geophysics Technical Report, no. 537. Austin, Texas.

De Boer, J. Z., M. J. Defant, R. H. Stewart, J. F. Restrepo, L. F. Clark, and A. H. Ramírez. 1988. Quaternary calc-alkaline volcanism and implication for the plate tectonic framework. *J. South Am. Earth Sci.* 1(3): 275–93.

Dengo, G. 1962. Tectonic-igneous sequence in Costa Rica. In *Petrologic studies: A volume to honor A. F. Buddington,* ed. A. E. J. Engel, H. L. James, and B. F. Edwards, 133–61. Boulder, Colo.: Geological Society of America.

———. 1985. Mid America: Tectonic setting for the Pacific margin from southern Mexico to northwestern Colombia. In *The ocean basin and margins,* ed. A. E. M. Nairn and F. G. Stehli, 7. New York: Plenum Press.

Donnelly, T. W. 1992. Geological setting and tectonic history of Mesoamerica. In *Insects of Panama and Mesoamerica: Selected studies,* ed. D. Quintero and A. Aiello, 1–13. Oxford: Oxford University Press.

Duque-Caro, H. 1972. Relaciones en la bioestratigrafía y la cronoestratigrafía en el llamado geosinclinal de Bolívar. *Ingeominas Bol. Geol.* 19(3): 25–68.

———. 1975. Los foraminíferos planctónicos y el Terciario de Colombia. *Rev. Esp. Micropaleontol.* 7(3): 403–27.

———. 1979. Major structural elements and evolution of northwestern Colombia. In *Geological and geophysical investigations of continental margins,* ed. J. S. Watkins, L. Montadert,and P. W. Dickerson, 329–51. American Association of Petroleum Geologists Memoir 29. Tulsa, Okla.

———. 1984. Structural style, diapirism, and accretionary episodes of the Sinu-San Jacinto terrane, southwestern Caribbean borderland. In *The South American–Caribbean Plate boundary and regional tectonics,* ed. W. E. Bonini, R. B. Hargraves, and R. Shagam, 303–16. Geological Society of America Memoir 162. Boulder, Colo.

———. 1990a. The Choco Block in the northwestern corner of South America: Structural, tectonostratigraphic, and paleogeographic implications. *J. South Am. Earth Sci.* 3(1): 71–84.

———. 1990b. Neogene stratigraphy, paleoceanography, and paleobiology in northwest South America and the evolution of the Panama Seaway. *Palaeogeogr., Palaeoclimatol., Palaeoecol.* 77:203–34.

Escalante, G. 1990. The geology of southern Central America and western Colombia. In *The Caribbean Region,* vol. H of *The geology of North America,* ed. G. Dengo and J. E. Case, 201–30. Boulder, Colo.: Geological Society of America.

Esso Exploration and Production, 1971. Final report: The exploration results related to the concession under contract no. 59, Darién Basin, Republic of Panama. Typescript.

Fischer, R. 1981a. El desarollo paleogeográfico del Mioceno de Costa Rica. *Anales II Cong. Lat.–Amer. Paleontol.* 2:565–79.

———. 1981b. *Die Herausformung des mittelamerikanischen Isthmus im Moizän Costa Rica.* 1(3–4): 210–21. Stuttgart: Ibe. Geologische Paläontologisches.

Fischer, R., and J. Franco. 1979. La Formación Coris, Mioceno, Valle Central, Costa Rica. *Inst. Geo. Nac., Info. Semestr.* (January–June): 15–17.

Franco, J. 1978. La Formación Coris (Mioceno, Valle Central, Costa Rica). B.S. thesis, Escuela Centroamericana de Geología, Universidad de Costa Rica, San José, Costa Rica.

Guendel, F. D. 1986. Seismotectonics of Costa Rica. An analytical view of the southern terminus of the Middle America Trench. Ph.D. diss., University of California, Santa Cruz, Calif.

Haffer, J. 1967. On the geology of the Uraba and northern Choco regions, northwestern Colombia. Ecopetrol Open-file Report 809(GR-357). Bogotá, Colombia.

―――. 1970. Geologic climatic history and zoogeographic significance of the Uraba region in northwestern Colombia. *Caldasia* 10(50): 603–36.

Haq, B. U., J. Hardenbol, and P. R. Vail. 1987. Chronology of fluctuating sea levels since the Triassic. *Science* 235:1156–67.

Howell, D. G., D. L. Jones, and E. R. Schermer. 1985. Tectonostratigraphic terranes of the Circum-Pacific Region. In *Tectonostratigraphic terranes of the Circum-Pacific Region*, ed. D. G. Howell, 1:3–30. American Association of Petroleum Geologists, Earth Science Series. Tulsa, Okla.

Jones, D. S., and P. F. Hasson. 1985. History and development of the marine invertebrate faunas separated by the Central American Isthmus. In *The Great American Biotic Interchange*, ed. F. G. Stehli and S. D. Webb, 325–55. New York: Plenum Press.

Keigwin, L. D., Jr. 1978. Pliocene closing of the Isthmus of Panama, based on biostratigraphic evidence from nearby Pacific Ocean and Caribbean Sea cores. *Geology* 6:630–34.

―――. 1982. Isotopic paleoceanography of the Caribbean and east Pacific: Role of Panama uplift Late Neogene time. *Science* 217:350–52.

Keller, G. 1986. Paleoceanographic implications of Middle Eocene to Pliocene deep-sea hiatuses. Paper presented at the Second International Conference on Paleoceanography (SICP), Woods Hole, Mass.

Keller, G., and J. A. Barron. 1983. Paleoceanographic implications of Miocene deep-sea hiatuses. *Geol. Soc. Am. Bull.* 94:590–613.

Knowlton, N., L. A. Weigt, L. A. Solorzano, D. K. Mills, and E. Bermingham, 1993. The divergence in protein, mitochondrial DNA, and reproductive compatibility across the Isthmus of Panama. *Science.* 260:1629–32.

Kolarsky, R. A., and P. Mann. 1995a. East Panama Deformed Belt: Structure, age, and neotectonic significance. In *geologic and tectonic development of the Caribbean Plate boundary in southern Central America*, ed. P. Mann, 111–30. Geological Society of America Special Paper 265. Boulder, Colo.

―――. 1995b. Structure and neotectonics of an oblique-subduction margin, southwestern Panama. In *Geologic and tectonic development of the Caribbean Plate boundary in southern Central America*, ed. P. Mann, 131–57. Geological Society of America Special Paper 265. Boulder, Colo.

Kolarsky, R. A., P. Mann, and W. Montero. 1995. Island arc response to shallow subduction of the Cocos Ridge, Costa Rica. In *Geologic and tectonic development of the Caribbean Plate boundary in southern Central America*, ed. P. Mann, 235–62. Geological Society of America Special Paper 265. Boulder, Colo.

Lonsdale, P., and K. D. Klitgord. 1978. Structure and tectonic history of the eastern Panama Basin. *Geol. Soc. Am. Bull.* 89:981–99.

Lowery, B. J. 1982. Sedimentology and tectonic implications of the Middle to Upper Miocene Curre Formation, southwestern Costa Rica. M.S. thesis, Louisiana State University, Baton Rouge, La.

Mann, P., and K. Burke. 1984. Neotectonics of the Caribbean. *Rev. Geophys. and Space Phys.* 22:309–62.

Mann, P., and J. Corrigan. 1990. Model for Late Neogene deformation in Panama. *Geology* 18:558–62.

Mann, P., C. Schubert, and K. Burke. 1990. Review of Caribbean neotectonics. In *The Caribbean Region*, vol. H of *The Geology of North America*, ed. G. Dengo and J. E. Case, 307–38. Boulder, Colo.: Geological Society of America.

Maury, R. C., M. J. Defant, H. Bellon, J. Z. de Boer, R. H. Stewart, and J. Cotton. 1995. Early Tertiary arc volcanics from eastern Panama. In *Geologic and tectonic development of the Caribbean Plate boundary in southern Central America*, ed. P. Mann, 29–34. Geological Society of America Special Paper 265. Boulder, Colo.

McBirney, A. R., and H. Williams. 1965. Volcanic history of Nicaragua. *Univ. Calif. Pub. Geol. Sci.* 55:1–65.

McCann, W. R., and W. D. Pennington. 1990. Seismicity, large earthquakes, and the margin of the Caribbean Plate. In *The Caribbean Region*, vol. H of *The geology of North America*, ed. G. Dengo and J. E. Case, 291–306. Boulder, Colo.: Geological Society of America.

Montero, P. W. 1974. Estratigrafía del Cenozoico del area de Turrucares, Provincia de Alajuela, Costa Rica. B.S. thesis, Escuela Centroamericana de Geología, Universidad de Costa Rica, San José, Costa Rica.

Mora, C. 1979. Estudio geológico de una parte de la región sureste del Valle del General, Provincia de Puntarenas, Costa Rica. M.S. thesis, Escuela Centroamericana de Geología, Universidad de Costa Rica, San José, Costa Rica.

Mora, C., P. O. Baumgartner, and L. Hottinger. 1989. Eocene shallow-water carbonate facies with larger Foraminifera in the Caño accretionary complex, Caño Island and Osa Peninsula, Costa Rica, Central America. Paper presented at the Twelfth Caribbean Geological Conference, St. Croix, U.S. Virgin Islands (abstr. 12).

Obando, J. A. 1986. Sedimentología y tectónica del Cretácico y Paleogeno de la Región de Golfito, Península de Burica y Península de Osa, Provincia de Puntarenas. M.S. thesis, Escuela Centroamericana de Geología, Universidad de Costa Rica, San José, Costa Rica.

Obando, J. A., and P. O. Baumgartner. 1986. Estratigrafía y tectónica de la cuenca de Golfo Dulce y Charco Azul. *J. Geol. C.R.* 2:15–16.

Obando, J. A., P. O. Baumgartner, G. Bottazi, and G. Gonzalez. 1990. Sedimentación y tectónica del Paleógeno Tardío y el Neógeno en La Cuenca del Golfo Dulce, Pacífico Sur de Costa Rica. Paper presented at the Seventh Geological Congress of Central America, San José, Costa Rica (abstr. 119, 19–23).

Obando, L. 1986. Estratigrafía de la Formación Venado y rocas sobreyacientes (Mioceno-Reciente) Provincia de Alajuela, Costa Rica. *Rev. Geol. Am. Central* 5:73–104.

Obando, L., G. Bottazi, and F. Alvarado. 1991. Sedimentología de algunas facies de la Formación Coris (Mioceno Medio, Mioceno Superior), Valle Central, Costa Rica, América Central. *Rev. Geol. Am. Central* 13:61–71.

Office of Interoceanic Canal Studies. 1966. Preliminary geologic report, route 17. Memorandum FD-5. Panama City, Report of Panama.

Olsson, A. A. 1922. The Miocene of northern Costa Rica with notes on its general stratigraphic relations. *Bulls. Am. Paleontol.* 9:181–92.

———. 1942. Tertiary and Quaternary fossils of the Burica Peninsula of Panama and Costa Rica. *Bulls. Am. Paleontol.* 27:1–106.

Pennington, W. D. 1981. Subduction of the eastern Panama Basin and seismotectonics of northwestern South America. *J. Geophys. Res.* 86:10753–70.

Phillips, P. J. 1983. Stratigraphy, sedimentology, and petrographic evolution of Tertiary sediments in southwestern Costa Rica. M.S. thesis, Louisiana State University, Baton Rouge, La.

Pindell, J. L., and J. F. Dewey. 1982. Permo-Triassic reconstruction of western Pangea and the evolution of the Gulf of Mexico/Caribbean Region. *Tectonics* 1:179–212.

Sen Gupta, B. K., L. R. Malavassi, and E. Malavassi. 1986. Late Miocene shore in northern Costa Rica: Benthic foraminiferal record. *Geology* 14:218–20.

Seyfried, H., A. Astorga, and C. Calvo. 1987. Sequence stratigraphy of deep and shallow water deposits from an evolving island arc: The Upper Cretaceous and Tertiary of Central America. *Facies* 17:203–14.

Shelton, B. J. 1952. Geology and petroleum prospects of Darién, south-eastern Panama. M.S. thesis, Oregon State University, Corvallis, Oreg.

Sigal, J. 1969. Quelques acquisitions récentes concernant la chronostratigraphie des formations sédimentaires de l'Équateur. *Rev. Esp. Micropaleontol.* 1: 205–36.

Silver, E. A., D. L. Reed, J. E. Tagudin, and D. J. Heil. 1990. Implications of the North and South Panama Deformed Belts for the origin of the Panama orocline. *Tectonics* 9:261–82.

Stehli, F. E., and S. D. Webb, eds. 1985. *The Great American Biotic Interchange.* New York: Plenum Press.

Stewart, R. H. 1966. The Rio Bayano Basin: A geological report. Memorandum PCC-4, Office of Interoceanic Canal Studies, Panama City, Republic of Panama.

Stewart, R. H., and J. L. Stewart. 1980. *Geologic map of the Panama Canal and vicinity, Republic of Panama.* U.S. Geological Survey Miscellaneous Investigations Series, map 1-1232. Reston, Va.

Terry, R. A. 1956. *A geological reconnaissance of Panama.* California Academy of Sciences Occasional Papers, no. 23. San Francisco.

Van Houten, F. B. 1976. Late Cenozoic volcaniclastic deposits, Andean Foredeep, Colombia. *Geol. Soc. Am. Bull.* 87:481–95.

Wadge, G., and K. Burke. 1983. Neogene Caribbean Plate rotation and associated Central American tectonic evolution. *Tectonics* 2:633–43.

Weiss, L. 1955. Planktonic index foraminifers of north-western Peru. *Micropaleontology* 1:301–19.

Woodring, W. P. 1957. Geology and paleontology of the Canal Zone and adjoining parts of Panama. *U.S. Geol. Surv. Prof. Pap.* 306-A:145.

———. 1970. Geology and paleontology of the Canal Zone and adjoining parts of Panama. Description of Tertiary mollusks. *U.S. Geol. Sur. Prof. Pap.* 306-D: 299.

———. 1982. Geology and Paleontology of the Canal Zone and adjoining parts of Panama. *U.S. Geol. Surv. Prof. Pap.* 306-F:542–845.

Woodring, W. P., and V. Malavassi. 1961. Miocene Foraminifera, mollusks, and a barnacle from the Valle Central, Costa Rica. *J. Paleontol.* 35:489–97.

Yuan, P. B. 1984. Stratigraphy, sedimentology, and geologic evolution of the eastern Terraba Trough, South-western Costa Rica. Ph.D. diss., Louisiana State University, Baton Rouge, La.

3

Graphic Correlation of Marine Deposits from the Central American Isthmus: Implications for Late Neogene Paleoceanography

Harry J. Dowsett and Mathew A. Cotton

INTRODUCTION

The Late Neogene stratigraphy in Panama and Costa Rica preserves an extensive record of the marine biologic and climatic changes associated with the rise and closure of the Central American Isthmus. Creating a high-resolution temporal framework, within which stratigraphic sections found on the isthmus can be interpreted, is fundamental to our understanding of the history and importance of these events.

Ironically, the planktonic zonations that have been the basis for refined temporal frameworks over the past few decades are now the primary stumbling block for future paleoceanographic work in ocean margin settings because of the implicit assumption of synchrony of fossil first and last occurrences.

In this chapter we point out the problems associated with these conventional methods of biochronology and discuss the advantages associated with graphic correlation (GC). We first describe the graphic correlation technique and summarize a GC model for the Late Neogene, and then apply it to specific sequences from the Caribbean coast of Panama.

CENTRAL AMERICAN BIOCHRONOLOGY

The Central American Isthmus contains rich marine macro- and microfaunas and floras on both the Pacific and Atlantic margins. Planktonic foraminifers and calcareous nannofossils are common and often well preserved (Coates et al. 1992). Preliminary analysis of these fossils reveals a rich sedimentary record spanning the Late Miocene to Pleistocene. Multivariate statistical analyses of these assemblages allow detailed interpretations of paleoenvironments from a region central to some of the most important climatic questions of the Late Neogene. Specifically, Dowsett and others (1992) argue for a significant increase in meridional ocean heat flux, relative to the modern flux in the Atlantic

Ocean, during the middle part of the Pliocene, with no sea-surface temperature change in the tropics. Developing additional data from shallow marine and terrestrial habitats of Central America is crucial to understanding the effects and controls of this phenomenon (see also Cronin and Dowsett, this volume, chap. 4), as well as the regional biological consequences of the development of the isthmus and associated environmental change (Jackson et al., this volume, chap. 9). All of these investigations require better chronostratigraphic control.

Calcareous microfossils from the Central American Isthmus region have been identified and studied by numerous investigators (Bandy 1970; Berrangé 1989; Cassell and Sen Gupta 1989; Corrigan et al. 1990; Cushman 1918; Duque-Caro 1990; Pizarro 1987). Until now, most of this work has involved age determinations based on correlation to conventional zonal schemes and calibration of portions of those zonations in areas remote from the Central American Isthmus. Although age determinations have been supported by radiometric control and paleomagnetic analyses in some areas, this is not the general case.

BIOSTRATIGRAPHIC METHODS

Calcareous Microfossil Zonations

Planktonic foraminifers and calcareous nannoplankton have been studied intensely over the past few decades and a number of zonations have been erected for different regions of the globe (fig. 3.1). These interval zonations, based upon the first or last occurrences (FOs and LOs) of individual taxa, can be used to subdivide geologic ages and provide first-order control for stratigraphic correlation. A biozone, by definition, conveys no concept of time (Bates and Jackson 1980), although many authors conceptualize interval zones as discrete intervals of time. The confusion is compounded by the calibration of many planktonic events to radiometric time scales (Backman and Shackleton 1983; Berggren et al. 1985). These calibrated events provide excellent regional time control (fig. 3.2). For example, calibration of biostratigraphic events (FOs and LOs) from paleomagnetic reversal stratigraphy in one or two deep-sea cores from a single basin can serve to calibrate all sites within that region, assuming the latitudinal spread of sites is not great.

Biostratigraphic zonation has been extremely useful in studies of long time periods (e.g., Cenozoic) and small areas (e.g., within single basins), where it provides an appropriate tool for constructing a framework within which one can solve geologic problems. However, for refined temporal correlation over more regional geographic scales, the

ZONE / AGE	Bolli (1970) Bolli & Bermudez (1965) Bolli & Premoli Silva (1973)	Banner & Blow (1965) Blow (1969)	Parker (1973)	Berggren (1973)	Jenkins (1975)	Kennet (1973)	Okada & Bukry (1980)	Martini (1971)
PLEI.	G. truncatulinoides	N22	V		G. truncatulinoides	G. truncatulinoides — - G. tosaensis	CN13a	NN19
PLIOCENE L	G. tosaensis	N21	IV	PL6	G. inflata	G. tosaensis	CN12d	NN18
							CN12c	NN17
PLIOCENE M	G. miocenica	N20	III	PL5	G. inflata	G. inflata	CN12b	NN16
							CN12a	
				PL4		G. crassaformis	CN11	NN15
				PL3				
PLIOCENE E	G. margaritae	N19	II	PL2	G. puncticulata	G. puncticulata	CN10c	NN14
								NN13
				PL1			CN10b	
		N18	I			G. margaritae	CN10a	NN12
Mio.	G. humerosa	N17			G. conomiozea	G. conomiozea		

Fig. 3.1 Correlation of major planktonic foraminifer and calcareous nannoplankton zonations for the Late Neogene.

fossil record is commonly diachronous, as has been shown by several authors for marine microfossils (Dowsett 1989a,b; Johnson et al. 1989; Hills and Thierstein 1989). Figure 3.3 shows a hypothetical example of a fossil first occurrence surface with a geographic center of origin and temporal first appearance datum. Later migration, delayed in some regions, creates a complex migration surface. Temporal calibration of this event in one region means very little in another. For example, Dowsett (1989a) found most Pliocene calcareous planktonic events to be diachronous by more than 0.2 my. This type of diachrony results in relatively low-resolution correlations, especially when paleoceanographic studies are analyzing sequences on Milankovitch time scales.

This is certainly the case in the Late Neogene of the Central American region. For example, many of the deposits analyzed by Coates and others (1992) can be placed in planktonic foraminiferal zone N19 (Early Pliocene). In order to answer detailed paleobiologic and paleoclimatic questions, more precise correlations between these sections are required, along with some indication of the duration of sedimentation represented.

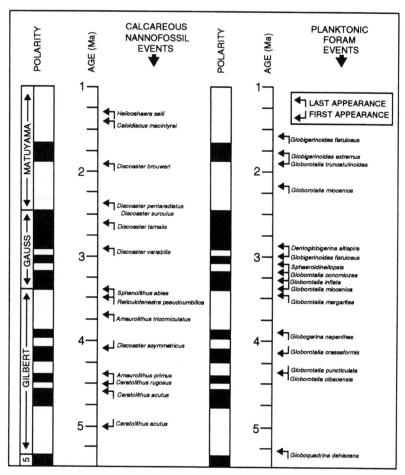

Fig. 3.2 Planktonic foraminifer and calcareous nannoplankton biochronologic events calibrated to the magnetic polarity time scale of Berggren et al. (1985). The integration of nannoplankton and foraminifers provides at least twenty-six events for Pliocene correlation.

In their recent reconnaissance study of units in Costa Rica and Panama, Coates and others (1992) used a series of planktonic events (fossil first and last occurrences of foraminifers and nannoplankton) calibrated to the magnetic polarity time scale and adjusted for known diachrony in the region. Further advances in this study await the development of a much more refined time scale against which paleobiologic responses to various types of environmental forcing can be examined. As climatic variability at periodicities close to known variations in obliquity and

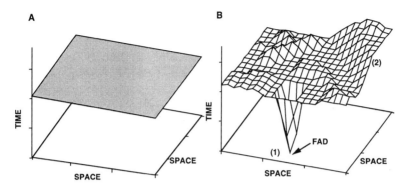

Fig. 3.3 Hypothetical first occurrence surfaces: (A) synchronous first occurrence surface; (B) diachronous first occurrence surface, where FAD (first appearance datum) (1) marks first appearance in time and the geographic center of origin in space, and (2) indicates the diachronous migration surface.

eccentricity of the Earth's orbit has been documented in the region (Dowsett and Poore 1991), we hope eventually to be able to resolve ages to the scale of about 10 ky.

Graphic Correlation Method

Graphic correlation utilizes first and last occurrences of fossil taxa in multiple stratigraphic sections to produce composite sections or biochronologic models (Shaw 1964; Miller 1977; Edwards 1984; Dowsett 1988, 1989a,b).

Two stratigraphic sections representing coincident or overlapping intervals of time are plotted against each other as the axes of a two-axis graph or Shaw Plot. Points plotted on such graphs (fig. 3.4) may represent first and last occurrences of fossil taxa, magnetostratigraphic boundaries, isotopic events, ash layers, and so forth. In general, any datum that can be uniquely identified in both sections (axes) may be plotted on a Shaw Plot.

As long as the two sections being correlated are in some way time-equivalent, a line of correlation (LOC) exists that can be used to link levels in one section with synchronous levels in the other section. Paleontological and other data are used to localize the probable position of the LOC.

Methods of positioning the LOC vary from worker to worker. The methodology followed here locates the LOC in accordance with several rules. First, the LOC is constrained to pass through isochronous datums (magnetostratigraphic boundaries, ash layers, unique isotopic events).

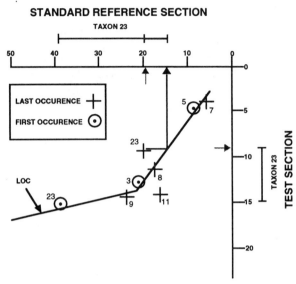

Fig. 3.4 Graphic correlation diagram (Shaw Plot) of hypothetical biostratigraphic data. Axes are scaled as depth in section. Line of correlation (LOC) is drawn in such a way as to separate fossil first and last occurrences. Numbers beside points identify taxa. The break in slope of the LOC indicates a relative change in the rate of rock accumulation. Short arrows near 20 m in the standard reference section and near 9 m in the test section represent the position of taxon 23 in these two sequences. The last occurrence of taxon 23 is revised in the standard reference section (to approximately 15 m or composite units) on the basis of the chosen LOC (long arrow).

Next, the LOC is positioned to put most of the first occurrences above the LOC and most of the last occurrences below the LOC. This configuration represents a situation in which taxon ranges along the reference-section axis are longer than the ranges on the test-section axis. Finally, the LOC should be drawn as one or more straight line segments (fig. 3.4).

In practice, one section is chosen as a standard reference section (SRS) on the basis of completeness, accumulation rate, fossil preservation, availability of magnetostratigraphy, sampling density, and similar factors. A second section (the test section) is plotted against the SRS to form a Shaw Plot. An LOC is determined using the procedures outlined above. Fossil first occurrences plotting earlier on the test-section axis than in the SRS and/or last occurrences plotting later on the test-section axis than in the SRS, based on the chosen LOC, are revised in the SRS (fig. 3.4). Once revisions have been made, the SRS becomes a composite standard reference section (CSRS) and is measured in com-

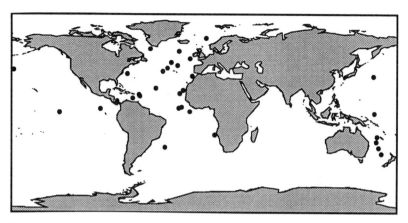

Fig. 3.5 Deep-sea cores and ocean margin sections used to construct the Pliocene composite model.

posite units. Composite units are units of accumulation expressed in terms of the depth scale of the section originally chosen as the standard.

Additional sections are plotted against the CSRS until geographic and stratigraphic coverage is sufficient to ensure that biostratigraphic events in the CSRS approximate the actual first and last appearance datums (FAD and LAD) as closely as these events can be sampled. This completes the first round of graphic correlation.

The procedure continues by plotting the first section (originally plotted against the SRS in round 1) against the end of round 1 composite (minus any information originally derived from that first section). The procedure continues with the plotting of the new composite against the second section (exclusive of information originally derived from that second section). After all sections have been replotted in a similar fashion, the procedure begins again with a third round of GC. After several rounds or cycles of GC, the positions of biostratigraphic events in the composite model tend to stabilize (Shaw 1964; Edwards 1984; Dowsett 1989a).

Stratigraphic information (first and last occurrences of planktonic foraminifers, calcareous nannofossils, and magnetostratigraphic events) from the Late Neogene of a large number of Deep Sea Drilling Project (DSDP) and Ocean Drilling Program (ODP) sites, as well as a series of land sections (fig. 3.5), were composited following the procedures described above to form a biochronological model (see also Dowsett 1989a,b).

The fit of geomagnetic reversal boundaries in the model to the mag-

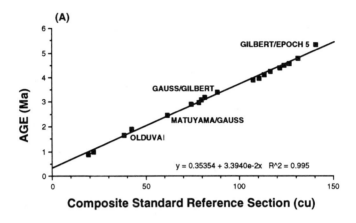

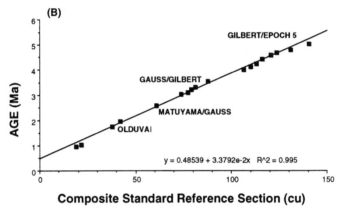

Fig. 3.6 Pliocene graphic correlation model calibrated to magnetic polarity time scales of (A) Berggren (Berggren et al. 1985) and (B) Cande and Kent (1992).

netic polarity time scales of Berggren (Berggren et al. 1985) and of Cande and Kent (1992) are nearly linear (fig. 3.6), indicating that composite units from different parts of the model represent equivalent amounts of time. The newly accepted time scale of Cande and Kent (1992) produces significantly older estimates for ages of magnetic reversals throughout the Pliocene (fig. 3.7). Because most of the available literature still uses the older time scale, ages derived in the following sections are referenced to the Berggren study (Berggren et al. 1985). Composite units of the model can be transformed to time (Ma) by the following equation:

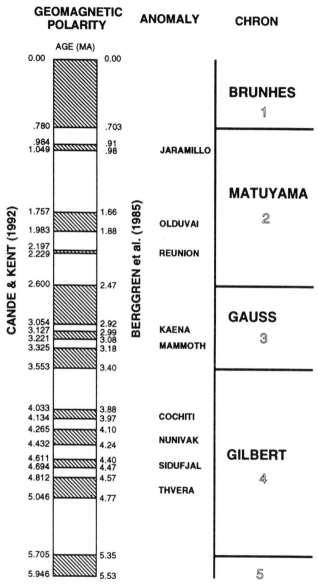

Fig. 3.7 Comparison of the ages estimated for geomagnetic polarity reversals in the Cande and Kent (1992) and Berggren (Berggren et al. 1985) time scales. Note that the difference between the two time scales increases through the Pliocene so that the offset at the top of Chron 5 is approximately 350 ky.

$$Ma = 0.354 + 0.034 \ CU, \qquad\qquad (1)$$

where $r^2 = .99$ and CU is the composite unit position in the model.[1] Therefore, ages of samples can be determined for any stratigraphic section that has been graphically correlated to the GC model.

This procedure has been employed in a number of cases where traditional methods of biochronology were insufficient and other independent stratigraphic data were either lacking or of poor quality. For example, DSDP Site 548 in the North Atlantic Ocean contains exceptionally warm Pliocene planktonic faunas and documents the transition from Middle Pliocene warmth to the cold surface conditions associated with the initiation of Northern Hemisphere glaciation near 2.4 Ma. The age model for this site previously was based upon planktonic foraminifers and, owing to the well-documented diachrony associated with planktonic biostratigraphic events at high latitude, was not very accurate. Dowsett and Loubere (1992) used GC to provide a new age model for the site, which enabled them to correlate much more precisely the changes in ocean circulation and sea surface temperature with other North Atlantic sites.

In another example, data on the beginning of upwelling off the coast of South Africa in the Middle Pliocene were rescaled to a GC age model by Dowsett (1989b), allowing more precise estimation of the timing of initiation of upwelling in this region as well as the first record of obliquity-driven $CaCO_3$ preservation cycles back to 4.8 Ma.

Graphic Correlation of selected sequences in Central America

We applied graphic correlation to two sequences in the Bocas del Toro Basin (Coates et al. 1992) to demonstrate the utility of this approach for overcoming the shortcomings of traditional biostratigraphic methods in the isthmian region. Earlier reconnaissance work produced thirty-six samples containing abundant identifiable microfossils. Analysis of the planktonic foraminifers and calcareous nannofossils shows a variable and discontinuous record with marginal preservation. The highest and lowest occurrences of each taxon were recorded, and the resulting information was plotted against the composite standard reference section outlined above.

Cayo Agua. Twenty-six samples, ranging through a little more than 300 meters of the Cayo Agua Formation (Bocas del Toro Group) on the

---

1. For reproduceability, numbers in this and other equations are carried out to three decimal places. This does not reflect the degree of precision.

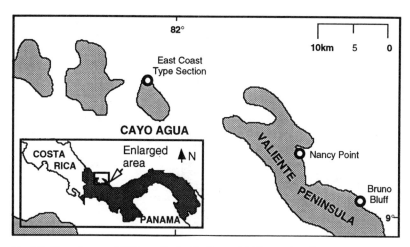

Fig. 3.8 Cayo Agua and Valiente Peninsula, Bocas del Toro region, Panama.

Island of Cayo Agua, were analyzed in this study (figs. 3.8, 3.9). Highest and lowest occurrences of five species of planktonic foraminifers and five species of nannoplankton were determined and plotted against their positions in the CSRS (fig. 3.10). The LOC determined from the distribution of data in figure 3.10 is constrained at the lower end by the first occurrences of *Globorotalia puncticulata* and *Ceratolithus rugosus* and the last occurrence of *Ceratolithus acutus*. Near the top of the Cayo Agua section a number of taxa—*Sphenolithus abies, Globorotalia margaritae, Sphaeroidinellopsis* spp., *Dentoglobigerina altispira*, and *Globigerinoides obliquus*—have last occurrences over a relatively short stratigraphic distance. This may be suggestive of a sudden and large decrease in relative accumulation rate or be due solely to diagenetic changes near the top of the section. Due to the lack of any first occurrences near the top of the section, the chosen LOC is placed conservatively to coincide with the stratigraphically lowest of these last occurrences (*Sphenolithus abies*). Based on the chosen LOC, the last occurrences of *Discoaster pentaradiatus* and *Reticulofenestra pseudoumbilica* fall significantly short of their positions in the CSRS. As both of these taxa are marker species in the common calcareous nannofossil zonations, traditional methods of correlation applied to this section would produce ages too young.

The equation of the chosen LOC is

$$\text{CSRS} = 137.5 - 0.16 \, (\text{Cayo Agua position}), \qquad (2)$$

which can be combined with equation (1) above to estimate age at the base of the Cayo Agua section as 5.03 Ma. The top of that part of the

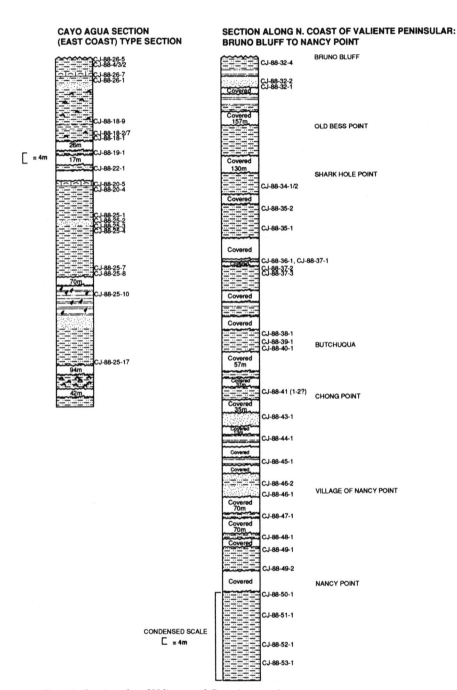

Fig. 3.9 Stratigraphy of Valiente and Cayo Agua sections.

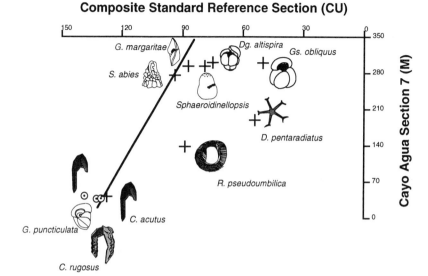

Composite Standard Reference Section (CU)

Fig. 3.10   Shaw Plot of Cayo Agua section and composite standard reference section of Pliocene graphic correlation model. Cayo Agua section (vertical axis) measured in meters above base of section. Composite standard reference section (horizontal axis) measured in composite units. Crosses (+) indicate last occurrences, and circles ⊙ indicate first occurrences. Dark line shows preferred line of correlation given by equation CSRS = 137.5 − 0.16 (Cayo Agua).

section which can be conservatively correlated to the CSRS is 3.51 Ma. The estimated time it took to (discontinuously) deposit approximately 280 m of the Cayo Agua section was just over 1.5 my, that is, a mean accumulation rate of 18.4 cm/ky.

Valiente Peninsula. Twenty-nine samples were selected from the exposed portions of the upper part of the Valiente Peninsula section (figs. 3.8, 3.9). In general planktonic foraminifer assemblages from the Shark Hole Point Formation have higher diversity than those of the Cayo Agua Formation. Highest and lowest occurrences of four species of planktonic foraminifers and three species of nannoplankton were determined and plotted against their positions in the CSRS (fig. 3.11). The LOC determined from the distribution of data in figure 3.11 is constrained at the lower end by the first and last occurrences of *Ceratolithus acutus* and last occurrence of *Discoaster quinqueramus*. The LOC extends up through the last occurrence of *Globigerina nepenthes* and *Globorotalia margaritae*. The equation of the chosen LOC is

$$CSRS = 86.13 + 0.147 \text{ (Valiente position)}, \tag{3}$$

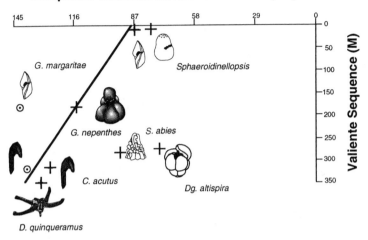

Composite Standard Reference Section (CU)

Fig. 3.11 Shaw Plot of Valiente sequence and composite standard reference section of Pliocene graphic correlation model. Valiente sequence (vertical axis) measured in meters below top of section. Composite standard reference section (horizontal axis) measured in composite units. Crosses (+) indicate last occurrences, and circles (⊙) indicate first occurrences. Dark line shows preferred line of correlation given by equation CSRS = 86.13 + 0.147 (Valiente position).

which can be combined with equation (1) above to estimate age at the top of the Valiente sequence as 3.28 Ma. The contact between the Shark Hole Point and Nancy Point Formations is projected to be at 5.68 Ma. Note that the last occurrence of *Dentoglobigerina altispira* falls well short of its last occurrence in the GC model and the Cayo Agua section. The GC method thus indicates that in the Valiente section *D. altispira* cannot be used to identify its usual level of 3 Ma. By displaying a large amount of stratigraphic information simultaneously, multiple correlation models can be tested and diachrony observed.

DISCUSSION

Coates and others (1992) determined the age at the top of the Cayo Agua section to be simply older than 2.9 Ma based upon the presence of *Dentoglobigerina altispira*, with a more definite age assignment below that of 3.6–3.5 Ma based upon the co-occurrence of *Sphenolithus abies* and *Pseudoemiliana lacunosa*. The graphic correlation analysis presented above refines the uppermost Cayo Agua to be 3.51 Ma and the base of

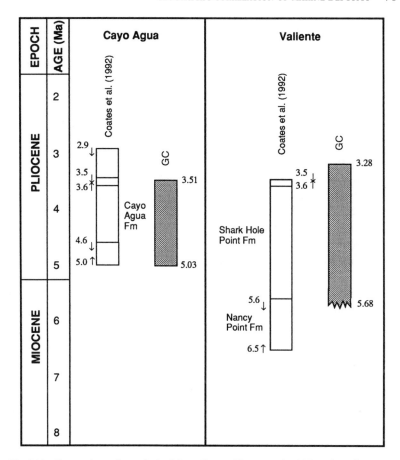

Fig. 3.12 Comparison of ages derived from Coates (Coates et al. 1992) and graphic correlation analysis (GC). Note that the Cayo Agua section is considered in our analysis to
include significantly less time than implied in the Coates study. The top of the Valiente
section is projected to be about 200 ky younger than previously thought.

the section to be 5.03 Ma (fig. 3.12). The fact that the section can be
correlated to the CSRS using one straight line segment, combined with
the knowledge that the CSRS has a nearly linear fit to absolute age,
suggests that sedimentation, while almost certainly composed of numerous discontinuous packages, happened at approximately the same
rate.

The top of the Valiente sequence was originally placed at 3.6–3.5 Ma
based upon the co-occurrence of *Sphenolithus abies* and *Pseudoemiliana
lacunosa*, while the base of the section was thought to be older than 5.6

Ma but younger than 6.5 Ma, based upon the occurrences of *Dicoaster berggrenii, Discoaster quinqueramus,* and *Globorotalia exilis* (Coates et al. 1992). The GC analysis above suggests that the top of the Valiente sequence is in fact younger and lies closer to 3.28 Ma. The original pick of older than 5.6 Ma for the contact between the Nancy Point and overlying Shark Hole Point is very close to the GC prediction of 5.68 Ma (fig. 3.12).

Actual verification or refinement of the age models developed for these sections will come from future paleomagnetic fieldwork in Central America. Based on the age models developed above, the position of magnetic reversal boundaries can be projected into each section. For example, the Valiente sequence should contain the Gilbert-Gauss boundary (3.40 Ma) at approximately 89.6 composite units (23.6 meters using equation [3]) below the top of the section.

In chapter 4 of this volume, Cronin and Dowsett argue for repeated closing of the Central American Straits and breaching of the isthmus between about 3.5 and 2.5 Ma. These changes are reflected in the open ocean planktonic faunas of the Caribbean Sea and the sedimentary record of the North American Atlantic Coastal Plain. One important conclusion from that chapter is that the magnitude and variability of Pliocene low-latitude SST (measured at Caribbean Sea Site 502) was very similar to present-day SST. Using the age models developed above, SST estimates based upon planktonic foraminiferal faunas from the Valiente and Cayo Agua sequences (Cotton and Dowsett 1991, unpublished data) can be used to corroborate the stability of SST in the region and suggest that it extends back to near 6.0 Ma.

The application of GC to these sequences in Panama is meant to serve as an example of the level of refinement time calibration can achieve. Eventually, with detailed, measured stratigraphic sections and systematic biostratigraphic analysis, we can build a temporal framework for all units in Central America. This will allow for precise correlations between widely separate areas with very different sedimentary environments, that can be expanded to other fossil occurrence data (benthic foraminifers, ostracodes, mollusks, etc.) composited to the CSRS to assess diachrony in those groups. Documenting the existence and magnitude of delayed migrations and local extinctions across the Central American Isthmus will be valuable data for all those interested in the paleobiologic and paleoceanographic effects of the formation of the Central American Isthmus.

ACKNOWLEDGMENTS

We thank Debra Willard, Scott Ishman, Anthony Coates, Ann Budd, and Jeremy Jackson for thoughtful reviews of this manuscript. Laurel Bybell generously provided preliminary occurrence data on calcareous nannoplankton without which this paper would not have been possible. We thank Thomas Cronin for numerous discussions of the Neogene stratigraphy and biochronology of the tropical Atlantic. Anthony Coates initiated our involvement in this project, made samples available, and provided many encouraging discussions. Emerson Polanco, Stephanie West, Marci Robinson, and Jean Self-Trail helped with various aspects of data reduction. This work is a product of the PRISM (Pliocene Research, Interpretation, and Synoptic Mapping) Project.

REFERENCES

Backman, J., and N. J. Shackleton. 1983. Quantitative biochronology of Pliocene and Early Pleistocene calcareous nannofossils from the Atlantic, Indian, and Pacific Oceans. *Mar. Micropaleontol.* 8:141–70.

Bandy, O. L. 1970. Upper Cretaceous Cenozoic paleobathymetric cycles, eastern Panama and northern Colombia. *Trans. Gulf Coast Assoc. Geol. Soc.* 20: 181–93.

Banner, F. T., and W. H. Blow. 1965. Progress in the planktonic foraminiferal biostratigraphy of the Neogene. *Nature* 208:1164–66.

Bates, R. L., and J. A. Jackson, eds. 1980. *Glossary of geology.* Falls Church, Va.: American Geologic Institute.

Berggren, W. A. 1973. The Pliocene time-scale: Calibration of planktonic foraminiferal and calcareous nannoplankton zones. *Nature* 243:391–97.

Berggren, W. A., D. V. Kent, et al. 1985. Neogene geochronology and chronostratigraphy. In *The chronology of the geological record,* 211–60. London: Blackwell Scientific Publications.

Berrangé, J. P. 1989. The Osa group: An auriferous Pliocene sedimentary unit from the Osa Peninsula, southern Costa Rica. *Rev. Geol. Am. Central* 10: 67–93.

Blow, W. H. 1969. Late Middle Eocene to Recent planktonic foraminiferal biostratigraphy. *Proc. Int. Conf. Planktonic Microfossils* (Geneva) 1:199–422.

Bolli, H. M. 1970. The Foraminifera of sites 23–31, leg 4. *DSDP* 4:577–643.

Bolli, H. M., and P. J. Bermudez. 1965. Zonation based on planktonic Foraminifera of middle Miocene to Pliocene warm-water sediments. *Boll. Inf. Asoc. Ven. Geol., Min., y Petrol.* 8(5):119–149.

Bolli, H. M., and I. Premoli-Silva. 1973. Oligocene to Recent planktonic Foraminifera and stratigraphy of the leg 15 sites in the Caribbean Sea. *Init. Repts. DSDP* 15:475–97.

Cande, S. C., and D. V. Kent. 1992. A new geomagnetic polarity time scale for the Late Cretaceous and Cenozoic. *J. Geophys. Res.* 97:13917–51.

Cassell, D. T., and B. K. Sen Gupta. 1989. Foraminiferal stratigraphy and paleoenvironments of the Tertiary Uscari Formation, Limón Basin, Costa Rica. *J. Foramin. Res.* 19(1):52–71.

Coates, A. G., J. B. C. Jackson, L. S. Collins, T. M. Cronin, H. J. Dowsett, L. M. Bybell, P. Jung, and J. A. Obando. 1992. Closure of the Isthmus of Panama: The near-shore marine record of Costa Rica and western Panama. *Geol. Soc. Am. Bull.* 104:814–28.

Cotton, M. A., and H. J. Dowsett. 1991. Quantitative environmental estimates from the Pliocene of the western Caribbean based on analyses of planktic foraminifers from Panama. *Geol. Soc. Am. Ann. Meeting, Abstr. Progr.* 22: A-365.

Corrigan, J., P. Mann, and J. C. Ingle. 1990. Forearc response to subduction of the Cocos Ridge, Panama–Costa Rica. *Geol. Soc. Am. Bull.* 102:628–52.

Cushman, J. A. 1918. The smaller Foraminifera of the Panama Canal zone. *U.S. Nat. Mus. Bull.* 103:45–87.

Dowsett, H. J. 1988. Diachrony of Late Neogene microfossils in the southwest Pacific Ocean: Application of the graphic correlation method. *Paleoceanography* 3:209–22.

———. 1989a. Application of the graphic correlation method to Pliocene marine sequences. *Mar. Micropaleontol.* 14:3–32.

———. 1989b. Improved dating of the Pliocene of the eastern South Atlantic using graphic correlation: Implications for paleobiogeography and paleoceanography. *Micropaleontology* 35(3): 279–92.

Dowsett, H. J., T. M. Cronin, R. Z. Poore, R. S. Thompson, R. C. Whatley, and A. M. Wood. 1992. Micropaleontological evidence for increased meridional heat transport in the North Atlantic Ocean during the Pliocene. *Science* 258:1133–35.

Dowsett, H. J., and P. Loubere. 1992. High resolution Late Pliocene sea-surface temperature record from the Northeast Atlantic Ocean. *Mar. Micropaleontol.* 20:91–105.

Dowsett, H. J., and R. Z. Poore. 1991. Pliocene sea surface temperatures of the North Atlantic Ocean at 3.0 Ma. *Quartern. Sci. Revs.* 10:189–204.

Dowsett, H. J., and L. B. Wiggs. 1992. Planktonic foraminiferal assemblage of the Yorktown Formation, USA. *Micropaleontology* 38(1): 75–86.

Duque-Caro, H. 1990. Neogene stratigraphy, paleoceanography, and paleobiogeography in northwest South America and the evolution of the Panama seaway. *Palaeogeogr., Palaeoclimatol., Palaeoecol.* 77:203–34.

Edwards, L. E. 1984. Insights on why graphic correlation (Shaw's Method) works. *J. Geol.* 92:583–97.

Hills, S. J., and H. R. Theirstein. 1989. Plio-Pleistocene calcareous plankton biochronology. *Mar. Micropaleontol.* 14:67–96.

Jenkins, D. G. 1975. Cenozoic planktonic foraminiferal biostratigraphy of the southwestern Pacific and Tasman Sea—DSDP leg 29. *DSDP* 29:449–67.

Johnson, D. A., D. A. Schneider, C. A. Nigrini, J. P. Caulet, and D. V. Kent. 1989. Pliocene-Pleistocene radiolarian events and magnetostratigraphic calibrations for the tropical Indian Ocean. *Mar. Micropaleontol.* 14:33–66.

Kennet, J. P. 1973. Middle and Late Cenozoic planktonic foraminiferal bio-stratigraphy of the Southeastern Pacific—DSDP leg 21. *DSDP* 21:575–640.

Martini, E. 1971. Standard Tertiary and Quaternary calcareous nannoplankton zonation. *Proc. II Planktonic Conf.* (Rome) 2:739–85.

Miller, F. X. 1977. The graphic correlation method. In *Concepts and methods of biostratigraphy*, ed. E. Kauffman and J. E. Hazel, 165–86. Stroudsburg, Pa.: Dowden Hutchinson and Ross.

Parker, F. L. 1973. Late Cenozoic biostratigraphy (planktonic Foraminifera) of tropical Atlantic deep-sea sections. *Rev. Esp. Micropaleontol.* 5:253–89.

Pizarro, D. 1987. Bioestratigrafía de la Formación Uscari con base en foramini-feros planctonicos. *Rev. Geol. Am. Central* 7:1–63.

Okada, H., and D. Bukry. 1980. Supplementary modification and introduction of code numbers to the low-latitude coccolith biostratigraphy zonation. *Mar. Micropaleontol.* 5(3):321–25.

Shaw, A. B. 1964. *Time in stratigraphy.* New York: McGraw Hill.

4

Biotic and Oceanographic Response to the Pliocene Closing of the Central American Isthmus

Thomas M. Cronin and Harry J. Dowsett

INTRODUCTION

The formation of the Central American Isthmus (CAI) during the Late Neogene closed an ocean gateway that had been open since the Mesozoic and simultaneously joined two long-isolated land masses. It was a key event for tropical biotic evolution, allowing the interchange of terrestrial species between North and South America (Marshall 1988) and isolating Pacific and Atlantic/Caribbean marine organisms (Jones and Hasson 1985; Stehli and Webb 1985).

The oceanographic and climatic effects of the closure of the isthmus have also been debated, in part because this event coincides with the initiation of major Northern Hemisphere glaciation and the amplification of Milankovitch climatic cycles about 2.5 Ma (Shackleton et al. 1984; Raymo et al. 1989). Berggren (1972) and Berggren and Hollister (1974) postulated, on the basis of early Deep Sea Drilling Project (DSDP) results, that isthmus closure may have had a profound effect on North Atlantic paleoceanography, and possibly on global climate, by diverting warm, saline water to high latitudes, causing Late Pliocene Northern Hemisphere ice build-up. Keigwin (1978, 1982) also suggested that as Pacific and Caribbean waters ceased to mix, significant changes in North Atlantic Ocean structure occurred, such as increased surface salinities in the Caribbean and the intensification of the Gulf Stream. Rind and Chandler (1991) argued, on the basis of a general circulation model, that relatively small changes in ocean heat flux, due to ocean circulation changes caused by isthmus closure, can substantially alter global climate.

In this chapter we examine the role of the CAI during the Pliocene in affecting tropical and extratropical oceanic and biotic events. We present evidence indicating there was near closure of the isthmus to surface water around 3.0–2.8 Ma, and perhaps again at about 2.0 Ma, which had major effects on marine paleobiogeography and altered

North Atlantic Ocean circulation to cause periods of Northern Hemispheric warmth. We will attempt to show, through the study of marine ostracodes and planktonic foraminifers from the North Atlantic/Caribbean, that the formation of the CAI directly or indirectly (1) increased North Atlantic oceanic heat flux from low to high latitudes at 3 Ma and possibly again at 2 Ma; (2) led to oceanic thermal gradients less steep than those today, thereby decreasing provinciality of marine organisms; (3) increased surface salinities but did not decrease sea surface temperatures in the Caribbean region.

MODERN NORTH ATLANTIC OCEANOGRAPHY

Modern North Atlantic Ocean thermohaline circulation is dominated by the northward flow of the warm Western Boundary Current (WBC), or the Gulf Stream–North Atlantic Drift, and the formation of cold, dense North Atlantic Deep Water (NADW) in the Norwegian Sea and adjacent seas. Broecker and others (1985) and Broecker and Denton (1989) postulated that this "conveyor belt" thermohaline circulation, in which warm, saline water from the south mixes with atmospherically cooled water in high latitudes, undergoes major reorganization during glacial-interglacial transitions. NADW formation, which ventilates the deep ocean with highly oxygenated water during the present interglacial, is turned off during cold periods, such as the last glacial maximum (LGM) or the Younger Dryas cool snap (11 Ka) and switched on during periods of climatic warming. Charles and Fairbanks (1992) called NADW the "primary amplifier" of glacial-interglacial climatic cycles in high latitudes 100 Ka. NADW is closely linked to surface water temperature and salinity; its formation should diminish when melting continental ice sheets discharge into the North Atlantic and should increase, due to increased evaporation, when heat flows from the tropics. Thus, although some aspects of the conveyor belt model have been questioned (Veum et al. 1992), there is little doubt that there were significant changes in the strength of NADW formation and deep ocean circulation during the Pliocene as well as the Pleistocene that were related to changes in surface salinities and oceanic heat transport.

The presence or absence of the Central American Isthmus may have influenced the degree of heat transported from low latitudes and thus played an important role in this complex system of oceanic circulation and NADW formation. For example, Maier-Reimer and others (1990) performed sensitivity tests of an oceanic general circulation model with both an open and closed isthmus. With the isthmus open, lower than modern surface salinities on the present Atlantic side caused a collapse of thermohaline circulation, decrease of NADW formation to near zero, and a weakening of poleward heat transport from the tropics. They suggest that closure of the deep water by the isthmus, which occurred during the Late Miocene, 10–7 Ma, could have initiated NADW formation but that more vigorous thermohaline circulation only occurred 4–3 Ma, after closure of surface waters occurred. We will address this problem of intensified heat flux between 4 and 2 Ma below.

NORTH ATLANTIC ZOOGEOGRAPHY

North Atlantic water temperatures strongly control the modern zoogeography of organisms living within the mixed ocean layer (surface to about 200 m)—in particular, planktonic foraminifers and shelf-dwelling ostracodes—making them sensitive indicators of changes in ocean temperature. Six major planktonic foraminiferal assemblages (polar, subpolar, gyre, transitional, subtropical, and tropical) reflect the control of sea surface temperatures (SST) on species' latitudinal distribution (Imbrie and Kipp 1971). Similarly, seven shallow marine ostracode assemblages (frigid, subfrigid, cold, mild temperate, warm temperate, subtropical, and tropical) inhabit various marine climatic zones. Zonal boundaries essentially reflect the effects of ocean currents on the distribution of temperature-sensitive species, limiting the poleward range expansion of thermophilic and equatorward expansion of cryophilic species (Hazel 1970). Figure 4.1 shows the major ocean currents influencing marine species along the U.S. Atlantic coast and in the Caribbean. The sharpest modern faunal boundary occurs where the Carolina Coastal and Virginia Currents converge at Cape Hatteras, defining the limits of the subtropical and mild temperate climatic zones. The modern distribution of ostracode species can be found in Hazel 1970, 1975, Valentine 1971, Cronin 1983, and papers cited therein.

PLIOCENE PALEOCEANOGRAPHY OF THE NORTH ATLANTIC

Pliocene paleoceanographic history of the North Atlantic involves changes in sea surface temperatures and surface salinities, heat flux from the North Atlantic Drift, and North Atlantic Deep Water formation that we only briefly review here. Shackleton and others (1984) determined that near 2.4 Ma (an age revised to 2.6 Ma by Curry and Miller 1989) major isotopic shifts occurred in the North Atlantic signifying

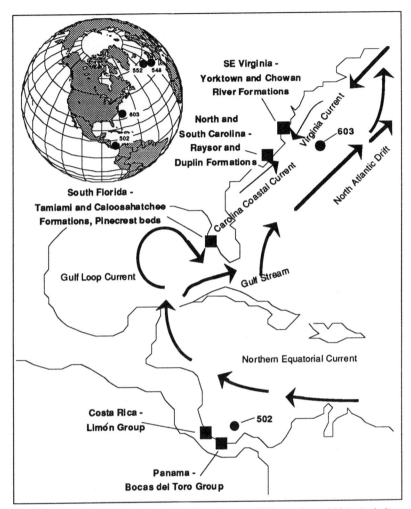

Fig. 4.1 Studied sections from Panama, Costa Rica, and Atlantic Coastal Plain, including deep-sea cores 502 and 603 and major currents. Inset shows all deep-sea cores discussed.

Northern Hemispheric ice build up. This Late Pliocene cooling, now widely recognized in many paleoclimatic records from around the world, was actually a progressive series of events (Dowsett and Poore 1990; Dowsett and Loubere 1992). For example, Raymo and others (1989) determined that high-amplitude 41,000-year Milankovitch cycles became more dominant during the Late Pliocene in the North Atlantic and that there were a series of sharp glacial events near 2.4–2.3 Ma (see also Sancetta et al. 1992). The Raymo study named these events oxygen isotope stages 100, 98, and 96, following the convention of naming Pleistocene glacial periods with even-numbered stages. From the history of ice-rafted detritus (IRD) in the Norwegian Sea, Jansen and others (1988) and Jansen and Sjøholm (1991) showed that although minor ice rafting occurred as far back as 5.5 Ma, a major increase in IRD occurred about 2.5 Ma, coinciding with the faunal and isotopic evidence for a large climatic transition. On the basis of carbon isotopic records on deep-sea foraminifers, Raymo and others (1992) also showed that between 2.5 and 2.3 Ma, suppression of NADW formation may have occurred coincident with these glacial events, but Late Pliocene NADW suppression was not nearly as strong as that which occurred during Late Pleistocene glacial periods.

The marine and glacial record of the Tjörnes Peninsula, northern Iceland, also records the high-latitude climate record of the North Atlantic region (Einarsson et al. 1967; Einarsson and Albertsson 1988; Eiriksson et al. 1990; Cronin 1991a). Following an extended period of relatively stable, warm water from 4.5 to 3.0 Ma, a series of Late Pliocene and Pleistocene glacial sediments were deposited, the first dated at about 2.5 Ma, alternating with glacio-marine and interglacial sediments. The shift from warm marine climates to frequent glacial-interglacial cycles near 2.5 Ma coincides with the shift in open ocean sea surface temperatures and increased IRD.

The warm period just preceding major ice buildup, 3.2–2.8 Ma, has been studied in detail (Dowsett and Poore 1991; Cronin and Dowsett 1991; Dowsett et al. 1992), with the conclusion that at 3.0 Ma, SSTs at higher latitudes were significantly warmer than those of today or any Pleistocene interglacial but that tropical SSTs were not significantly different from those of today.

PALEOENVIRONMENTAL HISTORY OF THE CARIBBEAN

Field investigations by the Panama Paleontology Project (Coates et al. 1992; Jackson et al. this volume, chap. 9) have resulted in a large collection of Pliocene marine sediments, many of which are well dated. Diverse microfaunas suitable for reconstructing the environmental history and biotic evolution of the western Caribbean are abundant. Our samples come from the Bocas del Toro Group of northwestern Panama and the Limón Group in eastern Costa Rica (fig. 4.2). Figure 4.3 adopted from Coates et al. 1992 shows the stratigraphy of these units. The samples (table 4.1) are representative of the varied lithofacies of the interval between 4 and 2 Ma found throughout the area.

To identify the broad trends in ostracode biofacies and to relate these

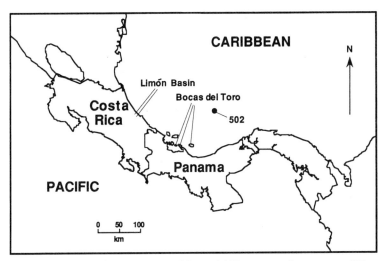

Fig. 4.2   Relationship of Limón Basin and Bocas del Toro area to deep-sea core 502.

patterns to age and paleoenvironment, R- and Q-mode cluster analyses using the Pearson correlation coefficient and complete linkage were carried out on 61 samples using 44 of the common ostracode species.[1] The results are shown in cluster diagrams in figures 4.4 and 4.5. Five primary groups of taxa can be identified (fig. 4.4) that, using the taxonomy and ecology for ostracode species and genera found in Teeter 1975 and Bold 1981, 1988, can be related to various tropical environments. Groups 1 and 4 are both carbonate platform assemblages, the former similar to inner-shelf assemblages described by Bold (1981), often found at water depths of 30–50 m. Some subgroups represent distinctive taxon groups, such as that comprised of *Loxocorniculum* and *Quadracythere* in group 4, which signifies very shallow, littoral environments. Group 2 is composed of an outer-neritic/upper-bathyal group of taxa; most notably, *Krithe* and *Bradleya* would indicate water depths greater than 200 m. Group 3 is comprised of the phytal-dwelling taxa *Loxoconcha* and *Paradoxostoma* and *Cativella* and *Cytheropteron*, which inhabit tropical lagoonal environments today. Group 5 consists of lagoonal/shelf genera whose species usually live on muddy substrates in modern environments. Bold (1981, 1988) found most group 5 taxa lived at shelf depths from about 30–100 m.

The most important result obtained from the ostracode data is that,

1. Ostracode species occurrence data for the Panisthmian and Florida material discussed here are available from Thomas Cronin and will be published in the *Panisthmian Compendium*.

Table 4.1 Pliocene Samples from Panama and Costa Rica

Sample Number	General Locale	Specific Locale	Formation
CJ 89-36-2	Limón, Costa Rica	Rte. 32 W. of Limón	Moin Fm.
CJ 89-38-2	Limón, Costa Rica	Rte. 32 W. of Limón	Moin Fm.
CJ 89-17-1	Limón, Costa Rica	Loma del Mar	Moin Fm.
CJ 89-17-8	Limón, Costa Rica	Loma del Mar	Moin Fm.
CJ 89-15-4	Limón, Costa Rica	Pueblo Nuevo Cem.	Moin Fm.
CJ 89-16-1	Limón, Costa Rica	Pueblo Nuevo Cem.	Moin Fm.
CJ 89-16-2	Limón, Costa Rica	Pueblo Nuevo Cem.	Moin Fm.
CJ 89-16-3	Limón, Costa Rica	Pueblo Nuevo Cem.	Moin Fm.
CJ 89-18-2	Limón, Costa Rica	Cangrejos Creek	Moin Fm.
CJ 89-18-7	Limón, Costa Rica	Cangrejos Creek	Moin Fm.
CJ 89-18-9	Limón, Costa Rica	Cangrejos Creek	Moin Fm.
CJ 89-18-11	Limón, Costa Rica	Cangrejos Creek	Moin Fm.
CJ 89-20-1	Bomba, Costa Rica	Rio Banano	Rio Banano Fm.
CJ 89-20-3	Bomba, Costa Rica	Rio Banano	Rio Banano Fm.
CJ 89-20-5	Bomba, Costa Rica	Rio Banano	Rio Banano Fm.
CJ 89-20-11	Bomba, Costa Rica	Rio Banano	Rio Banano Fm.
CJ 89-21-1	Bomba, Costa Rica	Rio Banano	Rio Banano Fm.
CJ 89-21-4	Bomba, Costa Rica	Rio Banano	Rio Banano Fm.
CJ 89-21-5	Bomba, Costa Rica	Rio Banano	Rio Banano Fm.
CJ 89-21-6	Bomba, Costa Rica	Rio Banano	Rio Banano Fm.
CJ 89-22-1	Bomba, Costa Rica	Rio Banano	Rio Banano Fm.
CJ 89-22-2	Bomba, Costa Rica	Rio Banano	Rio Banano Fm.
CJ 89-22-3	Bomba, Costa Rica	Rio Banano	Rio Banano Fm.
CJ 89-22-4	Bomba, Costa Rica	Rio Banano	Rio Banano Fm.
CJ 89-22-6	Bomba, Costa Rica	Rio Banano	Rio Banano Fm.
CJ 87-31-1	Bocas del Toro, Pan.	Cayo Agua	Cayo Agua Fm.
CJ 87-31-2	Bocas del Toro, Pan.	Cayo Agua	Cayo Agua Fm.
CJ 87-31-3	Bocas del Toro, Pan.	Cayo Agua	Cayo Agua Fm.
CJ 87-31-4	Bocas del Toro, Pan.	Cayo Agua	Cayo Agua Fm.
CJ 86-35-1	Bocas del Toro, Pan.	Cayo Agua	Cayo Agua Fm.
CJ 86-36-1	Bocas del Toro, Pan.	Cayo Agua	Cayo Agua Fm.
CJ 86-37-1	Bocas del Toro, Pan.	Cayo Agua	Cayo Agua Fm.
CJ 86-38-1	Bocas del Toro, Pan.	Cayo Agua	Cayo Agua Fm.
CJ 86-39-1	Bocas del Toro, Pan.	Cayo Agua	Cayo Agua Fm.
CJ 88-18-1	Bocas del Toro, Pan.	Cayo Agua	Cayo Agua Fm.
CJ 88-18-6	Bocas del Toro, Pan.	Cayo Agua	Cayo Agua Fm.
CJ 88-18-8	Bocas del Toro, Pan.	Cayo Agua	Cayo Agua Fm.
CJ 88-20-4	Bocas del Toro, Pan.	Cayo Agua	Cayo Agua Fm.
CJ 88-20-5	Bocas del Toro, Pan.	Cayo Agua	Cayo Agua Fm.
CJ 88-26-1	Bocas del Toro, Pan.	Cayo Agua	Cayo Agua Fm.
CJ 88-26-2	Bocas del Toro, Pan.	Cayo Agua	Cayo Agua Fm.
CJ 88-26-4	Bocas del Toro, Pan.	Cayo Agua	Cayo Agua Fm.
CJ 88-35-2	Bocas del Toro, Pan.	N. Coast Valiente	Shark Hole Point Fm.
CJ 88-51-1	Bocas del Toro, Pan.	N. Coast Valiente	Nancy Point Fm.
CJ 88-30-1	Bocas del Toro, Pan.	Escudo de Veraguas	Escudo de Verag. Fm.
CJ 88-30-3	Bocas del Toro, Pan.	Escudo de Veraguas	Escudo de Verag. Fm.
CJ 88-30-4	Bocas del Toro, Pan.	Escudo de Veraguas	Escudo de Verag. Fm.
CJ 88-30-5	Bocas del Toro, Pan.	Escudo de Veraguas	Escudo de Verag. Fm.

Table 4.1 *continued*

Sample Number	General Locale	Specific Locale	Formation
CJ 88-30-6	Bocas del Toro, Pan.	Escudo de Veraguas	Escudo de Verag. Fm.
CJ 88-30-7	Bocas del Toro, Pan.	Escudo de Veraguas	Escudo de Verag. Fm.
CJ 88-30-8	Bocas del Toro, Pan.	Escudo de Veraguas	Escudo de Verag. Fm.
CJ 88-30-9	Bocas del Toro, Pan.	Escudo de Veraguas	Escudo de Verag. Fm.
CJ 88-30-10	Bocas del Toro, Pan.	Escudo de Veraguas	Escudo de Verag. Fm.
CJ 88-30-11	Bocas del Toro, Pan.	Escudo de Veraguas	Escudo de Verag. Fm.
CJ 88-30-12	Bocas del Toro, Pan.	Escudo de Veraguas	Escudo de Verag. Fm.
CJ 87-9-1	Bocas del Toro, Pan.	Escudo de Veraguas	Escudo de Verag. Fm.
CJ 87-9-2	Bocas del Toro, Pan.	Escudo de Veraguas	Escudo de Verag. Fm.
CJ 87-10-5	Bocas del Toro, Pan.	Escudo de Veraguas	Escudo de Verag. Fm.
CJ 89-39-1	South of Limón, Costa Rica	Santa Rita	
CJ 89-39-2	South of Limón, Costa Rica	Santa Rita	
CJ 89-35-1	South of Limón, Costa Rica	Los Laureles	

whereas many distinct biofacies can be identified, all consist of species and genera that live in the tropics today and occur throughout the Caribbean in Neogene marine sediments. There is no indication of migration from extratropical areas, which would be expected if tropical SSTs had significantly decreased during this time. However, it is noteworthy that samples in group 3 from Cangrejos Creek, Limón Basin, Costa Rica, which we date at about 2.0 Ma on the basis of planktonic foraminifers, includes deeper water taxa such as *Bradleya* that often inhabit cooler waters below the thermocline. Their presence in sediments of the Moin Formation signifies either a migration at some localities onto the shelf, perhaps owing to upwelling, or that these facies of the Moin represent deeper water of upper-bathyal environments. We favor the former interpretation because of the occurrence of well-preserved shallow-water taxa, including some with eye tubercles, with the *Bradleya* assemblage. There are good modern analogs of such a situation of cool, deeper-water taxa occurring at shelf depths off Chile and elsewhere. Upwelling along the Caribbean side may be related to closure of the isthmus at this time (see Discussion).

The Q-mode cluster (figure 4.5) revealed 3 groups of samples. These groups can be dated on the basis of planktonic foraminifers and calcareous nannofossils (Coates et al. 1992). Group A is a Middle Pliocene (3.5–2.9 Ma) cluster consisting mostly of samples from the Rio Banano and Cayo Agua Formations. Group B is Late Pliocene (2.4–1.7 Ma) and includes several subgroups of samples from the Moin Formation at several localities, the upper part of the Escudo de Veraguas Formation, and two samples from the upper part of the Rio Banano. Group C is Early

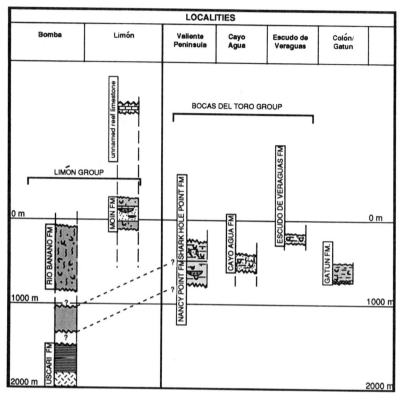

Fig. 4.3 Stratigraphy of Bocas del Toro and Limón Groups.

Pliocene (5.5–3.5 Ma) and includes samples from the Nancy Point Formation, the older part of the Cayo Agua Formation at Punta Norte, and several samples from the lower part of the Rio Banano. These 3 clusters illustrate differences among the ostracode assemblages related to the various tropical environments of deposition of each formation and, to a lesser degree, the extinction and origination of several species.

Outcrops of the Rio Banano Formation in the Limón Basin were examined to focus on Pliocene shallow marine environmental changes in the western Caribbean. Coates's study of the Rio Banano section (Coates et al. 1992) indicates it is 3.6–2.5 Ma. We studied the ostracode assemblages from twenty samples to establish paleoceanographic conditions represented by the Rio Banano Formation. Figure 4.6 shows the percentages of ostracode taxa from three sections exposed along the Rio Banano River. Assemblages from the lower part of the formation (section 89-21-1 through 89-21-6) have high percentages of *Puriana* and

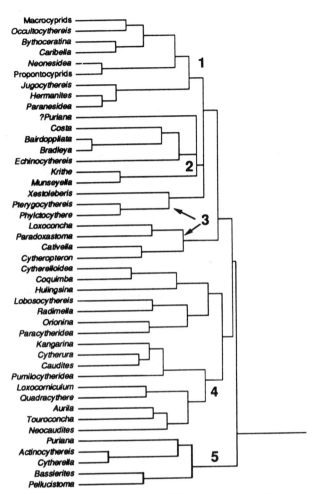

Fig. 4.4 Results of R-mode cluster analysis of Panisthmian ostracode samples. Groups 1–5 are clusters of species that characterize various tropical benthic habitats.

Xestoleberis with a spike in *Actinocythereis* in sample 89-21-6. These taxa signify inner-shelf environments. The middle part of the Rio Banano (samples 89-22-1 through 89-22-6) are dominated by *Cytherella, Touroconcha,* and *Basslerites,* which generally are characteristic of lagoonal and inner-shelf environments, often living in muddy and silty sediments. In the upper part, there are significant increases in *Loxocorniculum, Radimella, Orionina,* and *Caudites,* especially noteworthy between samples 89-20-6 and 89-20-5. These taxa are characteristic of carbonate platform environments throughout the modern Caribbean (Teeter 1975)

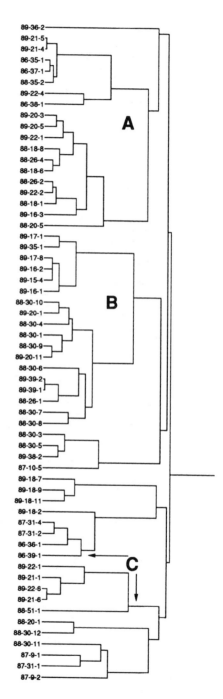

Fig. 4.5 Results of Q-mode cluster analysis of Panisthmian ostracode samples. Groups A, B, and C represent Middle, Late, and Early Pliocene samples, respectively.

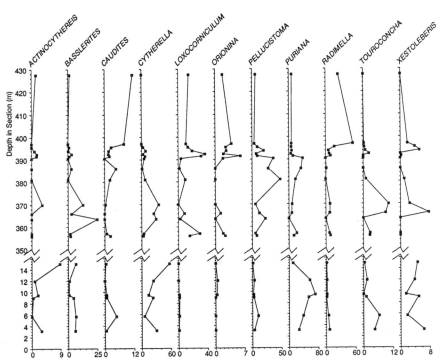

Fig. 4.6   Percentage of eleven environmentally diagnostic ostracode taxa, Rio Banano section, Costa Rica.

and areas of the Gulf of Mexico where scleractinian reefs commonly occur, for example, off Veracruz (Krutak 1982).

The faunal shift in ostracode assemblages may indicate the availability of more carbonate environments in the Limón Basin beginning about 3–2.5 Ma, based on the foraminiferal and nannofossil age data (Coates et al. 1992). Although additional age control is needed to better date this shift, it coincides with or precedes slightly the appearance of reefal and carbonate platform ostracode assemblages in the Late Pliocene Moin Formation that outcrops to the north of the Rio Banano sections and in the Escudo de Veraguas Formation in the Bocas del Toro region of Panama to the south.

PALEOCEANOGRAPHY OFF EASTERN NORTH AMERICA

The Neogene record of southern Florida provides an excellent opportunity to examine correlative Late Pliocene paleoceanographic and biotic events in a subtropical area near the Caribbean under the direct influence of the Gulf Stream. Today, southern Florida in the vicinity of the Florida Keys represents a transitional ostracode province between tropical faunas of the Caribbean and subtropical faunas living off the southeastern U.S. on the Atlantic side of the Florida peninsula and the northwest Gulf of Mexico on the west side (Bold 1977).

Florida stratigraphy. We studied ostracodes from outcrops of the Tamiami (including the Pinecrest beds), the Caloosahatchee and the Bermont Formations exposed at the APAC and Quality Aggregate (QA) pits near Sarasota (Jones et al. 1991), and the Desoto 5 pit in DeSoto County. Figure 4.7 and table 4.2 give the composite stratigraphy of the three pits, following Jones (Jones et al. 1991) and Ketcher (1993, personal communication). Figure 4.7 also shows the paleotemperature history on the basis of the ostracode assemblages.

We generally follow the age assignments of Jones (Jones et al. 1991), supplemented by ostracode biostratigraphic evidence from the Florida sequence (Willard et al. 1993), which allows us to correlate the Florida beds to the Yorktown Formation of Virginia and the Duplin and Raysor Formations of South and North Carolina (figure 4.1). These latter units have been well dated by planktonic foraminifers and calcareous nannofossils (see Dowsett and Wiggs 1992; Ward and Huddleston 1988). Allmon (1993) recently described the environment and mode of deposition of the Pinecrest beds.

"Unit 11" is an informal unit in the lowermost part of the Tamiami Formation (Ketcher 1993), most likely equivalent to the Sunken Meadows Member of the Yorktown Formation, deposited about 4 Ma. The ostracode fauna correlates with the *Pterygocythereis inexpectata* zone of Virginia (Hazel 1971a, 1977). Overlying unit 11, the lower Pinecrest beds of the Tamiami Formation represent a major transgression between about 3.5 and 2.8 Ma. These beds correspond to the Rushmere, Morgarts Beach, and probably the Moore House Members of the Yorktown. The ostracode assemblages correlate with the *Orionina vaughani* zone of Virginia (Hazel 1971a) and the *Murrayina barclayi* zone of the Carolinas (Cronin 1981). If the reversed paleomagnetic polarity of the Pinecrest beds obtained by Jones and others (1991) is correct, then at least part of the Pinecrest was deposited during the Kaena or Mammoth subchrons within the Gauss chron.

The upper Pinecrest beds are separated from the lower Pinecrest beds by a depositional hiatus indicated by the presence of a large Late Blancan vertebrate fauna, brackish-water ostracodes, pyritized microfossils, paleomagnetic data (reversed lower Matuyama Chron), lithologic changes, and ostracode biostratigraphy. These beds also contain a

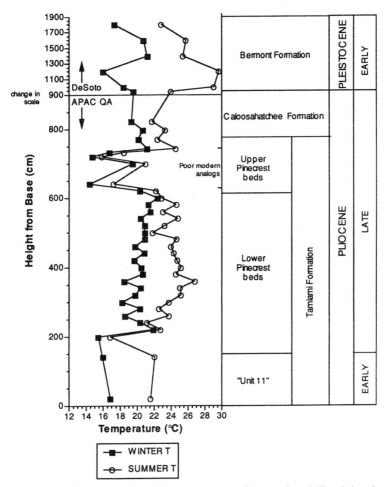

Fig. 4.7 Composite stratigraphy and water temperature history of south Florida based on analysis of ostracodes using the transfer function technique (Cronin and Dowsett 1990).

distinct ostracode fauna that may correlate with the *Puriana mesacostalis* zone of the Chowan River Formation of southeastern Virginia and northeastern North Carolina (Jones et al. 1991). The Caloosahatchee and Bermont Formations represent marine units of the latest Pliocene to Early Pleistocene age.

Florida paleotemperature history. We used the ostracode transfer function described by Cronin and Dowsett (1990) to estimate water temperatures during the deposition of these units (table 4.2; fig. 4.7). This technique was developed through Q-mode factor analysis of ostra-

Table 4.2 Estimated Winter and Summer Water Temperature (°C) for Floridian Late Neogene

			SAMPLE #			CM FROM BASE	COMMUNALITY	WINTER T	SUMMER T
	DeSoto		91TC19	Bermont Fm.		1800	0.427	17.30	22.86
			91TC18			1600	0.443	20.77	25.76
			91TC17			1400	0.797	21.22	25.46
			91TC16			1200	0.386	15.99	29.69
			91TC15			1000	0.845	18.39	29.08
			91TC14	Caloosa-hatchee Fm.		940	0.809	19.62	24.04
	APAC		K14			822	0.591	19.35	21.75
			K12			798	0.421	20.70	23.34
			K10			771	0.600	20.14	22.47
			K8	Upper Pinecrest		744	0.474	21.17	24.60
			K7			732	0.637	16.78	18.55
			K6			720	0.219	14.72	15.87
QUALITY AGGREGATES			QA-6-51			700	0.140	19.50	20.98
			QA-6-48			640	0.408	14.45	17.25
			QA-6-47			620	0.135	20.35	22.26
			QA-6-46			600	0.446	22.45	22.92
			QA-6-38			580	0.652	21.38	24.66
			QA-6-37			560	0.651	21.62	23.09
			QA-6-36		Tamiami Formation	540	0.611	20.45	24.81
			QA-6-35			520	0.604	20.93	23.27
			QA-6-34			500	0.350	20.91	21.85
			QA-6-33			480	0.249	20.91	24.66
			QA-6-32	Lower Pinecrest		460	0.616	19.75	24.04
			QA-6-24			440	0.391	20.83	24.31
			QA-6-23			420	0.542	19.64	24.72
			QA-6-22			400	0.454	20.49	25.19
			QA-6-21			380	0.570	20.66	24.63
			QA-6-20			360	0.425	18.51	26.81
			QA-6-19			340	0.732	20.43	25.08
			QA-6-18			320	0.694	19.79	25.22
			QA-6-17			300	0.686	18.28	23.80
			QA-6-16			280	0.569	20.34	22.60
			QA-6-15			260	0.614	18.57	23.72
			QA-6-14			240	0.524	20.35	21.15
			QA-6-13			220	0.452	21.96	22.76
			QA-6-12			200	0.378	15.43	16.85
			QA-6-8	Unit 11		140	0.264	15.97	22.08
			QA-6-2			20	0.549	16.88	21.61

code assemblages from 100 modern samples from the North Atlantic and Caribbean and multiple regression of factor analytic results against modern August and January bottom-water temperatures for each modern sample (see Imbrie and Kipp 1971; Cronin and Dowsett 1991). Pliocene August temperature estimates have an accuracy of +/−1.6°C and February temperature estimates of about 2.0°C.

Communalities provide a measure of the amount of variance explained by the assemblage and are thus a useful measure of the degree of confidence one should place in the temperature estimates. When the fossil assemblages are unlike those in the modern database in terms of the species proportions, communalities will be low; when similar, communalities will be higher. The Pinecrest beds and the Caloosahatchee Formation at DeSoto and the Bermont Formation yielded communalities associated with the paleotemperature estimate mostly between 0.55 and 0.84. Unit 11 had low communalities, signifying that there is no modern analog for this assemblage. Nonetheless, the ostracode assemblage includes *Murrayina, Bensonocythere,* and *Echinocythereis,* and lacks thermophilic species that appear in the Pinecrest beds, suggesting a mild temperate climatic zone during the Early Pliocene. Except for a noteworthy rise in summer temperatures in their lower part (200–400 cm, fig. 4.7), the overlying Pinecrest beds generally show relatively stable water temperatures during most of their depositional history. The lower Pinecrest beds indicate summer temperatures that increased from 21°C to almost 26°C and winter water temperatures that were 18–20°C. Then winter and summer temperatures stabilized at about 20°C and 25°C, respectively. In the upper 1.5 m, winter temperature was 20–21°C and summer temperature was 23–24°C.

In the upper Pinecrest beds, there are low communalities for the brackish-water facies and then a return to marine conditions. At both QA (samples QA-48, QA-51) and APAC (samples K6, K7), high percentages occur of species that dominate large bays and lagoons along the East Coast and Gulf of Mexico coast today. These environments were not included in the ostracode transfer function database, and thus these beds lack good modern analog assemblages. The apparent drop in temperatures at 640–732 cm (fig. 4.7; table 4.2) is an artifact of this non-analog situation (communalities were 0.1 to 0.4), and it is presently difficult to estimate water temperatures from these restricted environments. However, it is clear from the Late Pliocene stratigraphy in Florida and elsewhere in the Atlantic Coastal Plain that a major drop in sea level occurred at this time. Marine conditions returned during the deposition of the upper Pinecrest, when winter and summer temperatures ranged from about 17°C to 19°C, respectively (732 cm), and 19–21°C to 21–24°C (744–822 cm). These estimates represent Late Pliocene temperatures during the marine transgression that occurred about 2.4–2.0 Ma.

One Caloosahatchee (91TC14, table 4.2) and two Bermont samples (91TC15, 91TC17) had communalities of about 0.8–0.85 and winter and summer temperatures of 18–21°C and 24–29°C. Although these results are preliminary, they suggest that an increase in seasonality in marine environments occurred during the Pleistocene compared to the Middle and Late Pliocene.

In summary, water temperatures in southern Florida near 3.5 Ma were 18–20°C during the winter, but as low as 21–22°C during summer. Between 3.5 and 2.8 Ma, winter temperatures changed little, whereas summer temperatures increased to 24–26°C. Following a major regression, winter and summer temperatures during the subsequent transgression near 2.4–2.0 Ma were about 20°C and 21–24°C. This interval correlates with the period of deposition of the Moin Formation and the upper Escudo de Veraguas and the upper part of the Rio Banano Formations, discussed above.

At sometime during the Early Pleistocene, winter temperatures decreased slightly and summer temperatures increased. Today winter and summer water temperatures off southwest Florida, at depths comparable to those in which the Pinecrest beds were deposited, average about 16°C and 27°C (see Cronin and Dowsett 1990). Pliocene annual temperature ranges were narrower than those of today, with warmer winters and cooler summers. We have found no evidence from the ostracode faunas that true tropical conditions existed during the deposition of the Pinecrest beds or the Caloosahatchee. Cooler conditions than those of today existed during the Early Pliocene near 4.0 Ma and again near 3.5 Ma.

OPEN OCEAN PLIOCENE PALEOCEANOGRAPHY

The record of paleoceanographic changes within the North Atlantic Basin can be monitored through quantitative analysis of planktonic foraminiferal faunas from a series of Deep Sea Drilling Project (DSDP) cores (fig. 4.2). A factor analytic transfer function (GSF18) has been developed (Dowsett and Poore 1990; Dowsett 1991) specifically to monitor Pliocene to Recent sea-surface conditions in the North Atlantic Basin. GSF18 was constructed using the methodology pioneered by Imbrie and Kipp (1971) and Kipp (1976): factor analysis of core-top abundance of planktonic foraminifers and multiple regression of resulting factors against physical parameters such as SST or salinity (standard errors of estimate for SST are 1.5°C and 1.4°C for cold and warm seasons respectively; standard error for mean annual salinity estimates is approximately 0.4‰). The resulting equations are used to transform microfossil abundance data into SST estimates (Dowsett 1991).

Western Caribbean Sea (site 502). Site 502 (figs. 4.1, 4.2) is located in the Colombia Basin, due north of Colón, on a small fault block (3051 m) that rises above the surrounding seafloor (Prell et al. 1982). SST estimates (Dowsett and Poore 1991) based on the transfer function GSF18 (Dowsett and Poore 1990; Dowsett 1991) show high-frequency, low-amplitude fluctuations throughout the 4 to 2 Ma interval (fig. 4.8). During this interval, SST varied consistently about the modern mean of 27.9°C. In addition, spectral analysis of the faunal record shows the high frequency variability to be very close to the eccentricity, obliquity, and precession frequencies associated with Milankovitch forcing (Dowsett and Poore 1991). This "constancy" of SST in the tropics is representative of the low latitude North Atlantic based upon analysis of other deep-sea cores (Dowsett and Poore 1991).

Western Atlantic Margin (site 603). DSDP site 603 lies at the foot of the continental rise, about 550 km east of Cape Hatteras, North Carolina (fig. 4.1). The modern planktonic fauna at this site records conditions just east of the main path of the Gulf Stream Current. While planktonic faunas exhibit a moderate degree of dissolution throughout most of the record (which tends to decrease SST estimates through removal of fragile, warm-water taxa) there is an impressive difference from modern conditions occurring near 3 Ma (fig. 4.8). Winter SST estimates at that time are 5.4°C warmer than those of today, while summer estimates are indistinguishable from today's (Dowsett and Poore 1991). This indicates a dramatic decrease in seasonality (relative to today), which is corroborated by planktonic estimates from Yorktown Formation samples (Dowsett and Wiggs 1992) and the ostracode data from Florida presented above. Conditions similar to this are currently found south of Cape Hatteras where the Gulf Stream diverges from the North American continent. The records from the Yorktown Formation and site 603 suggest that the North Atlantic gyre may have been displaced to the north at that time. For much of the time between 2.5 and 2.0 Ma, a similar low seasonality warm winter situation seems to have existed. Immediately after 2.0 Ma, both winter and summer estimates from site 603 resemble modern conditions.

Northeastern Atlantic (sites 548 and 552). We have monitored surface conditions through analysis of planktonic foraminiferal assemblages at sites 552 and 548, located in the path of today's North Atlantic Drift (Dowsett and Poore 1990; Dowsett and Loubere 1992). Both records are remarkably similar to each other and to the temperature record obtained from analysis of ostracodes from Tjörnes, Iceland (Cronin 1991a). We will limit our discussion to the record at site 552, which has received much attention in recent years (Shackleton et al. 1984; Raymo et al. 1989; Curry and Miller 1989; Dowsett and Poore 1991). Factor analytic transfer functions show winter temperatures prior to 4 Ma were approximately 3°C cooler than current conditions in that region. Between 4 and 3.5 Ma, winter temperatures were considerably warmer than today. Between approximately 3.5 and 3.15 Ma, winter temperatures were again much cooler than the modern northeastern Atlantic. A warm interval between 3.15 and 2.85 Ma indicates winter or cold season temperatures about 3.6°C warmer and warm season temperatures more than 7°C warmer than modern conditions. From about 3 Ma on, the record shows a continuous change toward cooler conditions similar to today. These large-scale changes are superimposed on a record of high-frequency, low-amplitude "glacials" and "interglacials." However, during the large-scale intervals of warmth, "glacials" rarely reached SSTs as cool as the current interglacial.

In summary, analysis of these deep-sea cores provides some important insights into the paleoceanographic development of the North Atlantic during the Pliocene. First, while the tropics exhibited minor if any changes in surface temperature, the high-latitude regions show large-scale changes that we believe reflect the sensitivity of the extra-tropics to any kind of forcing. The evidence of warmer, more equable conditions along the North American coast and extreme warmth in the northeastern Atlantic suggest an enhanced Gulf Stream system leading up to 3 Ma. This trend culminated with a glacioeustatic high-stand just prior to 3 Ma (Dowsett and Cronin 1990), owing to the melting of substantial portions of Antarctic ice (Webb and Harwood 1991), which was apparently high enough to breach the recently emerging Central American Isthmus. The evidence for breaching consists of mean surface salinity estimates from the planktonic faunas at site 502 (fig. 4.8). While little confidence can be placed on the absolute values of the salinity estimates, the relative changes and trends in salinity are considered reliable and can be seen qualitatively as changes in assemblage composition. The estimates show a sustained trend toward decreasing salinity across the 3 my interval, which we interpret as indicating that eastern Pacific low-salinity water entered and mixed with Caribbean water. Afterward, a trend of increasing salinity from 2.8 to 2.5 Ma may indicate reclosure and possibly isolation of the Pacific and Caribbean at 2.8 Ma. While not as clear from the 502 paleosalinity record, based on gross trends in salinity and evidence from the Atlantic Coastal Plain and elsewhere in the North Atlantic (see below), another breach of the isthmus may have occurred just prior to 2 Ma.

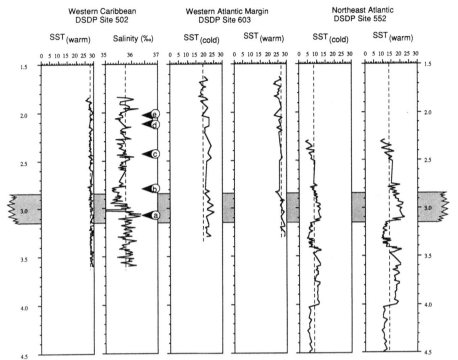

Fig. 4.8 Sea surface temperature (SST in °C) history of Caribbean (site 502) and North Atlantic (sites 603 and 552) based on planktonic foraminifer transfer function GSF18 (Dowsett and Poore 1991). Estimates of surface salinity are also given for site 502. Vertical dashed lines indicate modern conditions. Horizontal gray band indicates time slice (Ma) investigated by PRISM Project. Arrows on site 502 salinity panel give one interpretation of CAI closure history: prior to (a) CAI was closed and was breached at (a); CAI remained breached until (b), when it closed; CAI may have been breached again at (c?) and (d) and closed at (e) based upon salinity estimates.

DISCUSSION

Our studies have led us to develop the hypothesis that the closure of the Isthmus of Panama explains many observed paleoceanographic patterns in the Caribbean and North Atlantic Ocean. Table 4.3 presents the relationships between the isthmus, southwestern North Atlantic paleotemperatures, deep-sea salinity and temperatures, and sea level events between 4 and 2 Ma. Figure 4.9 represents the oceanographic history in the tropical and subtropical western North Atlantic based on the shallow marine record discussed above. Prior to 4 Ma, the isthmus was open, and relatively cool marine waters characterized the shelf off the

Table 4.3 Summary of Climatic Events Related to Closure of Central American Isthmus, 4–2 Ma

Time Interval (Ma)	CAI*	Southwestern North Atlantic Temperatures	Caribbean Salinity	Eustatic Sea Level	Atlantic Coastal Plain Marine Record
2.4–2.0	closed?	warm	low	high	Pulse 2
2.8–2.4	closing	cool	low → normal	low	
3.1–2.8	open	warm	decreasing	high	Pulse 1
3.5–3.1	closing	cool → warm	normal	high	Pulse 1
4.2–3.5	open	cool	low?	high	

*Central American Isthmus. Closing and opening of isthmus refers to surface water. Deeper waters were closed during Miocene.

eastern U.S., probably reflecting southward-flowing cool currents. There is no evidence from ostracode assemblages from the Atlantic Coastal Plain for warm Gulf Stream water during this period (see above). It is worth emphasizing that these cool water conditions along the southeastern U.S. occurred during a period of glacioeustatic high sea level and also during a period of warm climates in many parts of the world (see Gladenkov et al. 1991). Because this period preceded major Northern Hemisphere ice buildup, the Early Pliocene transgression along the Atlantic Coastal Plain was almost entirely due to reduced Antarctic ice volume (Webb and Harwood 1991). At this time, planktonic foraminiferal data indicate that Caribbean surface salinities were low.

Between 3.5 and 3.1 Ma, when global sea level was relatively high (Dowsett and Cronin 1990), water temperatures off the eastern U.S. began increasing (pulse 1 in figure 4.9). At the same time, Caribbean salinities were relatively high, suggesting the isthmus was beginning to close the surface-water connection between the Pacific and Atlantic. Increased transport of heat from low to high latitudes at this time would explain microfaunal evidence for warm oceanic and terrestrial Pliocene climates in high latitudes of the North Atlantic region and the Arctic Ocean (Cronin et al. 1993).

Eustatic sea level reached levels as high as 25–35 m above present-day levels by about 3.1–2.8 Ma (Dowsett and Cronin 1990), most likely owing to reduced Antarctic ice (Webb and Harwood 1991). The western North Atlantic ocean temperatures remained relatively high (Cronin 1991b; Dowsett et al. 1992), as did North Atlantic sea surface temperatures (Dowsett and Poore 1991). Raymo and others (1992) showed that by 3.0–2.6 Ma, there was strong NADW formation at DSDP site

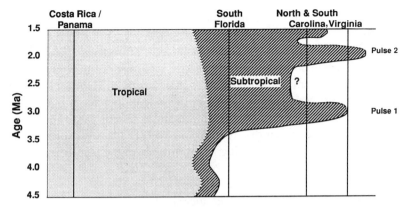

Fig. 4.9 Effects of the closing of the Central American Isthmus on shallow-water faunas and oceanic conditions off the eastern United States. Results based on sections from Costa Rica/Panama, south Florida, the Carolinas, and Virginia. Pulse 1 and pulse 2 refer to warm tropical water pulses caused by two separate closures, at about 2.8 Ma and 2 Ma. These events allowed the dispersal of thermophilic ostracode taxa to middle and high latitudes.

607, evidence consistent with the idea that there was enhanced transport of warm, saline surface waters from the tropics (Dowsett et al. 1992). These events seem to be related to the diminished exchange of surface water across the Isthmus of Panama, but it is difficult to be certain if the isthmus became completely closed at this time.

In fact, on the basis of the DSDP site 502 record (between points a and b in fig. 4.8), Caribbean salinities began decreasing markedly by 3.1–2.8 Ma, strongly suggesting a mixing of Pacific and Atlantic waters due to the breaching of the isthmus. Between 2.8 and 2.4 Ma, salinity rose again in the Caribbean (b–c in fig. 4.8). If this reversal in the salinity trend indicates the isthmus was closing near 2.8–2.4 Ma, then the closure would coincide with, and perhaps be related to, the time of major Northern Hemisphere ice buildup and major drops in eustatic sea level around 2.6–2.4 Ma. Based on isotopic evidence from deep-sea cores, Sarnthein and Tiedemann (1989) estimated that a sea level drop of about 50 m occurred at this time. This estimate is supported by evidence from the marine record of the Sea of Japan (Cronin et al. 1994) and from the major disconformity reflecting marine regression in the Atlantic Coastal Plain (Dowsett and Cronin 1990). Raymo and others (1992) documented moderate suppression of NADW formation in the North Atlantic Ocean at this time. The ostracode data presented above also indicate environmental changes were occurring in the Limón Basin of Costa Rica at this time, possibly related to oceanographic changes in the western Caribbean, but more detailed studies are needed to confirm the timing of these changes.

If our interpretations of these various lines of evidence are correct, then one can hypothesize that a negative feedback occurred in which the increased oceanic heat transport to high Northern Hemisphere latitudes near 3.0 Ma that resulted from the emerging isthmus led to reduced polar ice, high eustatic sea level, and a resumption of surface-water exchange between the Atlantic and Pacific. This exchange might have had the effect of decreasing heat transport to the high latitudes of the North Atlantic, diminishing NADW formation, and leading to the glacial events and drop in sea level near 2.6–2.3 Ma.

Between about 2.4 and 2.0 Ma, increased water temperatures along the western North Atlantic are in evidence during the major marine transgression that deposited the Chowan River Formation of Virginia and the upper Pinecrest beds of Florida (pulse 2, fig. 4.9). Hazel (1971b) found a peak in Pliocene water temperatures in Virginia and North Carolina in the Chowan River Formation (Hazel's lower *Puriana mesacostalis* zone). Although this transgression was not as extensive as that during the deposition of the Yorktown Formation, water temperatures along the eastern U.S. in some regions exceeded those of Yorktown. This transgression coincides with warm climates in high latitudes in evidence from the faunal and floral assemblages of the Kap Kobenhavn Formation, northern Greenland (Funder et al. 1985), and warm-water assemblages of the St. Erth beds of England (Maybury and Whatley 1988). It also coincides with reduced amplitude in high-frequency climatic cycles in the North Atlantic about 2.3–2.0 Ma and relatively strong NADW formation (Raymo et al. 1992), which one might expect if there was enhanced flow of warm saline surface water from the tropics. Thus, although there is not a strong drop in salinity at site 502 between 2.3 and 2.0 Ma as there was between 3.5 and 3.1 Ma, other indirect lines of evidence, including the possible upwelling in the Limón Basin discussed above, are consistent with the hypothesis that the isthmus closed a second, and possibly a final time, near 2.0 Ma. In summary, if we are correct that the emergence of the isthmus near 3.5–3.0 Ma directly or indirectly caused increased warming throughout the North Atlantic, then a second closing may also explain observed oceanographic evidence for relative warmth, high sea level, and strong NADW formation between 2.3 and 2.0 Ma.

If Late Pliocene pulses of warmth are the direct result of intensification of the Gulf Stream due to the CAI, it supports the hypothesis that short-term reorganization of ocean circulation due to the closure of a gateway can severely alter hemispheric climate. Enhanced heat flux near 3.0 and 2.0 Ma strongly influenced the northern North Atlantic and Arctic Oceans, leading to warm ocean temperatures at high latitudes (Cronin 1991b) and seasonally or even perennially ice-free conditions in parts of the Arctic (Scott et al. 1989; Brigham-Grette and Carter 1992; Cronin et al. 1993). Conversely, Early Pliocene warm climates between 4.5 and 4.0 Ma, known from many Northern and Southern Hemisphere sites, were not necessarily related to changes in the North Atlantic Drift system, but to separate factors, such as different atmospheric CO_2 levels.

We point out that this explanation only addresses the long-term pattern of oceanic changes and not the high-resolution changes occurring over 0.01–0.1 my and that these hypotheses must be tested with additional studies of Pliocene oceanography in the Caribbean. Moreover, there are other mechanisms operating during the Late Neogene that might lead to the observed climatic changes; notably, tectonic factors, such as the uplift of the Tibetan Plateau, many have strongly altered global climate (Raymo and Ruddiman 1992; see Raymo 1994 for a comprehensive review). Nonetheless, the potential influence of the isthmus in altering oceanic circulation should be considered one factor in explaining climatic events in the North Atlantic and Arctic Oceans and perhaps globally.

ACKNOWLEDGMENTS

We are very grateful to Jeremy B. C. Jackson and Anthony G. Coates for allowing us to participate in the Panama Paleontology Project and inviting us to contribute to this volume. Without their help, and that of Nancy Budd, and a host of project members in Panama, this study would not have been possible. We owe a great deal of thanks to Mathew Cotton for his assistance in sample preparation and work on Bocas del Toro and Limón foraminifers over the last four years of this project. Robert Ross helped with the analysis of Panisthmian ostracodes. Warren Allmon, Douglas Jones, Lynn Wingard, and Kathy Ketcher were invaluable in introducing Thomas Cronin to the stratigraphy of the Pinecrest beds. Debra Willard kindly provided preliminary pollen analyses from the Florida beds. Finally, special thanks are due to Andrew Shuckstes for his tireless work on the ostracodes from the Florida Neogene and to Thomas Holtz for his many computer analyses and graphical expertise. This chapter was originally part of the U.S. Geological Survey PRISM Project.

REFERENCES

Allmon, W. D. 1993. Age, environment, and mode of deposition of the densely fossiliferous Pinecrest Sand (Pliocene of Florida): Implications for the role of biological productivity in shell bed formation. *Palaios* 8.

Berggren, W. A. 1972. Late Pliocene–Pleistocene glaciation. *Initial repts. DSDP* 12:953–63.

Berggren, W. A., and C. D. Hollister. 1974. Paleogeography, paleobiogeography, and the history of circulation in the Atlantic Ocean. In *Studies in paleoceanography*, ed. W. W. Hay, 126–86. Society of Economic Paleontologists and Mineralogists, Special Publication 20.

Bold, W. A. van den. 1977. Distribution of marine podocopid ostracodes in the Gulf of Mexico and the Caribbean. In *Aspects of ecology and zoogeography of Recent and fossil Ostracoda*, ed. H. Loeffler and D. Danielopol, 175–86. The Hague: Dr. W. Junk.

———. 1981. Distribution of Ostracoda in the Neogene of Central Haiti. *Bulls. Am. Paleontol.* 79:1–136.

———. 1988. Neogene paleontology in the northern Dominican Republic. Part 7, The subclass Ostracoda (Anthropoda: Crustacea). *Bulls. Am. Paleontol.* 94:1–105.

Brigham-Grette, J., and L. D. Carter. 1992. Pliocene marine transgressions of Northern Alaska: Circumarctic correlations and paleoclimatic interpretations. *Arctic* 45 (1): 74–89.

Broecker, W. S., and G. H. Denton. 1989. The role of ocean-atmosphere reorganization in glacial cycles. *Geochim. Cosmochim. Acta* 53:2465–2501.

Broecker, W. S., D. M. Peteet, and D. Rind. 1985. Does the ocean atmosphere system have more than one stable mode of operation? *Nature* 315:21–26.

Charles, C. D., and R. G. Fairbanks. 1992. Evidence from Southern Ocean sediments for their effect of North Atlantic deep-water flux on climate. *Nature* 355: 416–19.

Coates, A. G., J. B. C. Jackson, L. S. Collins, T. M. Cronin, H. J. Dowsett, L. M. Bybell, P. Jung, J. A. Obando. 1992. Closure of the Isthmus of Panama: The near-shore marine record of Costa Rica and western Panama. *Geol. Soc. Am. Bull.* 104:814–28.

Cronin, T. M. 1981. Rates and possible causes of neotectonic vertical crustal movements of the emerged southeastern United States Atlantic Coastal Plain. *Geol. Soc. Am. Bull.* 92:812–33.

———. 1983. Bathyal ostracodes from the Florida-Hatteras Slope, the Straits of Florida, and the Blake Plateau. *Mar. Micropaleontol.* 8:89–119.

———. 1988. Evolution of marine climates of the U.S. Atlantic coast during the past four million years. *Phil. Trans. R. Soc. Lond.* B-318:661–78.

———. 1991a. Late Neogene marine Ostracoda from Tjörnes, Iceland. *J. Paleontol.* 65:767–94.

———. 1991b. Pliocene shallow water paleoceanography of the North Atlantic Ocean based on marine ostracodes. *Quatern. Sci. Revs.* 10:175–88.

Cronin, T. M., and H. J. Dowsett. 1990. A quantitative micropaleontologic method for shallow marine paleoclimatology: Application to Pliocene deposits of the western North Atlantic Ocean. *Mar. Micropaleontol.* 16:117–48.

Cronin, T. M., and H. J. Dowsett, eds. 1991. Pliocene climates. *Quatern. Sci. Revs.* 10:115–296.

Cronin, T. M., A. Kitamura, N. Ikeya, and T. Kamiya. 1994. Mid-Pliocene paleoceanography of the Sea of Japan. *Palaeogeogr., Palaeoclimatol., Palaeoecol.* 108:437–56.

Cronin, T. M., R. Whatley, A. Wood, A. Tsukagoshi, N. Ikeya, E. M. Brouwers, and W. M. Briggs, Jr., 1993. Microfaunal evidence for elevated mid-Pliocene temperatures in the Arctic Ocean. *Paleoceanography* 8:161–73.

Curry, W. B., and K. G. Miller. 1989. Oxygen and carbon isotopic variation in Pliocene benthic foraminifers of the equatorial Atlantic. *Proc. ODP, Sci. Results* 108:157–66.

Dowsett, H. J. 1991. The development of a long-range foraminifer transfer function and application to Late Pleistocene North Atlantic climatic extremes. *Paleoceanography* 6:259–73.

Dowsett, H. J., and T. M. Cronin. 1990. High eustatic sea level during the Middle Pliocene: Evidence from the southeastern U.S. Atlantic Coastal Plain. *Geology* 18:435–38.

Dowsett, H. J., T. M. Cronin, R. Z. Poore, R. S. Thompson, R. C. Whatley, and A. M. Wood. 1992. Micropaleontological evidence for increased meridional heat transport in the North Atlantic Ocean during the Pliocene. *Science* 258:1133–35.

Dowsett, H. J., and P. Loubere. 1992. High resolution Late Pliocene sea-surface temperature record from the Northeast Atlantic Ocean. *Mar. Micropaleontol.* 20:91–105.

Dowsett, H. J., and R. Z. Poore. 1990. A new planktic foraminifer transfer function for estimating Pliocene-Holocene paleoceanographic conditions in the North Atlantic. *Marine Micropaleontology* 16:1–24.

———. 1991. Pliocene sea surface temperatures of the North Atlantic Ocean at 3.0 Ma. *Quatern. Sci. Revs.* 10:189–204.

Dowsett, H. J., and L. B. Wiggs. 1992. Planktonic foraminiferal assemblage of the Yorktown Formation, Virginia, USA. *Micropaleontology* 38(1): 75–86.

Einarsson, T., and K. J. Albertsson. 1988. The glacial history of Iceland during the past three million years. *Phil. Trans. R. Soc. Lond.* B-318:637–44.

Einarsson, T., Hopkins, M. D., and R. R. Doell. 1967. The stratigraphy of Tjörnes, northern Iceland, and the history of the Bering Land Bridge. In *The Bering Land Bridge*, ed. D. M. Hopkins, 312–25. Stanford: Stanford University Press.

Eiriksson, J., A. I. Gudmundsson, L. Kristjansson, and K. Gunnarsson. 1990. Paleomagnetism of Pliocene-Pleistocene sediments and lava flows on Tjörnes and Flatey, North Iceland. *Boreas* 19:39–55.

Funder, S., N. Abrahamsen, O. Bennike, and R. W. Feyling-Hanssen. 1985. Forested arctic: Evidence from north Greenland. *Geology* 13:542–46.

Gladenkov, Yu. B., K. B. Barinov, A. E. Basilian, and T. M. Cronin. 1991. Stratigraphy and paleoceanography of Pliocene deposits of Karaginsky Island, eastern Kamchatka, U.S.S.R. *Quatern. Sci. Revs.* 10:239–45.

Hazel, J. E. 1970. Atlantic continental shelf and slope of the United States: Ostracode zoogeography in the southern Nova Scotian and northern Virginian faunal provinces. *U.S. Geol. Surv. Prof. Pap.* 529-E:1–21.

———. 1971a. Ostracode biostratigraphy of the Yorktown Formation (Upper Miocene and Lower Pliocene) of Virginia and North Carolina. *U.S. Geol. Surv. Prof. Pap.* 704:1–13.

———. 1971b. Paleoclimatology of the Yorktown Formation (Upper Miocene and Lower Pliocene) of Virginia and North Carolina. In *Paléoécologie ostracodes*, ed. H. J. Oertli, 361–75. Bull. Centre Rech. Pau, France: SNPA.

———. 1975. Ostracode biofacies in the Cape Hatteras, North Carolina, Area. *Bulls. Am. Paleontol.* 65:463–87.

———. 1977. Distribution of some biostratigraphically diagnostic ostracodes in the Pliocene and Lower Pleistocene of Virginia and North Carolina. *J. Res. U.S. Geol. Surv.* 5(3):373–88.

Imbrie, J., and N. G. Kipp. 1971. A new micropaleontological method for quantitative paleoclimatology: Application to a Late Pleistocene Caribbean core. In *The Late Cenozoic glacial ages*, ed. K. K. Turekian, 71–181. New Haven: Yale University Press.

Jansen, E., U. Bleil, R. Henrich, L. Kringstad, and B. Slettemark. 1988. Paleoenvironmental changes in the Norwegian Sea and the Northeast Atlantic during the last 2.8 m.y.: Deep Sea Drilling Project/Ocean Drilling Program sites 610, 642, 643, and 644. *Paleoceanography* 3:563–81.

Jansen, E., and J. Sjøholm. 1991. Reconstruction of glaciation over the past six Myr from ice-borne deposits in the Norwegian Sea. *Nature* 349:600–603.

Jones, D. S., and P. F. Hasson. 1985. History and development of the marine invertebrate faunas separated by the Central American Isthmus. In *The Great American Biotic Interchange*, ed. F. G. Stelhi and S. D. Webb, 325–56. New York: Plenum Press.

Jones, D. S., B. J. MacFadden, S. D. Webb, P. A. Mueller, D. A. Hodell, and T. M. Cronin. 1991. Integrated geochronology of a classic Pliocene fossil site in Florida: Linking marine and terrestrial biochronologies. *J. Geol.* 99: 637–48.

Keigwin, L. D., Jr. 1978. Pliocene closing of the Isthmus of Panama, based on biostratigraphic evidence from nearby Pacific Ocean and Caribbean Sea cores. *Geology* 6:630–34.

———. 1982. Isotopic paleoceanography of the Caribbean and east Pacific: Role of Panama uplift in Late Neogene time. *Science* 217:350–52.

Ketcher, K. M. 1993. Preliminary molluscan biozonation of Plio-Pleistocene fossiliferous deposits in South-Central Florida. *Geol. Sci. Am. Ann. Meeting, Abstr. Progr.* A-56.

Kipp, N. G. 1976. New transfer function for estimating past sea-surface conditions from sea-bed distribution of planktonic foraminiferal assemblages in the North Atlantic. In *Investigation of Late Quaternary paleoceanography and paleoclimatology*, ed. R. M. Cline and J. D. Hays, 3–41. Geological Society of America Memoir 145. Boulder, Colo.

Krutak, P. R. 1982. Modern ostracodes of the Veracruz–Anton Lizardo Reefs, Mexico. *Micropaleontology* 28(3): 258–88.

Maier-Reimer, E., U. Mikolajewicz, and T. Crowley. 1990. Ocean General Circulation Model sensitivity experiment with an open Central American Isthmus. *Paleoceanography* 5:349–66.

Marshall, L. G. 1988. Land mammals and the Great American Interchange. *Am. Sci.* 76:380–88.

Maybury, C., and R. C. Whatley. 1988. The evolution of high diversity in the ostracode communities of the Upper Pliocene faunas of St. Erth (Cornwall, England) and northwest France. In *Evolutionary biology of the Ostracoda: Its fundamentals and applications,* ed. T. Hanai, N. Ikeya, and K. Ishizaki, 569–96. Amsterdam: Elsevier.

Prell, W. L., J. V. Gardner, et al. 1982. Site 502: Colombia Basin, western Caribbean Sea. *Initial Repts. DSDP* 68:15–162.

Raymo, M. E. 1994. The initiation of Northern Hemisphere Glaciation. *Ann. Rev. Earth Planet. Sci.* 22:353–83.

Raymo, M. E., D. Hodell, and E. Jansen. 1992. Response of deep ocean circulation to initiation of Northern Hemisphere glaciation (3–2 Ma). *Paleoceanography* 7:645–72.

Raymo, M. E., and W. F. Ruddiman. 1992. Tectonic forcing of Late Cenozoic climate. *Nature* 359:117–22.

Raymo, M. E., W. F. Ruddiman, J. Backman, B. M. Clement, and D. G. Martinson. 1989. Late Pliocene variation in Northern Hemisphere ice sheets and North Atlantic deepwater circulation. *Paleoceanography* 4:413–46.

Raymo, M. E., W. F. Ruddiman, and P. N. Froelich. 1988. Influence of Late Cenozoic mountain building on ocean geochemical cycles. *Geology* 16: 649–53.

Rind, D., and M. Chandler. 1991. Increased ocean heat transports and warmer climate. *J. Geophys. Res.* 94:7437–61.

Sancetta, C., L. Heusser, and M. A. Hall. 1992. Late Pliocene climate in the Southeast Atlantic: Preliminary results from a multi-disciplinary study of DSDP site 532. *Mar. Micropaleontol.* 20:59–75.

Sarnthein, M., and R. Tiedemann. 1989. Toward a high-resolution isotope stratigraphy of the last 3.4 m.y., sites 658 and 659 off northwest Africa. *Proc. ODP, Sci. Results* 108:16–19.

Scott, D. P., P. J. Mudie, V. Baki, K. D. Mackinnon, and F. E. Cole. 1989. Biostratigraphy and Late Cenozoic paleoceanography of the Arctic Ocean: Foraminiferal, lithostratigraphic, and isotopic evidence. *Geol. Soc. Am. Bull.* 101: 260–77.

Shackleton, N. J., J. Backman, H. Zimmerman, D. V. Kent, M. A. Hall, D. G. Roberts, D. Schnitker, J. G. Baldauf, A. Desprairies, R. Homrighausen, P. Huddlestun, J. B. Keene, A. J. Kaltenback, K. A. O. Krumsiek, A. C. Morton, J. W. Murray, and J. Westberg-Smith. 1984. Oxygen isotopes calibration of the onset of ice-rafting and history of glaciation in the North Atlantic. *Nature* 307:620–27.

Stehli, F. G., and S. D. Webb, eds. 1985. *The Great American Biotic Interchange.* New York: Plenum Press.

Teeter, J. W. 1975. Distribution of Holocene Marine Ostracoda from Belize. In *Belize Shelf—carbonate sediments, clastic sediments, and ecology,* ed. K. F. Wantland, and W. C. Pusey, III, 400–498. American Association Petroleum Geologists, Studies in Geology, no. 2. Tulsa, Okla.

Valentine, P. C. 1971. Climatic implication of a Late Pleistocene ostracode assemblage from southeastern Virginia. *U.S. Geol. Surv. Prof. Pap.* 683-D: 1–28.

Veum, T., E. Jansen, M. Arnold, I. Beyer, J.-C. Duplessy. 1992. Water mass exchange between the North Atlantic and the Norwegian Sea during the past 28,000 years. *Nature* 356:783–85.

Ward, L. W., and P. F. Huddlestun. 1988. Age and stratigraphic correlation of the Raysor Formation, Late Pliocene, South Carolina. *Tulane Stud. Geol. Paleontol.* 21(2): 59–75.

Webb, P.-N., and D. M. Harwood. 1991. Late Cenozoic glacial history of the Ross Embayment, Antarctica. *Quatern. Sci. Revs.* 10:215–23.

Willard, D. A., T. M. Cronin, S. E. Ishman, and R. J. Litwin. 1993. Terrestrial and marine records of climatic and environmental changes during the Pliocene in subtropical Florida. *Geology* 21:679–82.

5

The Oxygen Isotopic Record of Seasonality in Neogene Bivalves from the Central American Isthmus

Jane L. Teranes, Dana H. Geary, and Bryan E. Bemis

INTRODUCTION

The formation of the Central American Isthmus during the Late Neogene isolated the previously contiguous Atlantic and Pacific Oceans. As recently as the Miocene, ocean currents passed directly over the area of today's isthmus and environmental differences between the southern Caribbean and the eastern equatorial Pacific were negligible (Holcombe and Moore 1977; Keigwin 1982a; Jones and Hasson 1985). Today, southern Caribbean and eastern Pacific environments differ in a number of important characteristics (Fuglister 1960; Bennett 1963; Wyrtki 1966, 1981; Glynn 1972, 1982; Robinson 1973, 1976; D'Croz et al. 1991). Annually, southern Caribbean waters average 2°C warmer and 1.5‰ (per mil) more saline than those of the eastern equatorial Pacific (Keigwin 1982a; Broecker 1989). This temperature and salinity contrast results from higher rates of evaporation on the Caribbean side and transport of the moisture-rich air to the Pacific side. Another striking environmental contrast is the strong seasonal upwelling that occurs at scattered localities along the Pacific coast during the late winter and early spring. This seasonal upwelling brings cool, nutrient-rich waters to the surface, driving large increases in primary production (Schaefer et al. 1958; Smayda 1965, 1966; Wyrtki 1966, 1981; Glynn 1972; Romine 1982; D'Croz et al. 1991). Seasonal upwelling is not an important environmental influence in most southern Caribbean waters.

Oxygen stable isotope profiles of mollusk shells contain a useful record of the environmental conditions in which the shell grew, including information on seasonal fluctuations (e.g., Horibe and Oba 1972; Killingley and Berger 1979; Wefer and Killingley 1980; Erlenkeuser and Wefer 1981; Williams et al. 1982; Jones et al. 1983; Jones et al. 1986; Krantz et al. 1987). We have analyzed the stable isotopic composition of well-preserved venerid bivalves selected from a sequence of fossiliferous Miocene to Pleistocene strata preserved along the coasts of Panama and

Costa Rica (Coates et al. 1992). The aim of our work is to highlight the potential of oxygen isotope profiles in understanding the Neogene paleoenvironmental history of the Central American Isthmus, particularly with respect to seasonality.

INTERPRETING PALEOENVIRONMENTS FROM OXYGEN ISOTOPIC DATA

Many authors have explored the relationship between the oxygen isotope ratio of shell carbonate and the environmental conditions in which the organism grew (e.g., Emiliani 1978; Fairbanks and Dodge 1979; Killingley and Berger 1979; Wefer and Berger 1980; Wefer and Killingley 1980; Erlenkeuser and Wefer 1981; Jones et al. 1983; Jones et al. 1986; Krantz et al. 1987). The ratio $^{18}O/^{16}O$ is usually reported as a δ (delta) value, defined as per mil deviation from a standard. The standard used in most paleotemperature work is Pee Dee Belemnite, PDB (Epstein and Mayeda 1953). $\delta^{18}O$ is mathematically defined as follows.

$$\delta^{18}O = \{[(^{18}O/^{16}O)_{sample} - (^{18}O/^{16}O)_{standard}] / (^{18}O/^{16}O)_{standard}\} \, 1000$$

The oxygen isotope ratio of mollusk shell carbonate is determined in large part by the temperature and isotopic composition of the ambient water (Urey 1947; Urey et al. 1951; Epstein and Mayeda 1953). Fortunately, physiological effects on mollusk oxygen isotopic composition are minimal (Epstein and Lowenstam 1953; Epstein and Mayeda 1953; Krantz et al. 1984; Jones 1985). Carbonate precipitates from seawater with a characteristic fractionation factor, α, a measurement of the differences in the chemical and physical behavior of the isotopes ^{16}O and ^{18}O. The value for α is inversely related to temperature, so that as the ambient water cools, proportionally more ^{18}O is concentrated in shell carbonate relative to the dissolved carbonate in the water. The temperature dependence of α makes the stable isotope ratio of shell carbonate a useful tool for calculating paleotemperatures. The following equation by Grossman and Ku (1986) gives the relationship between the isotopic composition of the water ($\delta^{18}O_{water \, [MOW]}$), the shell ($\delta^{18}O_{aragonite \, [PDB]}$), and the ambient temperature (T):

$$T(^{\circ}C) = 20.6 - 4.34 \, [\delta^{18}O_{aragonite \, (PDB)} - \delta^{18}O_{water \, (MOW)}],$$

where MOW [Mean Ocean Water] is equivalent to SMOW (Standard Mean Ocean Water) minus 0.2‰)

Seasonal variations in $\delta^{18}O_{water}$ occur in conjunction with salinity changes. Longer-term variations in $\delta^{18}O_{water}$ reflect the waxing and waning of continental glaciers. Because shell carbonate is precipitated in near isotopic equilibrium with seawater, its isotopic composition is correlated with salinity and global ice volume, in addition to the temperature dependence.

The oxygen isotope ratio of seawater is related to salinity through processes of evaporation and freshwater dilution (Craig and Gordon 1965; Broecker 1974). Evaporation enriches seawater in ^{18}O and salts, whereas freshwater influx contains a relatively higher proportion of ^{16}O. In shallow surface waters such as restricted lagoons, bays, and estuaries, $\delta^{18}O_{water}$ and salinity vary significantly and may correlate precisely (Mook 1971). The relationship between salinity and $\delta^{18}O_{water}$ differs from region to region because it depends on the influx of meteoric water (with highly variable $\delta^{18}O$ composition), the seawater $\delta^{18}O$, and regional circulation patterns. Fairbanks and others (1992) evaluated the $\delta^{18}O$-salinity relationship for modern waters of the eastern equatorial Pacific and the western equatorial Atlantic. The following are their empirically derived equations:

$$\delta^{18}O_{Pacific\ water\ (SMOW)} = 0.26\ (S) - 8.77$$
$$\delta^{18}O_{Atlantic\ water\ (SMOW)} = 0.19\ (S) - 5.97,$$

where S is salinity in ‰ units. Empirically derived equations can be determined for modern waters because salinity and $\delta^{18}O_{water}$ can be measured directly. Unfortunately, it is not possible to determine this relationship for ancient water without knowledge of the salinity and the $\delta^{18}O$ of seawater and meteoric water.

Early in the history of paleoenvironmental research, workers recognized that during glacial events shell $\delta^{18}O$ values recorded ice volume as well as paleotemperature (Shackleton and Opdyke 1973, 1976; Shackleton 1974). During glacial times, isotopically "lighter" water is preferentially stored in glacial ice, and ocean water is temporarily enriched in ^{18}O. Shell carbonate precipitated during glacial periods therefore exhibits correspondingly higher $\delta^{18}O$ values.

Estimates of the $\delta^{18}O$ of Pliocene seawater to within a few tenths of a per mil have been compiled by Krantz (1990) from benthic foraminiferal studies by various workers (e.g., Shackleton and Opdyke 1977; Shackleton and Cita 1979; Weissert et al. 1984; Shackleton and Hall 1985; Keigwin 1982b). During interglacial periods from 4.6–2.5 Ma, the offset in $\delta^{18}O_{water}$ relative to today varied between −0.2‰ and −0.6‰. Initiation of major Northern Hemisphere continental glaciation began approximately 2.5 Ma, resulting in higher $\delta^{18}O$ values

(Shackleton 1984; Keigwin 1987). From 2.5 until 1.6 Ma, the offset in $\delta^{18}O_{water}$ ranged from 0.0‰ to -0.2‰ relative to today.

The intrashell range of $\delta^{18}O$ is the most useful isotopic characteristic for interpreting seasonality, and is superimposed over any relative "whole-shell" enrichment in $\delta^{18}O$ (Krantz et al. 1987; Bemis 1992; Bemis and Geary n.d.). For example, Krantz and others (1987) found that oxygen isotope profiles from two Pleistocene specimens of the bivalve *Spisula* show a seasonal range of shell $\delta^{18}O$ superimposed on the shifts in oxygen isotope values attributed to ice-volume effects. Average oxygen isotope values from two specimens, dated at 70 Ka and 125 Ka, were 0.7‰ and 1.5‰ lighter, respectively, than those of modern shells, whereas the range of $\delta^{18}O$ within each shell (approximately 4‰) was attributed to seasonal environmental fluctuations (Krantz et al. 1987). Thus, we stress the relative magnitude of the $\delta^{18}O$ range across the shell as the important paleoenvironmental criterion, not the absolute values.

We evaluated the $\delta^{18}O$ range in the context of the overall pattern of $\delta^{18}O$ fluctuations within each shell. Unless otherwise noted, the reported range of $\delta^{18}O$ is contained within one overall cycle of increase and decrease and is interpreted as one year's record.

Carbon isotopes may contain much information about an organism and its environment, but biological influences on carbon isotopes are poorly understood and interfere with extracting environmental information. Shell $\delta^{13}C$ is determined in part by environmental conditions such as the ambient temperature of the water, phytoplankton productivity, and the dissolved inorganic carbon (DIC) content of the seawater, and in part by biological effects such as metabolically produced CO_2, growth rate, and ontogenetic age (Jones et al. 1983; Krantz et al. 1987). A recent isotopic study of modern venerid bivalves by Bemis (1992) concluded that it is difficult to determine environmental conditions from shell $\delta^{13}C$ profiles. Therefore, for this paper we base our paleoenvironmental interpretations exclusively on $\delta^{18}O$.

Mollusk isotope profiles provide a unique look at paleoseasonality, but are accompanied by serious sampling problems. Although each profile typically includes many analyzed samples, it represents at most only a few years time at a single locality. Particularly in shallow-water habitats, local environmental variation can be great, and these local effects may dominate the pattern recorded in a single shell. The "unusual" nature of any particular profile is difficult to assess without generating profiles from many specimens from the same or nearby localities or horizons. Unfortunately, such repeated sampling is generally unfeasible.

Thus far we have generated profiles for thirty-one fossil and modern shells from the isthmian region (Geary et al. 1992; Bemis 1992; Teranes 1993; Bemis and Geary, n.d.); these provide a solid basis for comparison and interpretation. We will continue to generate additional profiles to enhance geographic and temporal coverage, and to aid in the interpretation of existing profiles. In the meantime, no other such window on paleoseasonality exists.

MATERIALS AND METHODS

Shallow-marine sediments deposited from the Miocene to the Recent along the southern Central American Isthmus contain numerous well-preserved venerid bivalves. Fossil material from these sediments was collected, curated, and placed stratigraphically as part of the Panama Paleontology Project (Coates et al. 1992; Jackson et al. 1993). Table 5.1 outlines collection information and oxygen isotope data for analyzed specimens; localities are shown in figure 5.1.

Bivalves of the family Veneridae were chosen for this study for the following reasons: (1) They are sufficiently abundant to provide geographic and temporal coverage of environmental change during the formation of the isthmus. (2) Venerids precipitate a shell of aragonite, a metastable form of $CaCO_3$, from seawater (Cox 1969; Kobayashi 1969). We analyzed powder from all of the shells using x-ray diffraction and confirmed that fossil shell material is still aragonite. Therefore, the chance that diagenesis has reset the shell isotopic composition is minimal (Dodd and Stanton 1976). (3) Venerid shells are generally sturdy and moderate in size (analyzed shells range from 34 to 52 mm from beak to ventral margin), which facilitates the drilling procedure used to collect sample powder. (4) By restricting our analyses only to shells from the family Veneridae, biological (nonenvironmental) differences between shells are reduced. (5) Venerids live in shallow-marine shelf environments. Our data provide the shallow environmental complement to previous isotopic work on deep-ocean samples (Keigwin 1978, 1982a,b; Hodell et al. 1985).

Serial samples of material were obtained by drilling out short segments parallel to external growth laminations with a 0.5 mm dental burr and then collecting the carbonate powder (fig. 5.2). Care was taken to drill only from the outer shell layer. The average spacing of sampling grooves was approximately 1.0 mm; at least 0.5 mg of sample was collected from each segment. Initially twenty samples were drilled from

Table 5.1 Collection Localities and δ18O Data for Fossil and Modern Venerid Shells

Species and NMB ID Number	Paleodepth[a] (m)	Formation and Location	Age (Ma)	$\delta^{18}O_{PDB}$ (‰) Maximum	Minimum	Range
		Caribbean Side				
Chione (Chionopsis) tegulum 17649	30–50[b]	Lower Gaton Fm. near Colón, Panama	7.1–10.2	-1.2	-2.3	1.1
Liropbora (Panchione) mactropsis 17644	30–50[b]	Middle Gatun, Fm. near Colón, Panama	5.0–7.1	-1.6	-2.9	1.3
Liropbora sp. 17634	40–80	Lower Cayo Agua Fm. Cayo Agua, Panama	4.6–5.0	-0.6	-2.4	1.8
Ventricolaria sp. 17830	20–40	Upper Cayo Agua Fm. Cayo Agua, Panama	≥3.5	-1.2	-2.0	0.8
Ventricolaria sp. 17827	20–40	Upper Cayo Agua Fm. Cayo Agua, Panama	≥3.5	-0.3	-4.1	3.7
Liropbora (Panchione) mactropsis 17772	20–40	Rio Banano Fm. (Quitaria) near Limón, Costa Rica	3.5–3.6	-1.0	-2.2	1.2
Chione (Chionopsis) tegulum 17454	20–40	Rio Banano Fm. (Quitaria) near Limón, Costa Rica	3.5–3.6	-1.4	-2.9	1.5
Chione cf. *amathusia* 17447	20–40	Rio Banano Fm. (La Bomba) near Limón, Costa Rica	2.4–2.5	-1.1	-2.5	1.4

Species	Location					
Ventricolaria sp. 17834	Escudo de Veraguas Fm.	100–150	1.8–1.9	−0.2	−0.9	0.7
Callocardia cf. *catharia*	Escudo de Veraguas, Panama east of Panama Canal (non-upwelling)	24	modern	−1.4	−2.1	0.7
Callocardia cf. *catharia*	off Santa Marta, Colombia (weak upwelling)	16	modern	−0.8	−2.2	1.4
Pacific Side						
Chione cf. *tegulum* 18066	Peñita Fm. Burica Peninsula, Panama	uncertain[c]	≥3.5	−1.3	−3.1	1.8
Chione sp. 17441	Armuelles Fm. Burica Peninsula, Panama	80–400[d]	<1.9	−1.2	−2.6	1.4
Lirophora cf. *kellettii* 17443	Armuelles Fm. Burica Peninsula, Panama	80–400[d]	<1.9	−3.4	−6.3	2.9
Lirophora cf. *kellettii* 17736	Armuelles Fm. south Golfo Dulce, Costa Rica	28–80[d]	<1.9	−0.1	−1.9	1.8
Chione amathusia	Parita Bay, Gulf of Panama (upwelling)	20	modern	−0.4	−3.2	2.8
Megapitaria aurantiaca	Coiba Island, Gulf of Chiriquí (non-upwelling)	<10	modern	−2.3	−3.9	1.6

Note: See map for locations.

[a] Paleodepth information from benthic foraminiferal assemblages (Collins 1993; Collins et al. 1995).

[b] Preliminary paleodepth estimates for all Gatun Fm. material (L. Collins, personal communication).

[c] Collected from float.

[d] Preliminary paleodepth estimates (L. Collins, personal communication).

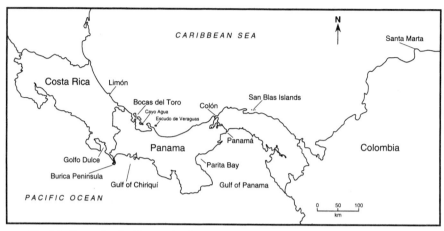

Fig. 5.1   Generalized collection localities in Panama and Costa Rica.

each shell. If significant variation in $\delta^{18}O$ was detected from the first twenty samples, then more samples were drilled and analyzed. (Refer to figures and figure captions for the location of each sample on individual shells and the length of each shell along the axis of maximum growth.)

Splits from a single sample groove may contain significant intra-sample isotopic variation (up to 1‰), presumably because the width of the dental burr is large enough to drill through several growth bands, each with a different isotopic composition. We believe that it is most instructive to analyze an average of each drilled band. Therefore, each sample powder was homogenized by grinding in a mortar and pestle before analysis. Analysis of splits from homogenized samples repro-duced $\delta^{18}O$ values to better than 0.2‰.

The isotopic composition of each sample powder was analyzed in the stable isotope lab of Dr. J. W. Valley at the University of Wisconsin, Madison. Shell material was roasted *in vacuo* for one hour at 350°C to remove any remaining organic material. Roasted samples were then re-acted with concentrated phosphoric acid (density approximately 1.92g/cm³) at either 50°C or 25°C. Necessary calculations were made to adjust for the kinetic fractionation effects from the different reaction tempera-tures. Prepared CO_2 was analyzed on a Finnigan/MAT 251 mass spec-trometer. Oxygen and carbon isotopic results are calculated in standard per mil notation relative to PDB. Analytical precision of the UW Calcite lab standard was typically 0.1‰.

Fig. 5.2 Ventricolaria sp. showing shell sampling technique. Grooves were drilled parallel to externally visible growth lines. A total of twenty-five samples were excavated and analyzed from this shell.

OXYGEN ISOTOPE PROFILES OF MOLLUSK SHELLS FROM MODERN ENVIRONMENTS

Analyses of modern mollusk shells from the eastern Pacific and the southern Caribbean demonstrate that oxygen isotope profiles reliably reflect regional environmental differences (Bemis 1992; Geary et al. 1992; Bemis and Geary n.d.). Locality information and oxygen isotope data from several modern venerid specimens are provided in table 5.1. Understanding the association between temperature, salinity, upwelling, and modern shell isotopic composition is key to accurate interpretation of fossil mollusk profiles.

Eastern equatorial Pacific upwelling environments exhibit pronounced, inversely correlated, seasonal temperature and salinity changes, averaging approximately 18–28°C and 24–34‰ (Bennett 1966; Robinson, 1973; D'Croz et al. 1991). These seasonal environmental swings produce characteristic $\delta^{18}O$ profiles in mollusk shells (Bemis 1992; Geary et al. 1992). Maximum shell $\delta^{18}O$ is precipitated during the late winter or early spring when strong upwelling brings cool waters of normal salinity to the surface. Minimum $\delta^{18}O$ is precipitated during the rainy season when warm, dilute, isotopically "light" meteoric water flows into the Gulf of Panama. For example, a specimen of the gastropod *Strombus gracilior* collected from the Gulf of Panama exhibits two annual cycles in oxygen isotopic composition (fig. 5.3A; Geary et al. 1992). The large range of $\delta^{18}O$ in this shell (4.5‰) reflects seasonal variation in its shallow habitat (probably less than 10 m).

Shell profiles constructed from venerid bivalves also reflect the strong seasonality of the Pacific side. Figure 5.3C presents the isotope profile from a specimen of *Chione amathusia* collected from 20 m depth in Parita Bay, Gulf of Panama. The range of $\delta^{18}O$ across this shell is 3.0‰; the profile records nearly two years of growth. Analyzed venerid bivalves exhibit smaller intrashell $\delta^{18}O$ ranges than do the strombid gastropods, probably because the venerids were collected from greater water depths and therefore experienced less intense seasonal temperature and salinity fluctuations. Furthermore, bivalves generally record a smaller temperature range than gastropods (Epstein and Lowenstam 1953).

In contrast to the upwelling environments of the Pacific, Caribbean coastal waters of northern Panama and northwestern Colombia experience no upwelling and very little seasonal variation in temperature and salinity. Mean seasonal surface-water temperatures at 20 m depth vary between 25.6°C and 26.7°C, and salinities vary between 32.6‰ and 33.8‰ (Levitus 1982). Also in contrast to Pacific waters, seasonal temperature and salinity changes are positively correlated in most Caribbean surface waters. Warmer temperatures are accompanied by increased evaporation and, hence, the saltiest water; cooler temperatures occur during the rainy season when precipitation exceeds evaporation. As a result of this positive correlation, the effects of temperature and salinity on shell $\delta^{18}O$ may partially cancel each other out in the southwestern Caribbean, for example, elevated temperature leads to decreased shell $\delta^{18}O$, whereas elevated salinity is correlated with increased shell $\delta^{18}O$ (Geary et al. 1992).

Reduced seasonal variation in temperature and salinity and the fact that the influences of these variables on shell $\delta^{18}O$ at least partially cancel each other out in the Caribbean produce characteristically flat oxygen isotope profiles. A specimen of *Strombus pugilis* collected from a few meters depth off the San Blas Islands (northern Panama) exhibits a relatively flat isotope profile, with a $\delta^{18}O$ range of 1.5‰ (fig. 5.3B). Figure 5.3D presents an isotope profile of the venerid bivalve *Callocardia* cf. *catharia* collected from 24 m water depth in the southern Caribbean. As with the *Strombus* in fig. 5.3B, the isotope profile is flat (range = 0.7‰), and annual cycles are obscured.

Neither the tropical eastern Pacific nor the southwestern Caribbean presents a uniform set of environments. There are exceptions to the generalizations discussed above, including eastern Pacific localities that

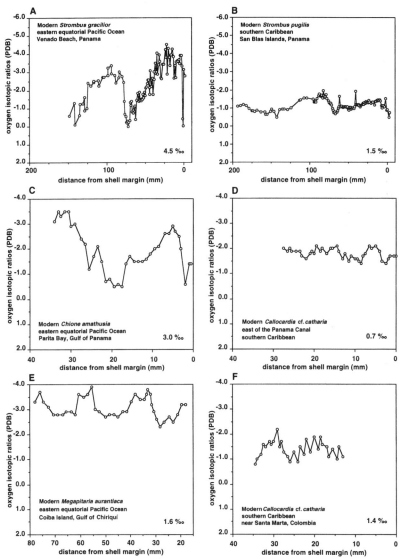

Fig. 5.3 δ¹⁸O profiles for modern specimens. Total intrashell δ¹⁸O range is given at lower right of each plot: (A) δ¹⁸O profile of a modern strombid gastropod from an area of strong upwelling in the eastern equatorial Pacific (after Geary et al. 1992); (B) δ¹⁸O profile of a modern strombid gastropod from area with no upwelling in the southern Caribbean (after Geary et al. 1992); (C) δ¹⁸O profile of a modern venerid bivalve from an area of strong upwelling in the eastern equatorial Pacific (after Bemis 1992); (D) δ¹⁸O profile of a modern venerid bivalve from an area with no upwelling in the southern Caribbean (after Bemis 1992); (E) δ¹⁸O profile of a modern venerid bivalve from an area with no upwelling in the eastern equatorial Pacific (after Bemis 1992); (F) δ¹⁸O profile of a modern venerid bivalve from an area of slight upwelling in the southern Caribbean (after Bemis 1992).

lack upwelling and Caribbean localities with upwelling. For instance, the Gulf of Chiriquí in the eastern Pacific exhibits no significant upwelling, relatively small average annual surface temperature variations (26.5–28.0°C), moderate salinity variations (29.5–33.5‰), and a roughly positive temperature-salinity correlation. As expected, shells from this region have $\delta^{18}O$ ranges intermediate between "typical" Pacific and "typical" Caribbean shells. The profile from a modern specimen of *Megapitaria aurantiaca*, collected from less than 10 m depth off of Coiba Island in the Gulf of Chiriquí is shown in figure 5.3E; the $\delta^{18}O$ range for this shell is 1.6‰.

The waters off northwestern Colombia and the northern Gulf of Venezuela are unusual for the southern Caribbean in that they experience moderate upwelling events (Kinder et al. 1985). In this region at 20 m depth there are moderate seasonal temperature variations (25.8–28.0°C) and a very small annual salinity range (35.5–36.3‰; Levitus 1982). As is characteristic of upwelling environments, temperature and salinity are inversely related. A representative oxygen isotope profile of *Callocardia* cf. *catharia*, collected from 16 m water depth off of Santa Marta, Colombia, is shown in figure 5.3F. Seasonal temperature variations produce two identifiable annual cycles of shell $\delta^{18}O$. The range of shell $\delta^{18}O$ is 1.4‰, smaller than that of Pacific-side upwelling shells.

In summary, seasonal hydrographic features are recorded in mollusk shells. Strong upwelling environments with large seasonal temperature and salinity variations result in large (several ‰) variations in $\delta^{18}O$. Such environments exist today in the Gulf of Panama and to a lesser extent in more localized environments of the southern Caribbean. Areas with no seasonal upwelling and positive temperature-salinity correlation, as in the northwestern Caribbean, produce consistently smaller ranges of shell $\delta^{18}O$.

FOSSIL VENERID SHELLS FROM THE CARIBBEAN COAST

A sequence of highly fossiliferous Neogene sediments is located on the Caribbean side of the isthmus. The rocks record very shallow to shallow inner-shelf (less than 200 m) and upper-slope environments (200–800 m; Coates et al. 1992; Collins 1993). Included in our study are specimens from the Miocene Gatun Formation, the Mio-Pliocene Bocas del Toro Group, and the Mio-Pliocene Limón Group.

Miocene Gatun Formation

The Late Miocene Gatun Formation consists mainly of burrowed, grey-green, tuffaceous litharenite, some siltstone, and abundant fossils, including scattered mollusks (Coates et al. 1992). We analyzed two shells from outcrops of the Gatun Formation near Colón, Panama: shell 17649 from the Lower Gatun Formation, 10.2–7.1 Ma (fig. 5.4A; Collins et al. 1995), and shell 17644 from the Middle Gatun Formation, 7.1–5.0 Ma (fig. 5.4B). The $\delta^{18}O$ ranges of these shells, 1.1‰ and 1.3‰, respectively, are approximately 0.5‰ greater than those characteristic of modern Caribbean (non-upwelling) shells (e.g., fig. 5.3D; Bemis and Geary n.d.). Our isotopic data support other paleontological and tectonic data (Duque-Caro 1990) that indicate a Miocene shallow-water connection between the eastern Pacific and the southwestern Caribbean.

Temperature and salinity patterns in the eastern Pacific and Caribbean would have been quite different without the isthmus. Today, the Caribbean experiences high rates of evaporation during the warm summer months. Moisture-rich air is transported by trade winds across the lowlands of Panama to the Pacific side. The net water transport is significant; the southern Caribbean is on average 1.5‰ more saline than the eastern Pacific (Broecker 1989; Maier-Reimer et al. 1990). Maier-Reimer and others (1990) simulated ocean circulation with an open isthmus region and found significant differences in salinity and thermohaline circulation in the absence of the isthmus. In particular, their simulation produced salinities in the North Atlantic that were diluted by more than 1.0‰ relative to those of today, owing to water exchange between the two oceans.

Our data offer evidence that seasonal salinity and temperature patterns were different in the Miocene, in addition to the net salinity differences. Isotope profiles in modern Caribbean shells are flat because temperature and salinity variations are relatively minor, and because these two variables are generally positively correlated in this region. The greater range of $\delta^{18}O$ in Miocene shells may be explained by one or more of the following scenarios: (1) Seasonal temperature variations were more pronounced. (2) The relative timing of salinity and temperature change was different from that of today, so that these seasonal fluctuations reinforced each other in the shell isotopic signature. In particular, such conditions may have existed if the Gatun Formation was deposited in a locally restricted environment, such as a large bay. (3) Seasonal salinity changes were positively correlated with temperature

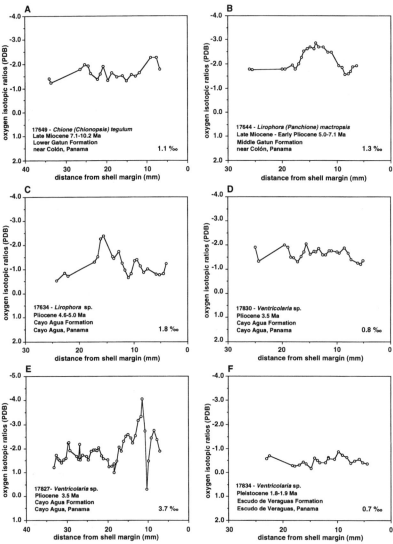

Fig. 5.4 δ¹⁸O profiles for fossil specimens from the Gatun Formation and the Bocas del Toro Basin. Total intrashell δ¹⁸O range is given at lower right of each plot: (A) δ¹⁸O profile of specimen 17649 from the Lower Gatun Formation (shell height is 48 mm measured from beak to ventral margin); (B) δ¹⁸O profile of specimen 17644 from the Middle Gatun Formation (shell height is 32 mm); (C) δ¹⁸O profile of specimen 17634 from the Cayo Agua Formation (shell height is 41 mm); (D) δ¹⁸O profile of specimen 17830 from the Cayo Agua Formation (shell height is 35 mm); (E) δ¹⁸O profile of specimen 17827 from the Cayo Agua Formation (shell height is 42 mm); (F) δ¹⁸O profile of specimen 17834 from the Escudo de Veraguas Formation (shell height is 41 mm).

but were less than those observed today and therefore did not mask the effect of temperature changes. Although it is impossible to determine the exact environmental factors that create a seasonal signal, we favor the third explanation (perhaps in combination with the second). Duque-Caro (1990) has argued that a shallow-water connection between the Pacific and Caribbean still existed in the Miocene, but was somewhat restricted. With a shallow-water connection, we would not expect the modern net evaporation and transportation of water, nor the modern salinity gradient between oceans. Furthermore, we expect that water circulation across the isthmus region would have reduced seasonal salinity variation on the Caribbean side relative to today, thereby unmasking the seasonal temperature signal.

Bocas del Toro Group

The Bocas del Toro Group from northern Panama consists of Late Miocene to latest Pliocene sediments, including the Cayo Agua Formation and the Escudo de Veraguas Formation. We analyzed four shells from the Bocas del Toro Group, including three from the Early Pliocene Cayo Agua Formation. Our oldest Cayo Agua specimen (17634, dated at 5.0–4.6 Ma), displays a $\delta^{18}O$ range of 1.8‰ (fig. 5.4C). The range of values and the shape of the shell profile suggest that the shell precipitated in waters with significant seasonal variations, more pronounced than those of the Miocene shells. This evidence supports the idea that from 5.0 to 4.6 Ma shallow waters still flowed between the Caribbean and the Pacific, and the modern positive correlation between temperature and salinity had not yet been established.

Specimens 17830 and 17827 are from the upper part of the Cayo Agua Formation, dated at approximately 3.5 Ma (Dowsett and Cotton, this volume, chap. 3), near the time of final closure. The profile for 17830 is relatively flat (fig. 5.4D); its $\delta^{18}O$ range of 0.8‰ is similar to that of modern Caribbean shells. From this specimen, then, it appears that modern Caribbean oceanic conditions (i.e., no seasonal upwelling events and a positive temperature and salinity correlation) were established by 3.5 Ma.

The oxygen isotope profile of 17827 (fig. 5.4E), however, is somewhat enigmatic. Fluctuations of approximately 1.8‰ characterize much of the shell, but a peak and valley separated by almost 4‰ occur across less than 5 mm near the ventral margin of the shell. Although shell margins are susceptible to diagenetic alteration (Krantz 1990), we do not believe the anomalous $\delta^{18}O$ values are a result of localized diagenesis

because (1) the composition of shell material at the margin is still aragonite, and there are no cracks, flaws, or other unusual characteristics; (2)
$\delta^{18}O$ values were reproduced with subsequent sampling along approximately the same growth bands, 1 cm laterally adjacent on the shell; (3)
we observe a wide fluctuation in $\delta^{18}O$, not the single isotopic overprint
expected from a diagenetic event.

The 4‰ range in this shell involves both a positive and a negative
pulse. No significant correlation exists between $\delta^{13}C$ and $\delta^{18}O$ across
the entire shell, but samples with extreme values of $\delta^{18}O$ near the shell
margin also exhibit respectively high and low values of $\delta^{13}C$ (total $\delta^{13}C$
range is 2.2‰). This "localized" correlation can aid in the interpretation of the $\delta^{18}O$ pulses. The negative pulse may have been caused by an
influx of freshwater, perhaps in combination with increased temperature. Freshwater typically carries isotopically lighter dissolved inorganic
carbon into a marine system, thus one would expect the accompanying
drop in $\delta^{13}C$. The positive excursion requires some combination of cool
and relatively saline waters. Although it brings relatively cool, saline
waters, upwelling cannot explain the positive $\delta^{18}O$ pulse in this shell
because it would also be accompanied by relatively light $\delta^{13}C$, not the
heavier (more positive) $\delta^{13}C$ values that accompany our positive $\delta^{18}O$
pulse. (One would expect that increased salinity caused by evaporation
in a shallow bay would be accompanied by increased rather than decreased temperatures.) Thus, we lack a straightforward explanation for
the positive pulse of $\delta^{18}O$ and $\delta^{13}C$ in this shell. None of our more than
thirty other modern or fossil shells exhibits a similar pattern.

Our youngest shell from the Bocas del Toro Group (17834 from the
Escudo de Veraguas Formation, 1.9–1.8 Ma) comes from a greater paleodepth than those of our other samples, probably 100–150 m (Collins
1993; Collins et al. 1995). The oxygen isotope profile across the shell is
relatively flat, with a range of only 0.7‰ (fig. 5.4F). As with shell 17830,
we interpret this small range to mean that modern oceanic circulation
was established in the Caribbean at this time. Values for $\delta^{18}O$ in this
shell are relatively high (maximum = −0.18‰, minimum = −0.87‰;
see table 5.1), which indicates deeper (cooler) water (Bemis and Geary
n.d.). Thus, our data are in agreement with paleodepth estimates based
on benthic foraminifers.

Limón Group

The Limón Group is represented by the Moin Formation and the Rio
Banano Formation (see Coates et al. 1992). Analyzed shells are from the
Rio Banano Formation, which is composed of blue-gray siltstones and

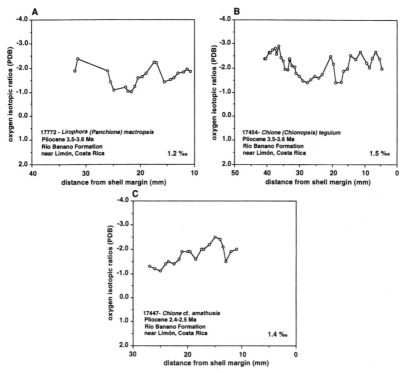

Fig. 5.5 δ¹⁸O profiles for fossil specimens from the Limón Basin, Rio Banano Formation. Total intrashell δ¹⁸O range is given at lower right of each plot: (A) δ¹⁸O profile of specimen 17772 from the Rio Banano Formation (shell height is 42 mm); (B) δ¹⁸O profile of specimen 17454 from the Rio Banano Formation (shell height is 45 mm); (C) δ¹⁸O profile of specimen 17447 from the Rio Banano Formation (dorsal portion of shell is broken; estimated shell height is approximately 45 mm).

volcanic litharenites that crop out along the Banano River near the city of Limón, Costa Rica (Coates et al. 1992).

Two specimens, 17772 and 17454 (fig. 5.5A,B), were collected from the Quitaría section of the Rio Banano Formation; both are dated at 3.6–3.5 Ma. The ranges of δ¹⁸O in the two shells are similar: 1.2‰ and 1.5‰, respectively. The third specimen, 17447 (fig. 5.5C), is from the type section of the Rio Banano Formation, dated at 2.5–2.4 Ma. This shell exhibits approximately the same oxygen isotope profile as the other Rio Banano shells (δ¹⁸O range = 1.4‰). The oxygen isotope profiles of all of the Limón Group fossil shells closely approximate, in terms of range and δ¹⁸O values, the profiles of the Gatun specimens (see fig. 5.3A, B), and exhibit consistently greater ranges than those of Bocas del

Toro Group specimens. Although additional sampling would be desirable, our data support the idea of environmental differences between the Limón and Bocas del Toro Basins (see Collins et al. 1995).

FOSSIL VENERID SHELLS FROM THE PACIFIC COAST

Obtaining usable specimens from the Pacific coast is difficult because it is an active margin. Most of the older sediments deposited during the emergence of the Central American Isthmus have been subducted, whereas younger Pliocene sediments were deposited in a deep-sea trench. (This paucity of available sediments will be relieved somewhat with upcoming collection of samples from the Darién region.) The original paleodepths of shells deposited in intraformational slumps within trench-slope environments are difficult to determine. Most Late Pliocene to Early Pleistocene deposits are from marginal-marine to inner-shelf settings (Coates et al. 1992). We have one Early Pliocene specimen and three Late Pliocene–Early Pleistocene specimens from the Pacific side.

Early Pliocene

Specimen 18066 (Peñita Formation; 3.5 Ma) exhibits a $\delta^{18}O$ range of 1.8‰ (fig. 6A). This Pacific-side specimen is similar in absolute value and slightly greater in range than the roughly contemporaneous Caribbean-side shells from the Quitaría locality of the Rio Banano Formation.

The ocean circulation model of Maier-Reimer (Maier-Reimer et al. 1990) indicated that upwelling in the eastern equatorial Pacific was present and similar to that of today even with an open isthmus. However, our oxygen isotope profile from this pre-isthmus bivalve does not indicate that the organism lived in a strong upwelling environment. Modern venerid bivalves from strong upwelling regions in the Gulf of Panama typically have shell $\delta^{18}O$ ranges of at least 2.2‰. Venerids from regions of weaker upwelling exhibit $\delta^{18}O$ ranges of 1.6‰ to 2.8‰ (Bemis and Geary n.d.). Our sample 18066 has a $\delta^{18}O$ range of 1.8‰, but unlike other profiles, this range is not confined to one observable cycle of increase and decrease. Additional sampling would be highly desirable here; nonetheless, existing data indicate that upwelling may have been weak in this part of the Pacific during the Early Pliocene, in contrast to the model results.

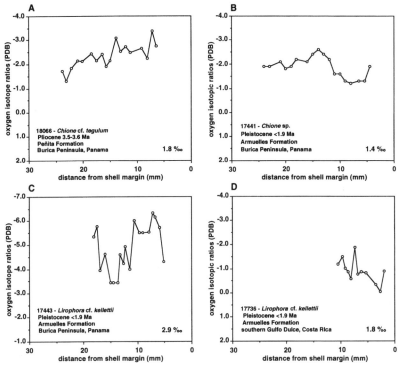

Fig. 5.6 δ^{18}O profiles for fossil specimens from the Pacific side. Total intrashell δ^{18}O range is given at lower right of each plot: (A) δ^{18}O profile of specimen 18066 from the Pacific side Peñita Formation (dorsal portion of shell is broken; estimated shell height is approximately 40 mm); (B) δ^{18}O profile of specimen 17441 from the Armuelles Formation (shell height is 33 mm); (C) δ^{18}O profile of specimen 17443 from the Armuelles Formation (shell height is 36 mm); (D) δ^{18}O profile of specimen 17736 from the Armuelles Formation (eight samples were lost during preparation owing to a malfunction of laboratory equipment, leaving only twelve samples to be analyzed; the shape of the oxygen isotope profile suggests that approximately one and a half years have been sampled and the δ^{18}O range still represents annual variation; shell height is 32 mm).

Plio-Pleistocene

Three shells were collected from the greenish-blue silts of the Pleistocene Armuelles Formation and dated as less than 1.9 Ma. Samples 17441 and 17736 (fig. 6B,D) exhibit moderate δ^{18}O ranges (1.4 and 1.8‰, respectively), indicating that upwelling was weak, if present. Sample 17443 (fig. 6C), although collected from the same section of the Armuelles Formation along the Rabo de Puerco River as specimen 17741, exhibits a considerably wider δ^{18}O range (2.9‰), and relatively

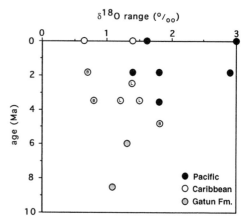

Fig. 5.7 $\delta^{18}O$ ranges for all fossil shells except the anomalous 17827 (see fig. 5.4E) and for characteristic modern venerids (fig. 5.3C–F): L = specimens from the Limón Group; B = specimens from Bocas del Toro Group.

low absolute values of $\delta^{18}O$ (maximum = −3.4‰, minimum = −6.3‰; note different values on the y-axis of fig. 6C). None of the modern shells analyzed by Bemis (Bemis and Geary n.d.) show such low $\delta^{18}O$. Ambient water temperature could not have been the primary control on shell $\delta^{18}O$ because water temperatures above 48°C (unrealistically warm for open marine waters or for shell growth) would be required to produce a value of −6.3‰ (using the Grossman and Ku equation [1986]). Localized freshwater runoff, perhaps in combination with elevated temperatures, is a more likely cause of the anomalously light shell $\delta^{18}O$ in this specimen. This interpretation is supported by the relatively low $\delta^{13}C$ that also characterizes this shell.

CONCLUSIONS

Figure 5.7 summarizes our data by plotting $\delta^{18}O$ range against age for each of our shells (excluding the anomalous 17827). "Typical" Caribbean non-upwelling and Pacific upwelling venerid ranges are represented by the specimens from figure 5.3C,D, top left and top right, respectively; we also include ranges from a Caribbean upwelling and a Pacific non-upwelling shell at top center (fig. 5.3E,F). We note the following points:

   1. Isotope profiles from Middle to Late Miocene shells from the Gatun Formation exhibit slightly greater shell $\delta^{18}O$ ranges than do typical (non-upwelling) modern Caribbean specimens. This suggests that seasonal salinity and temperature patterns were different in the Miocene, probably as a result of shallow-water

communication between the eastern Pacific and the southwestern Caribbean.

2. Trends toward the characteristic values of today are apparent in both oceans beginning approximately 3.5 Ma. The ongoing development of the isthmus and related changes in weather patterns (and possibly upwelling) resulted in an increase in seasonality on the Pacific side and a decrease on the Caribbean side. Divergence is well established by the end of the Pliocene.

3. At 3.5 Ma and again at 1.9 Ma, shells from the Bocas del Toro Basin suggest conditions similar to those of the modern Caribbean (excluding the enigmatic 17827). Three specimens from the Limón Basin (3.5–2.4 Ma) exhibit slightly greater $\delta^{18}O$ ranges, raising the possibility of different environmental conditions in these two basins (see Collins et al. 1995).

4. Oxygen isotope profiles of one Pliocene shell and two Pleistocene shells from the Pacific coast indicate substantially greater seasonality than in contemporaneous Caribbean habitats. The degree of upwelling present at these times is uncertain, but does not appear to have been as strong as in many Pacific localities today.

ACKNOWLEDGMENTS

We are grateful to J. W. Valley for use of the Stable Isotope Laboratory at the University of Wisconsin, Madison, and to M. J. Spicuzza for instruction and assistance in the laboratory. Special thanks to L. S. Collins for stratigraphic and paleoenvironmental information, and to P. Jung and N. Voss for assistance with specimens. A. F. Budd, C. W. Byers, L. S. Collins, J. B. C. Jackson, D. S. Jones, D. E. Krantz, and J. W. Valley made useful suggestions on the manuscript. This research was supported by grants from the National Science Foundation (EAR-9010086 and EAR-9106336) and the Petroleum Research Fund of the American Chemical Society (26239-AC8) to Dana H. Geary.

REFERENCES

Bemis, B. E. 1992. Stable isotopic records of venerid bivalve shells: Environmental information from the eastern Pacific Ocean and southern Caribbean Sea. M.S. thesis, University of Wisconsin, Madison, Wis.

Bemis, B. E., and D. H. Geary. N.d. The usefulness of bivalve stable isotope profiles as environmental indicators: Data from the eastern Pacific Ocean and southern Caribbean Sea. *Palaios.* In press.

Bennett, E. B. 1963. An oceanographic atlas of the eastern tropical Pacific Ocean, based on data from Eastropic Expedition, October–December 1955. *Bull. Inter-Am. Trop. Tuna Comm.* 8(2): 32–132.

———. 1966. Monthly charts of surface salinity in the eastern tropical Pacific Ocean. *Bull. Inter-Am. Trop. Tuna Comm.* 11(1): 1–41.

Broecker, W. S. 1974. *Chemical oceanography.* New York: Harcourt Brace Jovanovich.

———. 1989. The salinity contrast between the Atlantic and Pacific Oceans during glacial time. *Paleoceanography* 4:207–12.

Coates, A. G., J. B. C. Jackson, L. S. Collins, T. M. Cronin, H. J. Dowsett, L. M. Bybell, P. Jung, and J. A. Obando. 1992. Closure of the Isthmus of Panama: The near-shore marine record of Costa Rica and western Panama. *Geol. Soc. Am. Bull.* 104:814–28.

Collins, L. S. 1993. Neogene paleoenvironments of the Bocas del Toro Basin, Panama. *J. Paleontol.* 67:699–709.

Collins, L. S., A. G. Coates, J. B. C. Jackson, and J. A. Obando. 1995. Timing and rates of emergence of the Limón and Bocas del Toro Basin: Caribbean effects of Cocos Ridge subduction? In *Geologic and tectonic development of the Caribbean Plate boundary in southern Central America,* ed. P. Mann. Geological Society of America Special Paper 295. Boulder, Colo.

Cox, L. R. 1969. General features of Bivalvia. In *Treatise on invertebrate paleontology,* vol. 1, part N, ed. R. C. Moore, 2–129. Lawrence: Geological Society of America and University of Kansas.

Craig, H., and L. Gordon. 1965. Deuterium and ^{18}O variations in the ocean and the marine atmosphere. In *Stable isotopes in oceanographic studies and paleotemperatures,* no. 9, ed. E. Tongiorgi, 1–122. Pisa: Consiglio Nazionale delle Ricerche, Laboratorio di Geologica Nucleare.

Cubit, J. D., H. M. Caffey, R. C. Thompson, and D. M. Windsor. 1989. Meteorology and hydrography of a shoaling reef flat on the Caribbean coast of Panama. *Coral Reefs* 8:59–66.

D'Croz, L., J. B. Del Rosario, and J. A. Gomez. 1991. Upwelling and phytoplankton in the Bay of Panama. *Rev. Biol. Trop.* 39(2): 233–41.

Dodd, J. R., and R. J. Stanton. 1976. Paleosalinities within a Pliocene bay, Kettleman Hills, California: A study of the resolving power of isotopic and faunal techniques. *Geol. Soc. Am. Bull.* 87:51–64.

Duque-Caro, H. 1990. Neogene stratigraphy, paleoceanography and the paleobiogeography in northwest South America and the evolution of the Panama Seaway. *Palaeogeogr., Palaeoclimatol., Palaeoecol.* 77:203–34.

Emiliani, C. 1978. The cause of the ice ages. *Earth Planet. Sci. Lett.* 37:349–54.

Epstein, S., and T. K. Mayeda. 1953. Variations of ^{18}O content of waters from natural sources. *Geochim. Cosmochim. Acta* 4:213–24.

Epstein, S., and H. A. Lowenstam. 1953. Temperature-shell growth relations of Recent and interglacial Pleistocene shoalwater biota from Bermuda. *J. Geol.* 61:424–38.

Erlenkeuser, H., and G. Wefer. 1981. Seasonal growth of bivalves from Bermuda recorded in their ^{18}O profiles. *Proc. Fourth Int. Coral Reef Symp.* 4: 643–48.

Fairbanks, R. G., C. D. Charles, and J. D. Wright. 1992. Origin of global melt-water pulses. In *Radiocarbon after four decades*, ed. R. E. Taylor et al., 473–500. New York: Springer-Verlag.

Fairbanks, R. G., and R. E. Dodge. 1979. Annual periodicity of the $^{18}O/^{16}O$ and $^{13}C/^{12}C$ ratios in the coral *Montastrea annularis*. *Geochim. Cosmochim. Acta* 43:1009–20.

Fuglister, F. C. 1960. *Atlantic Ocean atlas of temperature and salinity profiles and data from the international geophysical year of 1957–1958*. Woods Hole, Mass.: Woods Hole Oceanographic Institution.

Geary, D. H., T. A. Brieske, and B. E. Bemis. 1992. The influence and interaction of temperature, salinity, and upwelling on the stable isotope profiles of strombid gastropod shells. *Palaios* 7:77–85.

Glynn, P. W. 1972. Observations of the ecology of the Caribbean and Pacific coasts of Panama. *Bull. Biol. Soc. Wash.*, no. 2:13–30.

————. 1982. Coral communities and their modifications relative to past and prospective Central American seaways. *Adv. Mar. Biol.* 19:91–132.

Grossman, E. L., and T. L. Ku. 1986. Oxygen and carbon isotope fractionation in biogenic aragonite: Temperature effects. *Chem. Geol., Iso. Geosci. Sec.* 59: 59–74.

Hodell, D. A., D. F. Williams, and J. P. Kennett. 1985. Late Pliocene reorganization of deep vertical water-mass structure in the western South Atlantic: Faunal and isotopic evidence. *Geol. Soc. Am. Bull.* 96:495–503.

Holcombe, T. L., and W. S. Moore. 1977. Paleocurrents in the eastern Caribbean: Geologic evidence and implications. *Mar. Geol.* 23:35–56.

Horibe, Y., and T. Oba. 1972. Temperature scales of aragonite-water and calcite-water systems. *Fossils* 23:69–79.

Jackson, J. B. C., P. Jung, A. G. Coates, and L. S. Collins. 1993. Diversity and extinction of tropical American mollusks and emergence of the Isthmus of Panama. *Science* 260:1624–26.

Jones, D. S. 1985. Growth increments and geochemical variations in the molluscan shell. In *Mollusks: Notes for a short course*, ed. T. W. Broadhead, 72–87. University of Tennessee Department of Geological Sciences Studies in Geology, no. 13, Knoxville, Tenn.

Jones, D. S., and P. F. Hasson. 1985. History and development of the marine invertebrate faunas separated by the Central American Isthmus. In *The Great American Biotic Interchange*, eds. F. G. Stehli and S. D. Webb, 325–55. New York: Plenum Press.

Jones, D. S., D. F. Williams, and M. A. Arthur. 1983. Growth history and ecology of the Atlantic surf clam, *Spisula solidissima* (Dillwyn), as revealed by stable isotopes and annual increments. *J. Exper. Mar. Biol. Ecol.* 73:225–42.

Jones, D. S., D. F. Williams, and C. S. Romanek. 1986. Life history of symbiont-bearing giant clams from stable isotope profiles. *Science* 231:46–48.

Keigwin, L. D., Jr. 1978. Pliocene closing of the Isthmus of Panama, based on biostratigraphic evidence from nearby Pacific Ocean and Caribbean Sea cores. *Geology* 6:630–34.

————. 1982a. Isotopic paleoceanography of the Caribbean and east Pacific: Role of Panama uplift in Late Neogene time. *Science* 217:350–52.

————. 1982b. Stable isotope stratigraphy and paleoceanography of sites 502 and 503. *Initial Repts. DSDP* 68:445–53.

————. 1987. North Pacific deep water formation during the latest glaciation. *Nature* 330:362–64.

Killingley, J. S., and W. W. Berger. 1979. Stable isotopes in a mollusk shell: Detection of upwelling events. *Science* 205:186–88.

Kinder, T. H., G. W. Heburn, and A. W. Green. 1985. Some aspects of Caribbean circulation. *Mar. Geol.* 68:25–52.

Kobayashi, T. 1969. Internal microstructure of the shell of bivalve molluscs. *Am. Zool.* 9:663–72.

Krantz, D. E. 1990. Mollusk-isotope record of Plio-Pleistocene marine paleoclimate, U.S. Middle Atlantic Coastal Plain. *Palaios* 5:317–35.

Krantz, D. E., D. F. Williams, and D. S. Jones. 1984. Growth rates of the sea scallop, *Placopecten magellanicus*, determined from the $^{18}O/^{16}O$ record in the shell carbonate. *Biol. Bull.* 167:186–99.

————. 1987. Ecological and paleoenvironmental information using stable isotopes from living and fossil molluscs. *Palaeogeogr., Palaeoclimatol., Palaeoecol.* 58:249–66.

Levitus, S. 1982. *Climatological atlas of the world ocean.* National Oceanic and Atmospheric Administration Professional Paper 13. Washington, D.C.

Maier-Reimer, E., U. Mikolajewicz, and T. Crowley. 1990. Ocean General Circulation Model sensitivity experiment with an open Central American Isthmus. *Paleoceanography* 5:349–66.

Mook, W. G. 1971. Paleotemperatures and chlorinites from stable carbon and oxygen isotopes in shell carbonate. *Palaeogeogr., Palaeoclimatol., Palaeoecol.* 9:245–63.

Robinson, M. K. 1973. *Gulf of Mexico and Caribbean Sea.* Atlas of monthly mean sea surface and subsurface temperature and depth of the top of the thermocline. Scripps Institution of Oceanography, University of California, San Diego. SIO Reference 73–78.

————. 1976. *Atlas of North Pacific Ocean monthly mean temperatures and mean salinities of the surface layer.* Naval Oceanographic Office, Reference Publication 2. Washington, D.C.

Romine, K. 1982. Late Quaternary history of atmospheric and oceanic circulation in the eastern equatorial Pacific. *Mar. Micropaleontol.* 7:163–87.

Schaefer, M. B., Y. M. M. Bishop, and G. V. Howard. 1958. Some aspects of upwelling in the Gulf of Panama. *Bull. Inter-Am. Trop. Tuna Comm.* 3(2):78–132.

Shackleton, N. J. 1974. Attainment of isotopic equilibrium between ocean water and the benthonic Foraminifera genus *Uvigerina*: Isotopic changes in the ocean during the last glacial. In *Les Méthodes quantitatives d'étude variations du climat au cours de Pleistocene. Cent. Natl. Rech. Sci. Colloq. Int.* 219:203–9.

Shackleton, N. J., and M. B. Cita. 1979. Oxygen and carbon isotope stratigraphy of benthic foraminifers at site 397: Detailed history of climate change during the Neogene. *Initial Repts. DSDP* 47:433–45.

Shackleton, N. J., and M. A. Hall. 1985. Oxygen isotope and carbon stratigraphy of Deep Sea Drilling Project hole 552A: Plio-Pleistocene glacial history. *Initial Repts. DSDP* 81:599–609.

Shackleton, N. J., and N. D. Opdyke. 1973. Oxygen isotope and paleomagnetic stratigraphy of equatorial Pacific core V28–238: Oxygen isotope temperatures and ice volumes on a 10^5 and 10^6 year scale. *Quatern. Res.* 3:9–55.

———. 1976. Oxygen isotope and paleomagnetic stratigraphy of the Pacific core V28–239, Late Pliocene to latest Pleistocene. In *Investigation of the Late Quaternary paleoceanography and paleoclimatology*, ed. R. M. Cline and J. D. Hays, 449–64. Geological Society of America Memoir 145. Boulder, Colo.

———. 1977. Oxygen isotope and paleomagnetic evidence for early Northern Hemisphere glaciation. *Nature* 270:216–19.

Smayda, T. J. 1965. A quantitative analysis of the phytoplankton of the Gulf of Panama. Part 2, On the relationship between the ^{14}C assimilation and the diatom standing crop. *Bull. Inter-Amr. Trop. Tuna Comm.* 9(7): 466–531.

———. 1966. A quantitative analysis of the phytoplankton of the Gulf of Panama. Part 3, General ecological conditions. *Bull. Inter-Amr. Trop. Tuna Comm.* 2(5): 354–83.

Teranes, J. L. 1993. Stable isotope records of seasonality in Neogene bivalves from the Central American Isthmus. M.S. thesis, University of Wisconsin, Madison, Wis.

Urey, H. C. 1947. The thermodynamics of isotopic substances. *J. Chem. Soc. Lond.*, 562–81.

Urey, H. C., H. A. Lowenstam, S. Epstein, and C. R. McKinney. 1951. Measurement of paleotemperatures and temperatures of the Upper Cretaceous of England, Denmark, and the southern United States. *Bull. Geol. Soc. Am.* 62: 399–416.

Wefer, G., and W. H. Berger. 1980. Stable isotopes in benthic Foraminifera: Seasonal variation in large tropical species. *Science* 209:803–5.

Wefer, G., and J. S. Killingley. 1979. Stable isotopes in a mollusk shell: Detection of upwelling events. *Science* 205:186–88.

———. 1980. Growth histories of strombid snails from Bermuda recorded in their ^{18}O and ^{13}C profiles. *Mar. Biol.* 60:129–35.

Weissert, H. J., and J. A. McKenzie, et al. 1984. Paleoclimate record of the Pliocene at Deep Sea Drilling Project sites 519, 521, 522, and 523 (central South Atlantic): *Initial Repts. DSDP* 73:701–15.

Williams, D. F., M. A. Arthur, D. S. Jones, and N. Healy-Williams. 1982. Seasonality and mean annual sea surface temperatures from isotopic and sclerochronological records. *Nature* 296:432.

Wyrtki, K. 1966. Oceanography of the eastern equatorial Pacific Ocean. *Oceanogr. Mar. Biol. Ann. Rev.* 4:33–68.

Wyrtki, K. 1981. An estimate of equatorial upwelling in the Pacific. *J. Phys. Oceanogr.* 11:1205–14.

6

Environmental Changes in Caribbean Shallow Waters Relative to the Closing Tropical American Seaway

Laurel S. Collins

INTRODUCTION

The complete closure of the oceanic seaway that had connected equatorial Atlantic and Pacific waters since the Permian breakup of Pangea is commonly considered the largest oceanographic event of Pliocene time (Haq 1984; Hay 1988). Before closure of this gateway, which I call the Tropical American Seaway, marine waters covered the southern Central American region, and marine environments on either side of the present-day isthmus were more similar. The seaway partially closed after Eocene time as a result of either (1) the northeastward movement of the Caribbean Plate from the eastern Pacific, which brought a protoisthmian island arc into collision with North America (Sykes et al. 1982), or (2) the relatively in situ formation of the island arc (but see Ross and Scotese 1988). As long-term, global sea level fell through Late Neogene time (Haq et al. 1987; Matthews 1988) and Central America experienced tectonic uplift, free mixing of Atlantic and Pacific equatorial waters became increasingly restricted, until the seaway was completely closed during the mid Pliocene. Today, marine waters differ markedly on either side of the isthmus. Tropical Pacific waters are silica-rich, less saline, and cooler on average from seasonal upwelling. Caribbean waters are carbonate-rich and are more saline because of a net flux of water vapor carried by warm trade winds from the Caribbean to the Pacific side (Weyl 1968; Maier-Reimer et al. 1990).

While emergence of the southern Central American isthmus evidently affected deep-sea circulation in middle Miocene time and surface-water changes had taken place by the middle Pliocene (Coates et al. 1992), the timing of shallow-water environmental change and initial evolutionary divergence of Caribbean and Pacific neritic faunas has remained unclear. This study uses the Late Neogene to Recent evolutionary and ecologic record of benthic foraminiferal species from upper bathyal to neritic waters (less than about 300 m deep) of Panama and Costa Rica to identify changes in Caribbean shallow marine environments through the interval of seaway closure. The first part of this chapter is a synthesis of the best evidence for the timing of seaway constriction and associated marine environmental changes. The second part examines the evolutionary record of common Caribbean benthic foraminifera, for the purpose of identifying ecological trends that tracked environmental changes in southwestern Caribbean waters of the closing seaway.

CONSTRICTION OF THE TROPICAL AMERICAN SEAWAY

History

During Paleogene time, the Caribbean Current flowed westward through the Tropical American Seaway, driven by northeast-east trade winds (Wüst 1964). An interoceanic, isthmian barrier had formed from North America southward to central Panama by Early Miocene time, as indicated by Late Oligocene, shallow-water benthic foraminifera of Costa Rica (Cassell and Sen Gupta 1989) and Panama (Cole 1952), and by the presence of diagnostic genera of a North American, Late Arikareean land mammal fauna (horses, rhinoceroses, oreodonts, and selenodont artiodactyls) in Early Miocene sediments (Tedford 1987) from Panama Canal excavations (Whitmore and Stewart 1965). Early to Middle(?) Miocene uplift associated with the collision of Panama and South America (Pindell and Barrett 1990), and increased volcanic activity 18–14 Ma (Donnelly et al. 1990) contributed to isthmian emergence.

Although shallow interoceanic passageways probably existed episodically among Miocene to Pliocene volcanic islands, significant mixing of Caribbean and Pacific deeper waters was most likely restricted to the area from central Panama to northwestern Colombia's Atrato River basin, resulting in partial deflection of the Caribbean Current northward (Berggren and Hollister 1974). A bathyal Caribbean-Pacific connection around 5 Ma is indicated by sediments from the Panama Canal basin, central Panama (Collins and Coates 1993), perhaps deposited during a large eustatic rise (Haq et al. 1987). The age of the youngest Atrato marine sediments is between 5.0 Ma (based on the highest occurrence of the planktonic foraminiferan *Globorotalia merotumida*, top of zone N18), and 1.9 Ma (based on co-occurring *Globigerina decoraperta*, *G. rubescens*, *Globoquadrina altispira*, and *Globorotalia conomiozea subconomio-*

zea, zone N21). Atrato sediments reflect a low-oxygen, neritic environment characteristic of a deep, borderland basin with restricted circulation (Duque-Caro 1990).

Patterns of deep-sea sedimentation indicate declining southern Caribbean circulation that began in Middle Miocene time. In the Venezuelan Basin, Late Cretaceous to Middle Miocene nondeposition ended, owing to the weakening of water circulation from seaway closures in Panama and the Lesser Antilles (Holcombe and Moore 1977). Formation of the lower Miami erosional terrace (Mullins and Neumann 1979) and the nature of west Florida slope sediments (Mullins et al. 1987) are ascribed to increased flow of the Gulf Stream from the addition of the deflected Caribbean Current in Middle Miocene time.

Distributions of benthic and planktonic foraminifera indicate an open, Early Miocene to earliest Pliocene seaway that was an interoceanic dispersal route for marine organisms. Cushman (1929), a taxonomist who tended to split species, noted that a Miocene fauna from Venezuela had "practically all the species" of an Early to Middle Miocene fauna from Manta, Ecuador. Hasson and Fischer (1986) observed that benthic foraminifera of the late Middle Miocene Borbón Formation, Ecuador (plankton zones N13–14), "resemble" those of Venezuela. Whittaker (1988) confirmed that Early and Middle Miocene faunas of the Manabí and Borbón Basins, Ecuador, are "very similar and have close affinities" with those of the Caribbean. These Early to Middle Miocene faunas are bathyal in distribution, indicating an open, unrestricted seaway to bathyal depths. (However, in contrast to previous studies, Duque-Caro 1990 suggested a late Middle Miocene Caribbean-Pacific barrier based on the observation that the Sierra Formation of Pacific Colombia and formations of Caribbean Colombia "bear different benthic foraminiferal assemblages.") Later, the bathyal fauna of the Esmeraldas Formation (3.5–3.2 Ma) of Ecuador "shows little similarity to the Pliocene faunas of the Caribbean, and has instead Californian affinities" (Hasson and Fischer 1986), indicating a Caribbean-Pacific barrier by 3.2 Ma. The occurrence in both Atlantic and Pacific waters of tropical planktonic foraminiferal species originating as late as Early Pliocene (e.g., *Globorotalia pertenuis, G. miocenica*) signifies an open-ocean seaway that existed until Early Pliocene time (Stainforth et al. 1975; Keigwin 1982b).

First occurrences of the tropical neritic benthic foraminiferan *Pararotalia magdalenensis* Lankford on either side of the Central American isthmus support the existence of a normal marine, Early Pliocene seaway. This species can be used to infer at least a shallow marine seaway because (1) its distribution today is restricted to tropical-subtropical waters, so high-latitude migration was unlikely; (2) it is an ecologic generalist, common in nearshore, sandy sediments, so it would not have been environmentally excluded by the near absence of carbonate facies in the eastern Pacific Ocean; (3) it occurs in abundance no deeper than about 40 m (Drooger and Kaasschieter 1958, as *Rotalia sarmientoi*; Golik and Phleger 1977, as *Ammonia decorata*), so its distribution provides evidence of a viable, inner- to middle-neritic connection between oceans. The earliest Caribbean occurrence of *P. magdalenensis* is in the Upper Gatun Formation, Panama Canal Basin, dated latest Miocene to earliest Pliocene (zones N17–18; Whittaker and Hodgkinson in Woodring 1982), or 7.1–5.0 Ma (time scale of Berggren et al. 1985). Its oldest Pacific occurrence is 5.3–3.0 Ma in the Jama Formation, Ecuador (zones N18–19, Whittaker 1988), which suggests migration through the seaway in earliest Pliocene time.

The most convincing marine evidence of complete seaway closure comes from stable isotopes. Keigwin's (1982a,b) study of the progressive divergence in stable isotope proportions between Caribbean and eastern Pacific basins suggests that deep bottom-water circulation was affected by the emerging isthmus by 6–5 Ma, and the Caribbean-Pacific salinity contrast of about 2‰ (Weyl 1968; Maier-Reimer et al. 1990) developed 4–3.5 Ma. Using measurements of water vapor flux in Panama City, Deffeyes (in Weyl 1968, 47) estimated that complete blockage of the net flux of water vapor from the Caribbean to the Pacific side for only 600 years would remove the present salinity contrast. Such a delicate balance between water vapor flux and salinity contrast implies that the Early Pliocene salinity contrast would have tracked isthmian emergence closely on a geologic time scale.

Sea-level changes also support increased emergence of the isthmus around 4–3.5 Ma. The eustatic curves of Haq and others (1987), constructed from the global record of coastal onlap, show a roughly 100 m drop in sea level 4–3.8 Ma. Matthews's eustatic curves (1988), based on oxygen isotopes from planktonic foraminifera, show a general trend of decreasing sea level 4–2.5 Ma.

Migrational patterns of terrestrial animals between North and South America provide an extreme upper age limit for complete seaway closure. The earliest unequivocal occurrence of land mammals that traversed the length of the Central American isthmus is that of a North American skunk and peccary preserved in the Argentinian Chapadmalal Formation 2.8–2.5 Ma (Marshall 1988; Marshall et al. 1992). Even if later sea-level rises breached the interoceanic isthmian barrier (Savin and Douglas 1985; Keller et al. 1989; Cronin and Dowsett, this volume, chap. 4), they apparently had little effect on large-scale oceanographic patterns (Keigwin 1982a,b) or on the continued evolutionary divergence of Caribbean and eastern Pacific marine faunas.

Shallow-water Environmental Consequences

The effects of constriction and closure of the Tropical American Seaway on Caribbean shallow-water environments have not been closely examined, but there are several hypotheses: (1) One idea is that typical Neogene shallow waters were colder, more turbid, nutrient-rich, and nonreefal before seaway closure, based upon a relict Neogene gastropod fauna from an upwelling zone along the Venezuelan coast (Vermeij 1978, 235; Petuch 1981; Vermeij and Petuch 1986). Development of carbonate-rich areas would have been inhibited by the cold temperatures and turbidity of coastal upwelling (Emiliani et al. 1972). (2) Another hypothesis is that closure may have caused warming of Caribbean surface waters. The predominant current direction of the Tethyan Seaway was westward, but colder, less saline equatorial Pacific waters also would have flowed into the Caribbean to create cooler surface waters than exist today (Emiliani et al. 1972), especially if Tethyan through-flow had been restricted in the area of Malaysia (Luyendyk et al. 1972). (3) A third hypothesis of environmental change, perhaps ultimately due to seaway closure, is mid-Pliocene to Recent cyclical cooling and warming, beginning in the North Atlantic 3.15 Ma (Ruddiman and Raymo 1988) to 3.4 Ma (Loubere and Moss 1986). Emiliani and others (1972) and Berggren and Hollister (1977) suggested that seaway closure may have raised surface temperatures in the western Atlantic by bringing warmer waters into higher latitudes, contributing to increasing precipitation through evaporation and cooling, and initiating a polar ice cap in the Northern Hemisphere.

The hypotheses of the development of a carbonate regime and significant Pleistocene cooling episodes in the southern Caribbean have specific predictions that are testable with benthic foraminiferal assemblages. Numerous tropical benthic foraminiferal taxa, such as many species of *Quinqueloculina*, are most abundant in biogenic carbonate sediments of shoal and reef areas, and significant increase in regional originations of these carbonate-associated taxa would suggest large-scale development of carbonate-rich areas. Plio-Pleistocene cyclic cold intervals have been associated with extinction of warm, shallow-water mollusks of the western Atlantic (Stanley 1986; Allmon 1992) and Caribbean (Jackson et al., this volume, chap. 9), and the appearance of the benthic foraminiferal cold-indicator species *Hyalinea balthica* in the Gulf of Mexico (Akers and Dorman 1964) to northern Caribbean (Bock 1970). Decreasing temperatures have been associated with extinction and taxonomic turnover in benthic foraminiferal assemblages (e.g., Woodruff et al. 1981). Significant cooling in the southwestern Caribbean would be expected to produce the appearance of *Hyalinea balthica* and increased extinction of benthic foraminifera.

ECOLOGIC AND EVOLUTIONARY TRENDS IN BENTHIC FORAMINIFERA

Ecologic Predictions

The utility of benthic foraminifera as environmental indicators is well established. They are abundant, widespread, and, particularly in the tropics, diverse, inhabiting virtually all marine to marginal marine sediments. Distributions of species (Murray 1991) and intraspecific morphologies (Boltovskoy et al. 1991; Collins 1991) have been correlated with environmental parameters such as temperature, salinity, carbonate solubility, water depth, nutrients, substratum, average sedimentary grain size, dissolved oxygen, illumination, pollution, water motion, and trace elements. In paleoenvironmental analyses, a working hypothesis is that fossils are environmentally analogous to their modern counterparts, that is, that physical tolerances of a taxon do not change with time. The further back in time this correlation is applied, the less it pertains, because environmental settings and species are increasingly different in progressively older deposits. Fortunately, about 90% of the species in this study are extant, and environmental inferences made with large assemblages can be applied with a fair degree of confidence.

Distributions of benthic foraminiferal species within a region such as the Caribbean are, relative to many other invertebrates, geographically wide and ecologically narrow. A compilation of species occurrences shows nearly all common taxa present from the northern to southern Caribbean, but inadequate sampling prevents comparisons of the eastern and western portions, and generalizations about rare taxa (Culver and Buzas 1982). Although ecological distributions are often given in terms of water depths, depth is probably not the limiting factor, but rather a proxy for the environmental parameters that change consistently with depth within and (less consistently) between regions. Environments generally change less with increasing depth, so paleodepth interpretations are more precise in shallower waters, to as little as a 10 m range.

Because of their strong environmental partitioning, benthic foraminiferal assemblages that show large changes through geologic time due to migration and evolution can be used to detect large-scale environmental changes. For example, a decrease in deep bottom-water temperature in the Middle Miocene interpreted from an enrichment in ^{18}O and ice-rafted debris, and ascribed to Southern Hemisphere glaciation and climatic cooling (Savin et al. 1975; Woodruff et al. 1981), has been correlated with major changes in Pacific deep-sea assemblages (Woodruff et al. 1981). Faunal turnover began with increasing $\delta^{18}O$, but lagged behind it into Late Miocene time, suggesting that conditions other than temperature also changed. Much of the Middle Miocene faunal change appeared to be origination and extinction, while the lagged change was largely due to migration (Douglas and Woodruff 1981).

If constriction of the Tropical American Seaway caused large environmental changes in shallow waters of the southern Central American isthmus, neritic (less than 200 m) benthic foraminiferal assemblages should have recorded unusually large faunal changes. Species time ranges in the Caribbean region should begin and end around a period of ecological restructuring, which would have happened before 4–3.5 Ma if environmental change preceded complete seaway closure. Since many environmental tolerances of shallow-water benthic foraminifera are known (Murray 1991), ecological trends in faunal turnover can provide information about the nature of environmental changes. If environmental change were great enough to exceed endemic species' tolerances, taxonomic turnover would reflect originations and extinctions (rather than migration), as would be the case if Pleistocene cooling intervals caused tropical extinctions of warm-water taxa. Here I address two predictions for Caribbean shallow waters:

1. If carbonate sediment production increased, first appearances of carbonate-associated species should have increased.

2. If Plio-Pleistocene cooling significantly affected southern Caribbean faunas, extinctions should have exceeded previous levels, and the cold-indicator species *Hyalinea balthica* should have migrated into the region.

Methods

To identify times of unusually great faunal change, I constructed a data set of species time ranges within the entire Caribbean region (app. 6.1). Common benthic foraminiferal taxa found in Late Miocene to latest Pliocene deposits of Caribbean Panama and Costa Rica comprise the

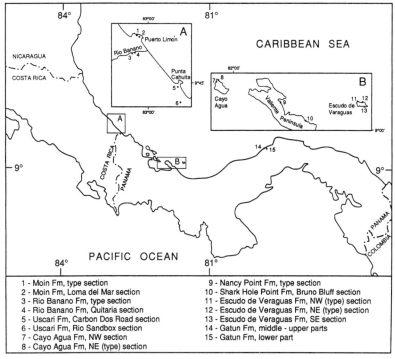

Fig. 6.1 Localities for Caribbean foraminiferal samples. Insets: (A) Limón Basin, Costa Rica; (B) Bocas del Toro Basin, Panama.

data set. The southern Central American subset of Caribbean species maximizes the possibility of detecting environmental change due to closure of the Tropical American Seaway, since effects of the event should be more pronounced proximal to the area of closure. Including only common species minimizes bias due to inadequate sampling.

Sediment samples were collected from fifteen stratigraphic sections (figs. 6.1, 6.2) that were biochronologically dated with calcareous nannoplankton and planktonic foraminifera (Coates et al. 1992; Collins 1993; Collins et al. 1995), using the geologic time scale of Berggren and others (1985). Neritic environments from Late Miocene to latest Pliocene strata are included, although upper bathyal environments are only represented in Late Miocene deposits, which bias is addressed in the ecologic analyses below. Micropaleontologic slides sampling total benthic foraminiferal faunas from eighty-three sediment samples were prepared using methods described by Collins (1993). Species having more than one occurrence per sample were identified with compara-

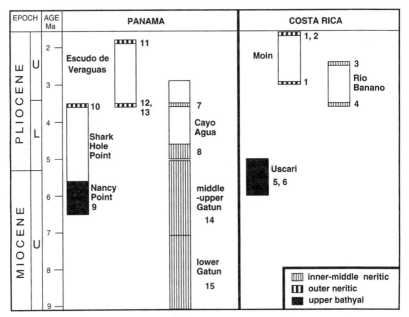

Fig. 6.2 Data set for Mio-Pliocene formations in Panama and Costa Rica. See figure 6.1 for numbered stratigraphic sections. Benthic foraminifera are from dated portions with patterns identifying general habitat. Water depths associated with general environments: inner neritic = 0–30 m; middle neritic = 30–100 m; outer neritic = 100–200 m; upper bathyal = 200–600 m.

tive collections housed at the U.S. National Museum (designated speci-men repository), American Museum of Natural History, University of California Museum of Paleontology, and a private collection of P. McLaughlin (1989). The data set includes all common species, defined here as those occurring in more than one sample with at least 1% abun-dance in at least one sample, for a total of 107 taxa (app. 6.1; Collins et al. 1995).

A data set of this size and environmental breadth should be represen-tative of the common southern Caribbean fauna. Ninety-four of the 107 taxa are alive today in the Caribbean region (Culver and Buzas 1982), from which 130 common and 1189 total species have been reported. Twenty-eight percent of the 94 taxa are common in today's Caribbean, compared with only 11% of living Caribbean species, so the fossilized species may be a more eurytopic subset of the fauna. Paleoenvironments sampled include nearshore, inner neritic, lagoon, carbonate shoal to reef, middle neritic, outer neritic, continental shelf edge, and upper bathyal (Collins 1993; Collins et al. 1995).

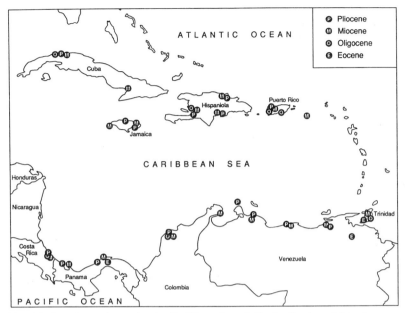

Fig. 6.3 Location and age of major Caribbean foraminiferal collections. More than one study is commonly associated with each area.

Determination of species time ranges in the Caribbean region entailed two procedures. First, taxonomic standardization of regional Caribbean and Gulf of Mexico collections resulted in the use of the same names for the same species from different times and places. Lack of taxonomic standardization is the greatest obstacle to understanding the evolution and paleoecology of benthic foraminifera (Culver et al. 1987). While benthic foraminiferal species are not particularly difficult to distinguish, their enormous diversity (about 40,000 have been described) and variable abundances have resulted in identifications endemic to specific time intervals and regions (Brolsma 1978).

I standardized species names used in this study with those of other foraminiferal studies from the Caribbean region (fig. 6.3). Published figures identified some potential synonymizations, which I commonly checked with comparative collections. Unpublished, non-type specimens stored under these species names, including those stored under different generic names, were also considered. In most cases (76%) of first and last occurrences, specimens were examined, and in the other cases I used figures judged to be of sufficient diagnostic quality. Names in a few taxonomic lists without figures or available specimens were excluded from this study because they could not be verified.

The Late Neogene history of the Caribbean region is generally one of tectonic uplift and emergence, so there is a bias toward land-based preservation of sediments from older, deeper waters and younger, shallower waters. To supplement these studies with collections from older, shallower sediments and younger, deeper sediments, I also included major works on Gulf Coast strata between Florida and Mexico. These are primarily works on deposits from pre-Late Miocene time, before steeper latitudinal temperature gradients were established (Savin and Douglas 1985), so mixing of Caribbean-Gulf of Mexico faunas would have been greater than present moderate levels.

Second, biochronologic ages of stratigraphic units that produced youngest or oldest occurrences of taxa were updated with the most recent biostratigraphic information available and the geologic time scale of Berggren and others (1985). Most major works on foraminifera and stratigraphy of the Caribbean and Gulf of Mexico regions were carried out in the middle part of this century (e.g., Renz 1948; Bermudez 1949; Redmond 1953), before the biostratigraphic frameworks in current use were established; in fact, many of those studies formed the basis of present geologic time scales. Compilations for the southeastern United States by Carter (1983, 1984), Carter and Wheeler (1983), and Carter and Rossbach (1989) provided the latest Gulf Coast biostratigraphies.

Biochronologic ranges in dates assigned to first and last species occurrences were the most constrained dates for a particular time interval. For example, a species that first occurred in both the Gurabo Formation of the Dominican Republic, 7.1–3.4 Ma (Saunders et al. 1986), and the upper part of the Gatun Formation, 7.1–5.0 Ma (Whittaker and Hodgkinson in Woodring 1982), such as *Bigenerina irregularis*, was assigned the Gatunian age because there is less uncertainty in that age range. Since the ages of the recently dated Panama-Costa Rica samples (Coates et al. 1992) have relatively short biochronologic uncertainties, I often used them in preference to other dates.

Results
Figure 6.4 shows 107 composite geologic ranges within the Caribbean region for species that were common to southern Central American waters at some point between 7.1 and 1.6 Ma. The chart is mostly a register of first occurrences, since no extinctions could occur before 7 Ma (with these data), few occurred thereafter, and species are generally long-ranging. The average taxon has lived in the region from Early Miocene time to the present, with a mean first occurrence of 21.4 Ma based on midpoints of uncertainties in biochronologic dates. This age

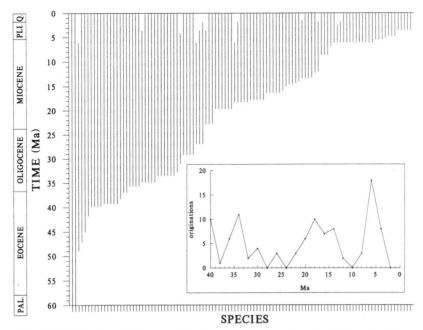

Fig. 6.4 Regional geologic ranges of 107 common benthic foraminiferal species. Time lines begin and end at midpoints of biochronologic ranges of the deposits that produced first and last occurrences. Inset: number of first occurrences per 2 my interval. Note that 88% of the species are still extant.

is comparable to the average species duration of 20 my for warm temperate, neritic taxa of the east coast of the United States determined by Culver and others (1987), based on midpoints of epochs. These 107 species first appeared in the region at a fairly steady rate from Late Eocene to Late Miocene time, although an Early Miocene increase around 20–17 Ma (fig. 6.4, inset) coincides with and may be in response to the end of a significant warming trend detected in Pacific low-latitude surface waters (Savin and Douglas 1985), related to development of the modern circum-Antarctic circulation and weakening of meridional heat transport between low and high latitudes.

The largest increase in species' regional first appearances occurred in latest Miocene to Early Pliocene time, 6–3.5 Ma, within the roughly 15 my interval spanned by estimates of seaway constriction and closure. Figure 6.5 shows ranges of the 47 species first or last occurring within this interval. Although these species occur in sediments in Panama and Costa Rica 7 Ma or younger (fig. 6.2), the latest Miocene pulse is not a sampling artifact dependent on first occurrences in these deposits.

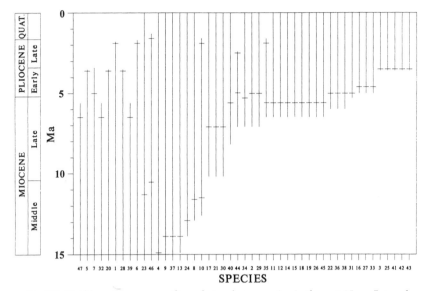

Fig. 6.5 Caribbean time ranges of taxa first or last occurring in the past 15 my. Intervals between cross bars and ends of range lines indicate uncertainties in dating based on planktonic microfossils. Numbers along *x*-axis refer to taxa in tables 6.1 and 6.2.

Sixty-nine percent (18/26) of first occurrences are comparably dated in other sediments of the Caribbean region (table 6.1) and 5 of the remaining 8 species are found as fossils elsewhere in the region. Even if the number of first occurrences in the last 7 my were reduced by 30%, it would still be a considerable increase over previous rates.

Are the first appearances in latest Miocene to earliest Pliocene time migrations into the region or the earliest evidence of evolutionary originations? Twenty-two of the 26 species (85%) first occurring in this interval are reported no earlier outside the region (table 6.1), and 11 have not been recorded in fossil or Recent sediments outside the western Atlantic or eastern Pacific regions. Thus, the latest Miocene-earliest Pliocene increase in first appearances appears to be an origination pulse, possibly in response to changing environmental conditions caused by seaway constriction.

An examination of the ecologies of neritic species first occurring during latest Miocene-earliest Pliocene time reveals accelerating diversification within carbonate environments. I divided the last 7.1 my into three approximately equal intervals of time (3.4–1.6, 5.3–3.4, 7.1–5.3 Ma) that correspond to stratigraphic intervals, which avoids methodologic difficulties in subdividing biochronologic intervals into smaller

Table 6.1 Benthic Foraminiferal Species with Regional First Occurrences since 7 Ma

Taxa	Regional Fossil Occurrences Other than SCA Samples	Regional First Occurrences with Dates Overlapping Those of SCA Samples	Global First Occurrences
2	x	Gurabo Fm., Dominican Rep.	x
3			x
11	x	Buff Bay deposits, Jamaica	
12	x	Gurabo Fm., Dominican Rep.	x
14	x		x
15	x	Upper Trinchera Fm., Dominican Rep.	x
16	x	Gurabo Fm., Dominican Rep.	x
18			x
19	x	Buff Bay deposits, Jamaica	x
22	x	Choctawhatchee Fm., (*Arca* zone), Fla.	
25			x
26	x		x
27	x		x
29	x	Jama Fm., Ecuador	x
31	x	Buff Bay deposits, Jamaica	x
33	x	Túbara Fm., Colombia	x
34	x	Choctawhatchee Fm. (*Arca* zone), Fla.	x
35	x	Encanto deposits, Mexico	x
36	x		
38	x	Marga de Las Hernández, Venezuela	
40	x	Cercado Fm., Dominican Rep.	x
41	x	Choctawhatchee Fm. (*Ecphora* zone), Fla.	x
42	x		x
43	x	Choctawhatchee Fm. (*Canc.* zone), Fla.	x
44	x	Pozón Fm., Venezuela	x
45	x	Trinchera Fm.(?), Dominican Rep.	x

Note: Dates of first occurrences are the midpoints of biochronologic ranges of deposits (see fig. 6.5). Table 6.2 gives names of taxa listed here by number. SCA = Southern Central America (Panama and Costa Rica samples).

units. Note that none of these taxa could have originated 1.6–0 Ma because the data set is composed of species alive before then. Figure 6.6 shows general bathymetric ranges of taxa (ordered by water depth within chronologic intervals) that first appear within the Early Pliocene 5.3–3.4 Ma, the latest Miocene planktonic foraminiferal zone N17, 7.1–5.3 Ma, and for comparison, between 15 and 7.1 Ma. None of the species first occur in the Caribbean in Late Pliocene time, 3.4–1.6 Ma, despite extensive sampling of that interval (Collins et al. 1995). More bathyal taxa *appear* to originate in the Late Miocene because no Pliocene bathyal sediments were collected, but there is no such sampling bias with regard to neritic taxa (fig. 6.2), which comprise the data set

Table 6.2 Common Caribbean Benthic Foraminifera
Originating or Becoming Extinct since 15 Ma

1. *Asterigerina pettersi*
2. *Bigenerina irregularis*
3. *Biloculinella eburnea*
4. *Bolivina alata*
5. *Bolivina byramensis*
6. *Bolivina imporcata*
7. *Bolivina isidroensa*
8. *Bolivina lowmani*
9. *Bolivina mexicana*
10. *Bolivina pozonensis*
11. *Bulimina aculeata*
12. *Bulimina marginata*
13. *Bulimina tessellata*
14. *Cassidulina minuta*
15. *Cassidulina norcrossi australis*
16. *Cymbaloporetta atlantica*
17. *Elphidium mexicanum*
18. *Eponides turgidus*
19. *Fursenkoina complanata*
20. *Hanzawaia isidroensa*
21. *Haynesina depressula*
22. *Lagena ornata*
23. *Lamarckina atlantica*
24. *Marginulinopsis marginulinoides*
25. *Miliolinella californica*
26. *Neoconorbina parkerae*
27. *Nodobaculariella cassis*
28. *Nonionina incisa*
29. *Pararotalia magdalenensis*
30. *Planulina ariminensis*
31. *Planulina foveolata*
32. *Planulina marialana*
33. *Quinqueloculina compta*
34. *Rosalina floridana*
35. *Rotalia garveyensis*
36. *Rotorbinella umbonata*
37. *Sagrina pulchella*
38. *Seabrookia earlandii*
39. *Siphogenerina lamellata*
40. *Sorites marginalis*
41. *Spiroplectammina floridana*
42. *Stetsonia minuta*
43. *Textularia foliacea occidentalis*
44. *Textularia panamensis*
45. *Textularia schencki*
46. *Trifarina eximia*
47. *Valvulineria haitiana*

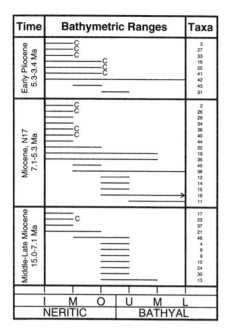

Fig. 6.6 Bathymetric ranges of twenty-six taxa first appearing in the Caribbean region within the last 15 my, using midpoints of biochronologic age ranges of the deposits that produced first and last occurrences. Species, numbered as in tables 6.1 and 6.2, are listed within intervals in bathymetric (not chronologic) order. C indicates exclusively neritic taxon strongly associated with carbonate environments; arrow on range line indicates extension of range beyond chart.

used in the ecologic analysis below. Neritic species show increasing originations of carbonate-associated taxa, that is, those species that are today found living most abundantly in carbonate shoal and reef environments, as reported in the literature on species distributions and ecologies. More exclusively neritic species associated with biogenic carbonate environments first appeared in latest Miocene zone N17 (4/9 = 44%) than in the previous 8 my (1/4 = 25%), and more first occurred in Early Pliocene time (6/7 = 86%) than in N17. A chi-square test of homogeneity yields a reasonable probability ($p < .100$) that proportions of carbonate-associated species are significantly different among these three intervals (even if partially neritic taxa are added to the exclusively neritic data set), but these results may be unreliable with this small number of observations (Noether 1976). This trend suggests an increasing proportion of carbonate-associated species originating in the southwestern Caribbean region, and supports a hypothesis of progressive, Late Miocene to mid-Pliocene development of a more carbonate-rich regime.

With respect to possible Pleistocene extinctions caused by glacial cooling cycles, figure 6.5 shows the opposite pattern. With one exception, *Trifarina eximia*, no extinctions of the 107 common species occurred during the Quaternary period. In contrast, 11% (12/107) became extinct during latest Miocene through Pliocene time. In addition, the well-known, latest Pliocene-Pleistocene, neritic-bathyal indicator species for cooler, glacial intervals, *Hyalinea balthica*, has not been found in Panama-Costa Rica deposits or any other southern Caribbean deposits of that age, despite its presence in Late Pliocene sediments of the northern Caribbean (Bock 1970) and Gulf of Mexico (Akers and Dorman 1964).

DISCUSSION

The trends in benthic foraminiferal assemblages suggest that shallow-water environmental changes preceded seaway closure by at least a few million years, and that carbonate-rich areas progressively developed from Late Miocene to mid-Pliocene time, when complete seaway closure occurred. The age ranges of common benthic foraminifera reported here show that 25% of exclusively neritic taxa first appearing between 15 and 7.1 Ma, 44% of neritic taxa first appearing in latest Miocene time, and 86% of neritic species originating in the Early Pliocene are associated with carbonate facies. This trend suggests an increase in the development of the carbonate regime from Late Miocene to Early Pliocene time. Vermeij (1978), Petuch (1981), Vermeij and Petuch (1986), and Allmon (1992) have proposed that shallow waters of the southern Caribbean may have been colder, more turbid, nutrient-rich, and nonreefal prior to the closure of the Tropical American Seaway. However, a greatly constricted seaway could have produced significant large-scale oceanographic changes preceding complete closure. A general circulation model with an open Tropical American Seaway, driven by present winds, atmospheric temperatures, and moisture fluxes, shows that near-closure would have been sufficient to have significantly altered circulation (Maier-Reimer et al. 1990; Mikolajewicz et al. 1993).

Caribbean sedimentation patterns support the hypothesis of environmental change and increasing carbonate deposition from Late Miocene seaway constriction. Measurements of carbonate content in Miocene to Pliocene sediments primarily come from the southern Caribbean deep-sea record. However, because continental shelves bypass a large proportion of sediment to deeper accumulation sites (Hay and Southam 1977), deep-water sediments can be used to track shallower trends. Based on depositional rates, a weakening of strong, westward circulation in the Middle-Late Miocene eastern Caribbean and later Miocene-Early Pliocene southwestern Caribbean was attributed to increasing constriction

(Holcombe and Moore 1977; Zimmerman 1982). Caribbean deep-sea environments became richer in carbonates, gradually increasing to present levels in the southern Caribbean from 7.5 to 3.8 Ma (Saunders et al. 1973; Gardner 1982; Keller and Barron 1983). Carbonate dissolution was insignificant in the cores examined (Gardner 1982); therefore, carbonate content must have increased because of (1) decreasing noncarbonate material, either from uplift away from terrigenous input (Gardner 1982; Saunders et al. 1973) or changing surface-current regimes due to seaway closure (Zimmerman 1982; Saunders et al. 1973), or (2) increased biogenic carbonate production (Saunders et al. 1973).

Biogenic carbonate sediments associated with shoals and reefs are known from many Early to Middle Miocene shallow-water deposits of the northern Caribbean to Gulf of Mexico, such as Florida (Cushman and Ponton 1932b), the Dominican Republic (Saunders et al. 1986) and Anguilla (Budd et al. 1989). However, very few Early-Middle Miocene, carbonate-rich sediments are reported from the southern Caribbean, an exception being the interbedded Early to Middle Miocene(?) clays and reefal limestones of Trinidad (Maurrasse 1990). Deposits of the Late Miocene Gatun Formation (fig. 6.1, sections 14–15) are clastic and lack the well-developed diverse miliolid facies characterizing later carbonates of southern Central America (Wantland 1967; Collins 1993; Collins et al. 1995) to South America (Redmond 1953). Early Pliocene neritic sediments of the southern Caribbean show a strong carbonate influence. Neritic sediments of the Magdalena Basin, Caribbean Colombia, dated as either Pliocene (Chorrera section) or some age between Late Miocene and Early Pliocene (Arroyo Juan de Acosta sections), contain taxa such as *Amphistegina lessonii* and species of *Quinqueloculina* and *Vertebralina* (= *Nodobaculariella*) (Bordine 1974), which are closely associated with carbonate-rich sediments. Early Pliocene benthic foraminifera from Túbara, northern Colombia (Redmond 1953), and Aruba (Drooger 1953) indicate proximal reefs to shoals.

The Late Miocene-Early Pliocene increases in originations of carbonate-associated benthic foraminifera and carbonate content of deep-sea sediments are in accord with evolutionary trends in reef corals (Budd et al., this volume, chap. 7). The diversity of Caribbean reef corals, which are major producers of carbonate sediments, rose in Late Miocene time owing to increased origination and decreased extinction. Oceanographic conditions may have become more conducive to the establishment of new reef-coral species and thus promoted expansion of carbonate production in Late Miocene Caribbean shallow waters.

The ideas of the development of carbonate-rich areas and progressively warmer waters in the southern Caribbean Sea are complementary. Emiliani and others (1972) reasoned that colder Pacific water flowed into the Caribbean Basin prior to seaway closure, and increasing temperatures and salinities could have promoted an increase in shallow-water carbonate sediments. Although temperature is only one factor influencing carbonate production, if other factors are equal, a change in climate can cause a change in the position of the latitudinal carbonate/noncarbonate boundary (30°N and 30°S today) in nearshore sediments (Rodgers 1957; Sanders and Friedman 1969). Higher temperatures and salinities apparently control nearshore carbonate distribution; carbonates are more abundant where nearshore salinities exceed 34‰ (Trask 1937) and where annual temperatures do not fall below 18.5°C (Newell and Rigby 1957).

The hypothesis of a progressive warming of Caribbean waters with seaway constriction is somewhat difficult to test by conventional paleotemperature methods using stable oxygen isotopes or planktonic foraminiferal migrations. The proportion of stable isotopes of oxygen preserved in foraminiferal tests is commonly used to detect past temperature change, although $\delta^{18}O$ is a mixed temperature/salinity signal. For example, Keigwin (1982a,b) interpreted increasing values in Caribbean $\delta^{18}O$ through the Early to Late Pliocene as a progressive salinity increase in surface waters rather than gradual cooling, because endemic, tropical Atlantic species of the planktonic foraminiferan *Globorotalia* originated in the Early Pliocene. Additional support for increasing salinity with no surface-water cooling in the Pliocene comes from Caribbean planktonic foraminiferal assemblages, which remained relatively stable during the Pliocene (Cotton and Dowsett 1991; Cronin and Dowsett, this volume, chap. 4). However, warming surface waters would have damped increasing $\delta^{18}O$. Using Keigwin's (1982a,b) isotopic values in a paleotemperature equation for Caribbean surface water (Imbrie et al. 1973, 34), waters could have warmed a few degrees and still remained within the isotopic values for a salinity increase sufficient to account for modern salinity. Stable isotope analyses of Neogene mollusks from Caribbean Panama and Costa Rica have yielded somewhat equivocal results; although Early Pliocene values indicate conditions different from today's, specific changes in temperatures and salinities cannot be deciphered, and Late Pliocene results from different areas conflict (Teranes et al., this volume, chap. 5).

Changing proportions of warm- and cold-water species in planktonic foraminiferal assemblages are also used to identify temperature changes, but for tropical assemblages this method is reduced to a one-way argument of the migration of subtropical and perhaps temperate species. Thus, it may be difficult to identify temperature increases if equatorial assemblages are already within a relatively warm interval—from where would warmer-water species emigrate? In addition to this problem, the usefulness of tropical Atlantic planktonic foraminifera in conjunction with transfer functions for estimating past sea surface temperatures has been called into question because faunal assemblages of the present tropical Atlantic are not well correlated with sea surface temperature (Ravelo et al. 1990; Emiliani 1992; Guilderson et al. 1994).

The Caribbean region experienced cyclical Pleistocene cooling (Emiliani 1971; Imbrie and Kipp 1971; Rögl and Bolli 1973), but the magnitude of temperature change was not large enough in the southern Caribbean to cause either extinctions in common benthic foraminifera or migration of the well-known Plio-Pleistocene cold-indicator species *Hyalinea balthica* southward from northern Caribbean waters (Bock 1970). The Quaternary period experienced only one extinction of the 107 common benthic foraminiferal species, fewer than the five extinctions of the Late Pliocene and four during the Early Pliocene (fig. 6.5). In contrast, Late Pliocene-Pleistocene Caribbean mollusks experienced faunal turnover (Jackson et al. 1993; Jackson et al. this volume, chap. 9), and Caribbean reef-coral diversity was dramatically reduced by extinction (Budd et al., this volume, chap. 7). Although Caribbean surface waters apparently did not cool during Pliocene time (Cronin and Dowsett, this volume, chap. 4), glacial intervals became colder in the Pleistocene and probably affected equatorial waters, as evidenced by studies of Caribbean planktonic foraminiferal assemblages (Emiliani 1971, 1992; Imbrie and Kipp 1971; Rögl and Bolli 1973; Emiliani and Ericson 1991) and stable isotope and Sr/Ca measurements of corals (Beck et al. 1992; Guilderson et al. 1994), all of which indicate Pleistocene intervals of surface-water temperatures cooler by about 5°C.

If the Late Pliocene to Pleistocene was a time of turnover in Caribbean mollusk and reef-coral taxa, as suggested in chapters 7 and 9 of this volume, why did benthic foraminifera show the opposite pattern of decreased extinction? Assuming cooling or related changes in marine conditions during glacial intervals caused the evolutionary changes in corals and mollusks, a possible reason for the difference in benthic foraminifera is my use of only common species, that is, those with greater than 1% abundance in at least one sample. This method assumes that the sampling bias introduced by including rare species would be greater than the bias introduced by using only common species. Common taxa should be more eurytopic than rare taxa, and thus more resistant to fluctuating temperature. Higher extinction rates in Early and Late Pliocene time affected mostly older, deeper-water taxa such as *Bolivina byramensis*, *Planulina mexicana*, and *Hanzawaia isidroensa*, and may have been due to changing bottom-water conditions. Speculation that benthic foraminifera would generally not be sensitive to temperature change is not supported by previous studies that have used them to detect temperature change (e.g., Woodruff et al. 1981), and by the strong association of some species such as *Hyalinea balthica* with cooling episodes (Bock 1970). In any case, Plio-Pleistocene temperature fluctuations were not large enough to have caused extinctions in common Caribbean benthic foraminifera of southern Central America, 90% of which are living in Caribbean waters today.

CONCLUSIONS

1. The Tropical American Seaway was constricted by Early Miocene time, as evidenced by shallow-water benthic foraminifera in Costa Rica to Panama and the migration of North American land mammals as far south as the Panama Canal basin.

2. In the Late Miocene, an increase in the rate of first occurrences of common Caribbean benthic foraminifera, most of which are the earliest records, suggests that environmental changes caused by seaway constriction promoted increased origination.

3. Progressively more of the new, neritic taxa first appearing during Late Miocene to Early Pliocene time are species that are most abundant in Caribbean carbonate shoal and reef environments today. This suggests expansion of carbonate-producing environments in the southern Caribbean during that time. Late Miocene-Early Pliocene development of a Caribbean carbonate regime is supported by increasing carbonate content in deep-sea sediments through this period and a Late Miocene diversification of reef corals, which are major producers of carbonate sediment in shallow waters.

4. The changes in benthic foraminiferal assemblages, and by inference, environmental changes, happened a few million years before complete seaway closure around 3.5 Ma, after which time there apparently was little speciation or extinction.

5. Extinction in common Caribbean benthic foraminifera decreased during the Quaternary, in contrast to patterns of extinction or taxonomic turnover seen in mollusks and reef corals. Since independent evi-

dence indicates Pleistocene cooling in the tropics, the near lack of extinction in benthic foraminifera may be due to the nature of the data set, which includes only common, presumably more eurytopic taxa.

ACKNOWLEDGMENTS

Helena Fortunato and Magnolia Calderon prepared foraminiferal specimens, and Laurel Bybell and Harry Dowsett provided the biochronologic dates published by Coates and others (1992). Laurie Anderson, Mercedes Arosemena, Anthony Coates, Timothy Collins, Mathew Cotton, Thomas Cronin, Dana Geary, Jeremy Jackson, Peter Jung, Elena Lombardo, Rene Panchaud, and David West assisted in field logistics or collecting. Catherine Badgely, Ann Budd, Martin Buzas, Harry Dowsett, Jeremy Jackson, Lisa Sloan, and Geerat Vermeij gave much useful advice. This work was supported primarily by the National Science Foundation, Smithsonian Tropical Research Institute, and National Geographic Society.

APPENDIX 6.1: FIRST AND LAST REGIONAL OCCURRENCES OF COMMON BENTHIC FORAMINIFERA

Name	Event	Reference	If Different, Recorded as	Stratigraphic Unit	Epoch	Zone/Stage/ Subepoch	Age (Ma)
Ammonia beccarii	F	Poag 1966		Paynes Hammock Fm.	Oligocene	N2–3, NP24	30.4–28.0
Amphistegina gibbosa	F	Nuttall 1932	*Amphistegina lessonii*	Alazan Shale	Oligocene	Vicksburgian	36.9–30.1
Asterigerina carinata	F	USNM, CC18552		Chipola Fm.	Miocene	N7–8	17.6–15.2
Asterigerina pettersi	F	USNM, CC43247	*Asterigerina bracteata*	Anahuac Fm.	Oligocene	Late	30.0–23.7
Asterigerina pettersi	L	CJ88-30-1		Escudo de Veraguas Fm.	Pliocene	*	1.9– 1.8
Bigenerina irregularis	F	CJ86-14-1		Gatun Fm., upper part	Miocene	N17–18	7.1– 5.0
Biloculinella eburnea	F	CJ87-9-1		Escudo de Veraguas Fm.	Pliocene	*	3.6– 3.5
Bolivina alata	F	Cushman & Renz 1947	*Bolivina alata*	St. Croix Fm.	Miocene	N9	15.2–14.9
Bolivina barbata	F	Cassell 1986		Uscari Fm.	Miocene	N5–6	21.8–17.6
Bolivina byramensis	F	Cushman 1923	*Bolivina caelata byramensis*	Byram Marl	Oligocene	NP22	35.1–34.6
Bolivina byramensis	L	CJ88-32-1		Shark Hole Point Fm.	Pliocene	*	3.6– 3.5
Bolivina imporcata	F	Blow 1959		Tocuyo (= San Lorenzo) Fm.	Miocene	N6	19.0–17.6
Bolivina imporcata	L	CJ89-17-4		Moin Fm.	Pliocene	*	1.9– 1.7
Bolivina isidroensis	F	Todd & Low 1976		Juana Diaz Fm.	Oligocene	N1–2	34.0–28.2
Bolivina isidroensis	L	Kohl 1985		Encanto Fm. or "Biozone"	Pliocene	N19	5.0– 3.4
Bolivina lowmani	F	McLaughlin 1989		Sombrerito Fm.	Miocene	N12	12.9–11.6
Bolivina mexicana	F	Cassell 1986		Uscari Fm.	Miocene	N9–10	15.2–13.9

Species		Reference		Formation	Epoch	Zone	Age
Bolivina paula	F	Cushman & McGlamery 1938		Chickasawhay Fm.	Oligocene	NP24, N2–3	30.3–28.1
Bolivina pozonensis	F	Blow 1959		Pozón Fm.	Miocene	*G. fohsi robusta*	12.6–11.5
Bolivina pozonensis	L	CJ89-18-12		Moin Fm.	Pliocene	*	1.9– 1.6
Bolivina striatula	F	Cushman 1923	*Bolivina cf. B. punctata*	Mint Spring Fm.	Oligocene	NP21–22	36.8–34.6
Bolivina subexcavata	F	Cassell 1986		Uscari Fm.	Miocene	N5–6	21.8–17.6
Bolivina subspinescens	F	Cassell 1986	*Rectobolivina spinescens*	Uscari Fm.	Miocene	N6–7	19.0–16.6
Bolivina tortuosa	F	USNM, W.H. Monroe #9		Byram Fm.	Oligocene	NP22	35.1–34.6
Buccella hannai	F	Cassell 1986		Uscari Fm.	Miocene	N5–6	21.8–17.6
Bulimina aculeata	F	CJ88-51-1		Nancy Point Fm.	Miocene	*	6.5– 5.6
Bulimina inflata mexicana	F	Coryell & Embich 1937	*Bulimina inflata*	Tranquilla Shale	Eocene	Jacksonian	42.2–36.3
Bulimina marginata	F	CJ88-49-1		Nancy Point Fm.	Miocene	*	6.5– 5.6
Bulimina tessellata	F	McLaughlin 1989		Sombrerito Fm.	Miocene	N10	14.9–13.9
Buliminella bassendorfensis	F	Cassell 1986		Uscari Fm.	Miocene	N6	19.0 17.6
Buliminella elegantissima	F	USNM, Saline Co., Ark.		Midway	Paleocene	NP2–6	65.9–60.4
Cancris sagra	F	Nuttall 1932		Alazan Shale	Oligocene	Vicksburgian	36.9 30.1
Cassidulina carapitana	F	Bandy 1949	*Cassidulina alabamensis*	Shubuta Clay, Yazoo Fm.	Eocene	NP19/20–21, P16–17	37.8–36.2
Cassidulina laevigata	F	Nuttall 1932		Alazan Shale	Oligocene	Vicksburgian	36.9–30.3
Cassidulina minuta	F	CJ88-49-1		Nancy Point Fm.	Miocene	*	6.5– 5.6
Cassidulina norcrossi australis	F	CJ88-49-1		Nancy Point Fm.	Miocene	*	6.5– 5.6
Cassidulina subglobosa	F	Bandy 1949		Shubuta Clay, Yazoo Fm.	Oligocene	NP19/20–21, P16–17	37.8–36.2
Cibicides pachyderma	F	USNM, CC37596		Ocala Ls.	Eocene	P14?–16	42.6–37.1

APPENDIX 6.1: *continued*

Name	Event	Reference	If Different, Recorded as	Stratigraphic Unit	Epoch	Zone/Stage/subepoch	Age (Ma)
Cornuspira planorbis	F	USNM, CC17935	*Cornuspira involvens*	Chipola Fm.	Miocene	N7–8	17.6–15.2
Cymbaloporetta atlantica	F	CJ86-36-1		Cayo Agua Fm.	Pliocene	*	5.0– 4.6
Discorbina patelliformis	F	Cushman 1923		Byram Fm.	Oligocene	NP22	35.1–34.6
Discorbis bulbosa	F	Cassell 1986	*Rosalina bulbosa*	Uscari Fm.	Miocene	N6–7	19.0–16.6
Dyocibicides biserialis	F	Cushman & Ponton 1932b		Chipola Fm.	Miocene	N7–8	17.6–15.2
Elphidium discoidale	F	Cushman & McGlamery 1938		Chickasawhay Fm.	Oligocene	NP24	30.3–28.1
Elphidium excavatum	F	Cushman & McGlamery 1938	*Elphidium* cf. *E. poeyanum*	Chickasawhay Fm.	Oligocene	NP24	30.3–28.1
Elphidium mexicanum	F	CJ86-3-1		Gatun Fm., lower part	Miocene	N16	10.2– 7.1
Epistominella vitrea	F	Cassell 1986	*Epistominella exigua*	Uscari Fm.	Miocene	N5–6	21.8–17.6
Eponides turgidus	F	CJ88-51-1		Nancy Point Fm.	Miocene	*	6.5– 5.6
Fissurina laevigata	F	Cushman 1922b	*Lagena laevigata*	Mint Spring Fm.	Oligocene	NP21–22	36.8–34.6
Fursenkoina complanata	F	CJ88-51-1		Nancy Point Fm.	Miocene	*	6.5– 5.6
Fursenkoina pontoni	F	Blacut & Kleinpell 1969		Culebra Fm.	Miocene	N6	19.0–17.6
Gyroidina regularis	F	Cassell 1986		Uscari Fm.	Miocene	N5–6	21.8–17.6
Gyroidina umbonata	F	Becker & Dusenbury 1958	*Gyroidina parva*	Goajira Fm.	Miocene	Aquitanian	23.7–21.8
Hanzawaia concentrica	F	USNM, USGS loc. 6452	*Truncatulina americana*	Mint Spring Fm.	Oligocene	NP21–22	36.8–34.6

Species	F/L	Reference	Synonym	Formation	Epoch	Zone	Age
Hanzawaia isidroensa	F	Cushman & Ellisor 1945	*Cibicides isidroensis*	Anahuac Fm.	Oligocene	Late	30.0–23.7
Hanzawaia isidroensa	L	CJ88-32-1		Shark Hole Point Fm.	Pliocene	*	3.6– 3.5
Hauerina fragillissima	F	Cushman 1922a		Byram Fm.	Oligocene	NP22	35.1–34.6
Haynesina depressula	F	CJ86-1-6		Gatun Fm., lower part	Miocene	N16	10.2– 7.1
Hoeglundina elegans	F	Coryell & Embich 1937	*Epistomina elegans*	Tranquilla Shale	Eocene	Jacksonian	42.2–36.3
Lagena ornata	F	CJ89-42-2		Uscari Fm., uppermost	Miocene	*	6.0– 5.0
Lamarckina atlantica	F	Cushman & Ponton 1932b		Shoal River Fm.	Miocene	N8–13	16.6–11.3
Lenticulina calcar	F	Coryell & Embich 1937	*Robulus alato-limbatus*	Tranquilla Shale	Eocene	Jacksonian	42.2–36.3
Marginulinopsis marginulinoides	F	Sansores 1972		Deposito Fm.	Miocene	N11	13.9–12.9
Melonis barleeanum	F	Bandy 1949	*Nonion nicobarensis*	Lisbon Fm.	Eocene	P11–14	49.0–41.3
Miliolinella californica	F	CJ87-9-2		Escudo de Veraguas Fm.	Pliocene	*	3.6– 3.5
Neocomorbina parkerae	F	CJ88-51-1		Nancy Point Fm.	Miocene	*	6.5– 5.6
Neoeponides antillarum	F	Coryell & Embich 1937	*Eponides jacksonensis*	Tranquilla Shale	Eocene	Jacksonian	42.2–36.3
Nodobacullariella cassis	F	CJ86-38-1		Cayo Agua Fm.	Pliocene	*	5.0– 4.6
Nonionella altantica	F	Cassell 1986	*Pseudononion grateloupi*	Uscari Fm.	Miocene	N6–7	19.0–16.6
Nonionella incisa	F	Becker & Dusenbury 1958		Goajira Fm.	Miocene	Aquitanian	23.7–21.8
Nonionella incisa	L	CJ88-32-1		Shark Hole Point Fm.	Pliocene	*	3.6– 3.5

APPENDIX 6.1: *continued*

Name	Event	Reference	If Different, Recorded as	Stratigraphic Unit	Epoch	Zone/Stage/ subepoch	Age (Ma)
Nonionella jacksonensis	F	Cushman 1933		Ocala Ls.	Eocene	P14–16	42.6–37.2
Oridorsalis umbonatus	F	Nuttall 1932	*Eponides umbonata*	Alazan Shale	Oligocene	NP21–23	36.9–30.3
Pararotalia magdalenensis	F	CJ86-7-1		Gatun Fm., upper part	Miocene	N17–18	7.1– 5.0
Peneroplis carinatus	F	USNM, CC17677	*Peneroplis proteus*	Chipola Fm.	Miocene	N7–8	17.6–15.2
Planorbulina acervalis	F	USNM, W. H. Monroe #18	*Planorbulina mediterranensis*	Byram Fm.	Oligocene	NP22	35.1–34.6
Planulina ariminensis	F	van Morkhoven et al. 1986		Eureka core	Miocene	N16	10.2– 7.1
Planulina foveolata	F	van Morkhoven et al. 1986		Caribbean	Pliocene	N18	5.3– 5.0
Planulina marialana	L	CJ88-53-1		Nancy Point Fm.	Miocene	*	6.5– 5.6
Planulina marialana	F	Hadley 1934		?	Oligocene	Late	30.1–23.7
Pullenia bulloides	F	Cushman & Renz 1948		Navet Fm.	Eocene	Early–Late	57.8–36.6
Quinqueloculina compta	F	CJ86-36-1		Cayo Agua Fm.	Pliocene	*	5.0– 4.6
Quinqueloculina costata	F	USNM, CC17808		Chipola Fm.	Miocene	N7–8	17.6–15.2
Quinqueloculina lamarckiana	F	Nuttall 1935		Pauji Shale	Eocene	Late	40.0 36.6
Rectobolivina advena	F	Cassell 1986		Uscari Fm.	Miocene	N6–7	19.0–16.6
Reussella spinulosa	F	Cassell 1986		Uscari Fm.	Miocene	N6–7	19.0–16.6
Rosalina concinna	F	Cassell 1986	*Neocomorbina terquemi*	Uscari Fm.	Miocene	N8	16.6–15.2

Species	Status	Reference	Synonym	Formation	Epoch	Zone	Range
Rosalina floridana	F	Cushman & Ponton 1932b	*Discorbis floridana*	Choctawhatchee Fm.	Miocene	*Area zone*, N17	7.1– 5.3
Rosalina floridensis	F	Palmer 1941	*Discorbis bertheloti floridensis*	Cojimar Fm.	Miocene	Middle	16.3–10.5
Rosalina subaraucana	F	USNM, CC36954	*Discorbis subaraucana*	Ocala Ls.	Eocene	P14–16	42.6–37.1
Rotalia garveyensis	L	CJ89-18-12		Moin Fm.	Pliocene	*	1.9 –1.6
Rotalia garveyensis	F	CJ88-53-1		Nancy Point Fm.	Miocene	*	6.5 –5.6
Rotorbinella umbonata	F	CJ89-46-1		Uscari Fm., uppermost	Miocene	*	6.0– 5.0
Sagrina pulchella	F	Cassell 1986		Uscari Fm.	Miocene	N9–10	15.2–13.9
Seabrookia earlandii	F	CJ89-42-3		Uscari Fm., uppermost	Miocene	*	6.0– 5.0
Sigmoilina tenuis	F	Nuttall 1932		Alazan Shale	Oligocene	Vicksburgian	36.9–30.1
Sigmoilopsis schlumbergeri	F	Cushman & Stainforth 1945	*Sigmoilina schlumbergeri*	Cipero Fm.	Miocene	N10–12	14.9–11.6
Siphogenerina lamellata	F	Renz 1948		Pozón Fm.	Miocene	N6	19.0–17.6
Siphogenerina lamellata	L	CJ88-53-1		Nancy Point Fm.	Miocene	*	6.5– 5.6
Siphonina pulchra	F	Howe & Wallace 1932	*Siphonina darvillensis*	Jacksonian	Eocene	P14–16	42.6–37.2
Sorites marginalis	F	Bermudez 1949		Cercado Fm.	Miocene	NN11	8.2– 5.6
Sphaeroidina bulloides	F	van Morkhoven et al. 1986		Eureka core	Oligocene	P19	34.0–31.6
Spirillina vivipara	F	Cushman & Todd 1942		Naheola Fm.	Paleocene	NP5–6, P3	62.0–61.0
Spiroloculina antillarum	F	USNM, USGS loc. 6452		Mint Spring Fm.	Oligocene	NP21–22	36.8–34.6
Spiroplectammina floridana	F	CJ87-9-2		Escudo de Veraguas Fm.	Pliocene	*	3.6– 3.5
Stetsonia minuta	F	CJ88-30-12		Escudo de Veraguas Fm.	Pliocene	*	3.6– 3.5

APPENDIX 6.1: *continued*

Name	Event	Reference	If Different, Recorded as	Stratigraphic Unit	Epoch	Zone/Stage/ subepoch	Age (Ma)
Textularia foliacea occidentalis	F	CJ89-21-1		Rio Banano Fm.	Pliocene	*	3.6– 3.5
Textularia panamensis	F	CJ86-5-1		Gatun Fm., middle part	Miocene	N17–18	7.1– 5.0
Textularia panamensis	L	CJ89-20-11		Rio Banano Fm.	Pliocene	*	2.5– 2.4
Textularia schencki	F	CJ88-53-1		Nancy Point Fm.	Miocene	*	6.5– 5.6
Trifarina carinata	F	Becker & Dusenbury 1958		Goajira Fm.	Miocene	Aquitanian	23.7–21.8
Trifarina eximia	L	Akers & Dorman 1964	*Angulogerina* cf. *A eximia*	Louisiana well cores	Pleistocene	Nebraskan	1.6– 1.3
Trifarina eximia	F	Palmer 1941	*Angulogerina danvillensis*	Cojimar Fm.	Miocene	Middle	16.3–10.5
Trifarina occidentalis	F	Cushman 1946		Cocoa Sand	Eocene	P15–16	41.3–37.2
Triloculina trigonula	F	USNM		Moodys Fm.	Eocene	P14,NP17	42.3–41.3
Uvigerina laevis	F	Blow 1959	*Uvigerina auberiana attentuata*	Tocuyo (= San Lorenzo) Fm.	Miocene	N6	19.0–17.6
Uvigerina peregrina	F	Nuttall 1932	*Uvigerina pigmaea*	Alazan Shale	Oligocene	NP21–23	36.9–30.3
Valvulineria baitiana	L	CJ88-52-1		Nancy Point Fm.	Miocene	*	6.5– 5.6
Valvulineria baitiana	F	Bermudez 1949		Abuillot Fm.	Eocene	Early–Middle	57.8–40.0
Valvulineria palmerae	F	Cassell 1986		Uscari Fm.	Miocene	N6–7	21.8–17.6

Note: Reference is either a work cited in the reference list, a sample (CJ no.), or a specimen at the U.S. National Museum (USNM no.) documenting first or last occurrence. F = first regional occurrence; L = last regional occurrence; asterisks indicate biochronologic zonation found in Coates et al. 1992, Collins 1993, or Collins et al. 1995.

REFERENCES

Akers, W. H., and J. H. Dorman. 1964. Pleistocene foraminifera of the Gulf Coast. *Tulane Stud. Geol.* 3:1–93.

Allmon, W. D. 1992. Role of temperature and nutrients in extinction of turritelline gastropods: Cenozoic of the northwestern Atlantic and northeastern Pacific. *Palaeogeogr., Palaeoclimatol., Palaeoecol.* 92:41–54.

Andersen, H. V. 1961. Genesis and paleontology of the Mississippi River mudlumps, lower Mississippi River delta. *La. Geol. Soc. Bull.* 35:1–208.

Ayala-Castañares, A. 1963. Sistemática y distribución de los foraminíferos recientes de la Laguna de Términos, Campeche, Mexico. *Inst. Geol. Univ. Nac. Auton. México, Bol.* 67:1–130, 11 plates.

Bandy, O. L. 1949. Eocene and Oligocene Foraminifera from Little Stave Creek, Clarke County, Alabama. *Bulls. Am. Paleontol.* 32:1–211.

———. 1954. Distribution of some shallow-water Foraminifera in the Gulf of Mexico. *U.S. Geol. Surv. Prof. Pap.* 254-F:124–42, plates 28–31.

———. 1956. Ecology of Foraminifera in northeastern Gulf of Mexico. *U.S. Geol. Surv. Prof. Pap.* 274-G:179–204, plates 29–31.

Beck, J. W., R. L. Edwards, E. Ito, F. W. Taylor, J. Recy, F. Rougerie, P. Joannot, and C. Henin. 1992. Sea-surface temperature from coral skeletal strontium/calcium ratios. *Science* 257:644–47.

Becker, L. E., and A. N. Dusenbury, Jr. 1958. *Mio-Oligocene (Aquitanian) Foraminifera from the Goajira Peninsula, Colombia.* Cushman Foundation for Foraminiferal Research, Special Publication 4. Bridgewater, Mass.

Berggren, W. A., and C. D. Hollister. 1974. Paleogeography, paleobiogeography, and the history of circulation in the Atlantic Ocean. In *Studies in paleooceanography*, ed. W. W. Hay, 126–86. Society of Economic Paleontologists and Mineralogists, Special Publication 20. Tulsa, Okla.

———. 1977. Plate tectonics and paleocirculation: A commotion in the ocean. *Tectonophysics* 38:11–48.

Berggren, W. A., D. V. Kent, J. J. Flynn, and J. A. van Couvering. 1985. Cenozoic geochronology. *Geol. Soc. Am. Bull.* 96:1407–18.

Bermudez, P. J. 1949. *Tertiary smaller Foraminifera of the Dominican Republic.* Cushman Foundation for Foraminiferal Research, Special Publication 25. Sharon, Mass.

Blacut, G., and R. M. Kleinpell. 1969. A stratigraphic sequence of benthonic smaller foraminifera from the La Boca Formation, Panama Canal Zone. *Contrib. Cushman Found. Foram. Res.* 20:1–22.

Blow, W. H. 1959. Age, correlation, and biostratigraphy of the Upper Tocuyo (San Lorenzo) and Pozón Formations, Eastern Falcón, Venezuela. *Bulls. Am. Paleontol.* 39:1–251.

Bock, W. D. 1970. *Hyalinea baltica* and the Plio-Pleistocene boundary in the Caribbean Sea. *Science* 170:847–48.

Boltovskoy, E., D. B. Scott, and F. S. Medioli. 1991. Morphological variations of benthic foraminiferal tests in response to changes in ecological parameters: A review. *J. Paleontol.* 65:175–85.

Bordine, B. W. 1974. Neogene biostratigraphy and paleoenvironments, Lower

Magdalena Basin, Colombia. Ph.D. diss., Louisiana State University, Baton Rouge, La.

Brolsma, M. J. 1978. Benthonic foraminifera. In *Micropaleontological counting methods and techniques—an exercise on an eight metres section of the Lower Pliocene of Capo Rossello, Sicily,* 47–80. *Utrecht Micropaleontology Bulletin,* vol. 17. Utrecht: State University of Utrecht.

Budd, A. F., K. G. Johnson, and J. C. Edwards. 1989. Miocene coral assemblages in Anguilla, B.W.I., and their implications for the interpretation of vertical succession on fossil reefs. *Palaios* 4:264–75.

Carter, J. G. 1983. Bibliography and index of Gulf and Atlantic coastal plain biostratigraphy. *Biostrat. Newslett.* (Chapel Hill, N.C.), supp. 1 (October): 1–118.

———. 1984. Additions to the newsletter bibliography. *Biostrat. Newslett.* (Chapel Hill, N.C.) 2 (January): 10–38.

Carter, J. G., and T. J. Rossbach. 1989. Summary of lithostratigraphy and biostratigraphy for the coastal plain of the southeastern United States. *Biostrat. Newslett.* (Chapel Hill, N.C.) 3:1.

Carter, J. G., and W. H. Wheeler. 1983. Correlation of the coastal plain formations of the southeastern United States, exclusive of peninsular Florida. *Biostrat. Newslett.* (Chapel Hill, N.C.) 1 (March): 1–44.

Cassell, D. T. 1986. Neogene foraminifera of the Limón Basin of Costa Rica. Ph.D. diss., Louisiana State University, Baton Rouge, La.

Cassell, D. T., and B. K. Sen Gupta. 1989. Foraminiferal stratigraphy and paleoenvironments of the Tertiary Uscari Formation, Limón Basin, Costa Rica. *J. Foram. Res.* 19:52–71.

Coates, A. G., J. B. C. Jackson, L. S. Collins, T. M. Cronin, H. J. Dowsett, L. M. Bybell, P. Jung, and J. A. Obando. 1992. Closure of the Isthmus of Panama: The near-shore marine record in Costa Rica and western Panama. *Geol. Soc. Am. Bull.* 104:814–28.

Cole, W. S. 1927. A foraminiferal fauna from the Guayabal Formation in Mexico. *Bulls. Am. Paleontol.* 14:1–47.

———. 1931. The Pliocene and Pleistocene Foraminifera of Florida. *Fla. Geol. Surv. Bull.* 6:1–79.

———. 1938. Stratigraphy and micropaleontology of two deep wells in Florida. *Fla. Geol. Surv. Bull.* 16:1–73.

———. 1952. Eocene and Oligocene larger Foraminifera from Panama. *J. Paleontol.* 27:332–37.

Collins, L. S. 1991. Regional versus physiographic effects on morphologic variability within *Bulimina aculeata* and *B. marginata. Mar. Micropaleontol.* 17: 155–70.

———. 1993. Neogene paleoenvironments of the Bocas del Toro Basin, Panama. *J. Paleontol.* 67:699–710.

Collins, L. S., and A. G. Coates. 1993. Marine paleobiogeography of Caribbean Panama: Last Pacific influences before closure of the Tropical American Seaway. *Geol. Soc. Am. Abstr. Prog.* 25:A428–29.

Collins, L. S., and A. G. Coates, J. B. C. Jackson, and J. Obando. 1995. Timing and rates of emergence of the Limón and Bocas del Toro Basins: Caribbean effects of Cocos Ridge subduction? In *Geologic and tectonic development of the Caribbean Plate boundary in southern Central America*, ed. P. Mann, 263–89. Geological Society of America Special Paper 295. Boulder, Colo.

Coryell, H. N., and J. R. Embich. 1937. The Tranquilla Shale (Upper Eocene) of Panama and its foraminiferal fauna. *J. Paleontol.* 11:289–305.

Cotton, M. A., and H. J. Dowsett. 1991. Quantitative environmental estimates from the Pliocene of the western Caribbean based on analyses of planktic foraminifers from Panama. *Geolog. Soc. Am. Abstr. Prog.*: 22:A336.

Culver, S. J., and M. A. Buzas. 1981. *Distribution of Recent benthic foraminifera in the Gulf of Mexico.* Smithsonian Contributions to Marine Science, vol. 8. Washington, D.C.

———. 1982. *Distribution of Recent benthic foraminifera in the Caribbean region.* Smithsonian Contributions to Marine Science, vol. 14. Washington, D.C.

Culver, S. J., M. A. Buzas, and L. S. Collins. 1987. On the value of taxonomic standardization in evolutionary studies. *Paleobiology* 13:169–76.

Cushman, J. A. 1918. Some Pliocene and Miocene Foraminifera of the coastal plain of the United States. *U.S. Geol. Surv. Bull.* 676:1–100.

———. 1920. Lower Miocene Foraminifera of Florida. *U.S. Geol. Surv. Prof. Pap.* 128-B:67–75.

———. 1922a. The Foraminifera of the Byram Calcareous Marl at Byram, Mississippi. *U.S. Geol. Surv. Prof. Pap.* 129-E:87–123.

———. 1922b. The Foraminifera of the Mint Spring calcareous marl member of the Marianna Limestone. *U.S. Geol. Surv. Prof. Pap.* 129-F:123–53.

———. 1922c. Shallow-water Foraminifera of the Tortugas region. *Carnegie Inst. Pub.* 311:1–85, plates 1–14.

———. 1923. The Foraminifera of the Vicksburg Group. *U.S. Geol. Surv. Prof. Pap.* 133:11–71.

———. 1926. Some new Foraminifera from the Upper Eocene of the southeastern coastal plain of the United States. *Contrib. Cushman Lab. Foram. Res.* 2:29–38, plates 4–5.

———. 1927a. New and interesting Foraminifera from Mexico and Texas. *Contrib. Cushman Lab. Foram. Res.* 3:111–19.

———. 1927b. Some characteristic Mexican fossil Foraminifera. *J. Paleontol.* 1:147–72.

———. 1929. A late Tertiary fauna of Venezuela and other related regions. *Contrib. Cushman Lab. Foram. Res.* 5:76–107.

———. 1930a. Common Foraminifera of the east Texas greensands. *J. Paleontol.* 4:33–41.

———. 1930b. The Foraminifera of the Choctawhatchee Formation of Florida. *Fla. Geol. Surv. Bull.* 4:1–92.

———. 1933. New Foraminifera from the Upper Jackson Eocene of the southeastern Coastal Plain region of the United States. *Contrib. Cushman Lab. Foram. Res.* 9:1–21.

———. 1946. *A rich foraminiferal fauna from the Cocoa Sand of Alabama.* Cushman Laboratory for Foraminiferal Research Special Publication 16. Sharon, Mass.

———. 1947. New species and varieties of Foraminifera from off the southeastern coast of the United States. *Contrib. Cushman Lab. Foram. Res.* 23:86–92.

Cushman, J. A., and E. D. Cahill. 1933. Miocene Foraminifera of the coastal plain of the eastern United States. *U.S. Geol. Surv. Prof. Pap.* 175-A:1–51.

Cushman, J. A., and A. C. Ellisor. 1939. New species of Foraminifera from the Oligocene and Miocene. *Contrib. Cushman Lab. Foram. Res.* 15:1–14.

———. 1945. The foraminiferal fauna of the Anahuac Formation. *J. Paleontol.* 19:545–72.

Cushman, J. A., and W. McGlamery. 1938. Oligocene Foraminifera from Choctaw Bluff, Alabama. *U.S. Geol. Surv. Prof. Pap.* 189-D:103–19.

———. 1939. New species of Foraminifera from the Lower Oligocene of Alabama. *Contrib. Cushman Lab. Foram. Res.* 15:45–49, plate 9.

———. 1942. Oligocene Foraminifera near Millry, Alabama. *U.S. Geol. Surv. Prof. Pap.* 197-B:64–84.

Cushman, J. A., and G. M. Ponton. 1932a. An Eocene foraminiferal fauna of Wilcox age from Alabama. *Contrib. Cushman Lab. Foram. Res.* 8:51–72.

———. 1932b. Foraminifera of the Upper, Middle, and part of the Lower Miocene of Florida. *Fla. Geol. Surv. Bull.* 9:1–147.

Cushman, J. A., and H. H. Renz. 1947. *The foraminiferal fauna of the Oligocene, Ste. Croix Formation, of Trinidad, B.W.I.* Cushman Laboratory for Foraminiferal Research, Special Publication 22. Sharon, Mass.

———. 1948. *Eocene Foraminifera of the Navet and Hospital Hill Formations of Trinidad, B.W.I.* Cushman Laboratory for Foraminiferal Research, Special Publication 24. Sharon, Mass.

Cushman, J. A., and R. M. Stainforth. 1945. *The Foraminifera of the Cipero Marl Formation of Trinidad, British West Indies.* Cushman Laboratory for Foraminiferal Research, Special Publication 14. Sharon, Mass.

Cushman, J. A., and R. Todd. 1941. Species of *Uvigerina* occurring in the American Miocene. *Contrib. Cushman Lab. Foram. Res.* 17:43–52, plates 11–14.

———. 1942. The Foraminifera of the type locality of the Naheola Formation. *Contrib. Cushman Lab. Foram. Res.* 18:23–46.

———. 1946. A foraminiferal fauna from the Byram Marl at its type locality. *Contrib. Cushman Lab. Foram. Res.* 22:76–102.

Donnelly, T. W., D. Beets, M. J. Carr, T. Jackson, G. Klaver, J. Lewis, R. Maury, H. Schellenkens, A. L. Smith, G. Wadge, and D. Westercamp. 1990. History and tectonic setting of Caribbean magmatism. In *The Caribbean Region*, vol. H of *The Geology of North America*, ed. G. Dengo and J. E. Case, 339–74. Boulder, Colorado: Geological Society of America.

Douglas, R. G., and F. Woodruff. 1981. Deep sea benthic foraminifera. In *The Sea*, vol. 7, ed. C. Emiliani, 1233–1327. New York: Wiley-Intersci.

Drooger, C. W. 1953. Miocene and Pleistocene Foraminifera from Oranjestad, Aruba (Netherlands Antilles). *Contrib. Cushman Found. Foram. Res.* 4:116–47.

Drooger, C. W., and J. P. H. Kaasschieter. 1958. *Foraminifera of the Orinoco-Trinidad-Paria shelf.* Vol. 4 of *Reports of the Orinoco Shelf Expedition.* Verhandl. Konink. Ned. Akad. Wetenschap., Natuurk., Eerste Reeks 22. Amsterdam.

Duque-Caro, H. 1990. Neogene stratigraphy, paleoceanography, and paleobiogeography in northwest South America and the evolution of the Panama Seaway. *Palaeogeogr., Palaeoclimatol., Palaeoecol.* 77:203–34.

Ekdale, S. F. 1974. Recent foraminiferal associations from northeastern Quintana Roo, Mexico. M.A. thesis, Rice University, Houston, Texas.

Emiliani, C. 1971. The amplitude of Pleistocene climatic cycles at low latitudes and the isotopic composition of glacial ice. In *The Late Cenozoic Glacial Ages*, ed. K. K. Turekian, 183–97. New Haven: Yale University Press.

———. 1992. Pleistocene paleotemperatures. *Science* 257:1462.

Emiliani, C., and D. B. Ericson. 1991. The glacial/interglacial temperature range of the surface water of the oceans at low latitudes. In *Stable isotope geochemistry: A tribute to Samuel Epstein*, ed. H. P. Taylor, Jr., J. R. O'Neil, and I. R. Kaplan, 223–28. Geochemical Society Special Publication 3. San Antonio, Texas.

Emiliani, C., S. Gartner, and B. Lidz. 1972. Neogene sedimentation on the Blake Plateau and the emergence of the Central American Isthmus. *Palaeogeogr., Palaeoclimatol., Palaeoecol.* 11:1–10.

Gardner, J. V. 1982. High-resolution carbonate and organic-carbon stratigraphies for the Late Neogene and Quaternary from the western Caribbean and eastern equatorial Pacific. *Init. Repts. DSDP* 68:347–64.

Golik, A., and F. B. Phleger. 1977. Benthonic foraminifera from the Gulf of Panama. *J. Foram. Res.* 7:83–99.

Guilderson, T. P., R. G. Fairbanks, and J. L. Rubenstone. 1994. Tropical temperature variations since 20,000 years ago: Modulating interhemispheric climate change. *Science* 263:663–65.

Hadley, W. H., Jr. 1934. Some Tertiary Foraminifera from the north coast of Cuba. *Bulls. Am. Paleontol.*

Haq, B. U. 1984. Paleoceanography: A synoptic overview of 200 million years of ocean history. In *Marine Geology and Oceanography of Arabian Sea and Coastal Pakistan*, ed. B. U. Haq and J. D. Milliman, 201–31. New York: Van Nostrand Reinhold Co.

Haq, B. U., J. Hardenbol, and P. R. Vail. 1987. Chronology of fluctuating sea levels since the Triassic. *Science* 235:1156–67.

Hasson, P. F., and A. G. Fischer. 1986. Observations on the Neogene of northwestern Ecuador. *Micropaleontol.* 32:32–42.

Hay, W. W. 1988. Paleoceanography: A review for the Geological Society of America Centennial. *Geo. Soc. Am. Bull.* 100:1934–56.

Hay, W. W., and J. R. Southam. 1977. Modulation of marine sedimentation by the continental shelves. In *The Fate of Fossil Fuel CO$_2$ in the oceans*, ed. M. R. Anderson and A. Malahoff, 569–604. New York: Plenum Press.

Holcombe, T. L., and W. S. Moore. 1977. Paleocurrents in the eastern Caribbean: Geologic evidence and implications. *Mar. Geol.* 23:35–56.

Howe, H. V. 1930. Distinctive new species of Foraminifera from the Oligocene of Mississippi. *J. Paleontol.* 4:327–31.

Howe, H. V., and W. E. Wallace. 1932. Foraminifera of the Jackson Eocene at Danville Landing on the Ouachita, Catahoula Parish, Louisiana. *La. Dept. Conserv. Geol. Bull.* 2:1–118, 15 plates.

Imbrie, J., and N. G. Kipp. 1971. A new micropaleontological method for quantitative paleoclimatology: Application to a Late Pleistocene Caribbean core. In *The Late Cenozoic glacial ages*, ed. K. K. Turekian, 71–181. New Haven: Yale University Press.

Imbrie, J., J. van Donk, and N. G. Kipp. 1973. Paleoclimatic investigation of a Late Pleistocene Caribbean deep-sea core: Comparison of isotopic and faunal methods. *Quat. Res.* 3:10–38.

Jackson, J. B. C., P. Jung, A. G. Coates, and L. S. Collins. 1993. Diversity and extinction of tropical American mollusks and emergence of the Isthmus of Panama. *Science* 260:1624–26.

Keigwin, L. D., Jr. 1982a. Isotopic paleoceanography of the Caribbean and East Pacific: Role of Panama uplift in Late Neogene time. *Science* 217: 350–53.

———. 1982b. Stable isotope stratigraphy and paleoceanography of sites 502 and 503. *Init. Repts. DSDP* 68:445–53.

Keller, G., and J. A. Barron. 1983. Paleoceanographic implications of Miocene deep-sea hiatuses. *Geol. Soc. Am. Bull.* 94:590–613.

Keller, G., C. E. Zenker, and S. M. Stone. 1989. Late Neogene history of the Pacific-Caribbean gateway. *J. South Am. Earth Sci.* 2:73–108.

Kohl, B. 1985. Early Pliocene benthic foraminifers from the Salina Basin, southeastern Mexico. *Bulls. Am. Paleontol.* 88:1–173.

Kornfeld, M. M. 1931. Recent littoral Foraminifera from Texas and Louisiana. *Contrib. Dept. Geol., Stanford Univ.* 1(3): 76–101.

Lankford, R. R. 1959. Distribution and ecology of Foraminifera from east Mississippi Delta margin. *Am. Assoc. Petrol. Geol. Bull.* 43:2068–99.

Loubere, P., and K. Moss. 1986. Late Pliocene climatic change and the onset of Northern Hemisphere glaciation as recorded in the northeast Atlantic Ocean. *Geol. Soc. Am. Bull.* 97:818–28.

Luyendyk, B. P., E. Forsyth, and J. D. Phillips. 1972. Experimental approach to the paleocirculation of the oceanic surface waters. *Geol. Soc. Am. Bull.* 83: 2649–64.

Maier-Reimer, E., U. Mikolajewicz, and T. Crowley. 1990. Ocean General Circulation Model sensitivity experiment with an open Central American isthmus. *Paleoceanography* 5:349–66.

Marshall, L. G. 1988. Land mammals and the Great American Interchange. *Am. Scient.* 76:380–88.

Marshall, L. G., C. C. Swisher III, A. Lavenu, R. Hoffstetter, and G. H. Curtis. 1992. Geochronology of the mammal-bearing Late Cenozoic on the northern Altiplano, Bolivia. *J. South Am. Earth Sci.* 5:1–19.

Matthews, R. K. 1988. Technical Comment on sea level history. *Science* 241: 597–99.

Maurrasse, F. J.-M. R. 1990. Stratigraphic correlation for the Circum-Caribbean Region. In *The Caribbean Region*, vol. H of *The Geology of North America*, ed. G. Dengo and J. E. Case, plates 4, 5A, and 5B. Boulder, Colo.: Geological Society of America.

McLaughlin, P. 1989. Neogene basin evolution in the southwestern Dominican Republic: A foraminiferal study. Ph.D. diss., Louisiana State University, Baton Rouge, La.

Mikolajewicz, U., E. Maier-Reimer, T. J. Crowley, and K.-Y. Kim. 1993. Effect of Drake and Panamanian gateways on the circulation of an ocean model. *Paleoceanography* 8:409–26.

Mullins, H. T., A. F. Gardulski, S. W. Wise, Jr., and J. Applegate. 1987. Middle Miocene oceanographic event in the eastern Gulf of Mexico: Implications for seismic stratigraphic succession and Loop Current/Gulf Stream circulation. *Geol. Soc. Am. Bull.* 98:702–13.

Mullins, H. T., and A. C. Neumann. 1979. Geology of the Miami Terrace and its paleoceanographic implications. *Mar. Geol.* 30:205–32.

Murray, J. W. 1991. *Ecology and palaeoecology of benthic foraminifera.* New York: Longman Scientific and Technical and John Wiley and Sons.

Newell, N. D., and J. K. Rigby. 1957. Geological studies on the Great Bahama Bank. In *Regional aspects of carbonate deposition,* ed. R. J. LeBlanc and J. G. Breeding, 15–72. Society of Economic Paleontologists and Mineralogists, Special Publication 5. Tulsa, Okla.

Noether, G. E. 1976. *Introduction to statistics: A nonparametric approach.* 2d ed. Boston: Houghton Mifflin.

Nuttall, W. L. F. 1932. Lower Oligocene Foraminifera from Mexico. *J. Paleontol.* 6:3–35.

———. 1935. Upper Eocene Foraminifera from Venezuela. *J. Paleontol.* 9: 121–31.

Palmer, D. K. 1941. Foraminifera of the Upper Oligocene Cojimar Formation of Cuba. Part 4. *Mems. Soc. Cubana Hist. Nat.* 15:181–200.

Parker, F. L., F. B. Phleger, and J. F. Peirson. 1953. *Ecology of Foraminifera from San Antonio Bay and environs, southwest Texas.* Cushman Foundation for Foraminiferal Research, Special Publication 2. Washington, D.C.

Petuch, E. J. 1981. A relict Neogene caenogastropod fauna from northern South America. *Malacologia* 20:307–47.

Pflum, C. E., and W. E. Frerichs. 1976. *Gulf of Mexico deep-water foraminifers.* Cushman Foundation for Foraminiferal Research, Special Publication 14. Lawrence, Kansas.

Phleger, F. B., and F. L. Parker. 1951. *Ecology of Foraminifera, northwest Gulf of Mexico.* Part 2, *Foraminifera species.* Geological Society of America Memoir 46. Boulder, Colo.

Pindell, J. L., and S. F. Barrett. 1990. Geological evolution of the Caribbean Region: A plate-tectonic perspective. In *The Caribbean Region,* vol. H of *The Geology of North America,* ed. G. Dengo and J. E. Case, 405–32. Boulder, Co.: Geological Society of America.

Poag, C. W. 1966. Paynes Hammock (Lower Miocene?) foraminifera of Alabama and Mississippi. *Micropaleontology* 12:393–440.

———. 1981. *Ecologic atlas of benthic foraminifera of the Gulf of Mexico.* Woods Hole, Mass.: Marine Science International.

Poag, C. W., and R. C. Tresslar. 1981. Living foraminifers of West Flower Garden Bank, northernmost coral reef in the Gulf of Mexico. *Micropaleontology* 27:31–70.

Puri, H. S. 1953. Contribution to the study of the Miocene of the Florida panhandle. *Fla. Geol. Surv. Bull.* 36:1–213.

Ravelo, A. C., R. G. Fairbanks, and S. G. H. Philander. 1990. Reconstructing tropical Atlantic hydrography using planktonic foraminifera and an ocean model. *Paleoceanography* 5:409–31.

Redmond, C. D. 1953. Miocene Foraminifera from the Tubara beds of northern Colombia. *J. Paleontol.* 27:708–33.

Renz, H. H. 1948. *Stratigraphy and fauna of the Agua Salada Group, state of Falcón, Venezuela.* Geological Society of America Memoir 32. Boulder, Colo.

Rodgers, J. 1957. The distribution of marine carbonate sediments: A review. In *Regional aspects of carbonate deposition,* ed. R. J. LeBlanc and J. G. Breeding, 2–13. Society of Economic Paleontologists and Mineralogists, Special Publication 5. Tulsa, Okla.

Rögl, F., and H. M. Bolli. 1973. Holocene to Pleistocene planktonic foraminifera of leg 15, site 147 (Carioco Basin [Trench], Caribbean Sea) and their climatic interpretation. *Init. Repts. DSDP* 15:553–99.

Ross, M. I., and C. R. Scotese. 1988. A hierarchical tectonic model of the Gulf of Mexico and Caribbean Region. *Tectonophysics* 155:139–68.

Ruddiman, W. F., and M. E. Raymo. 1988. Northern Hemisphere climate regimes during the past 3 Ma: Possible tectonic connections. *Phil. Trans. Roy. Soc. London* B318:411–30.

Sanders, J. E., and G. M. Friedman. 1969. Position of regional carbonate/noncarbonate boundary in nearshore sediments along a coast: Possible climatic indicator. *Geol. Soc. Am. Bull.* 80:1789–96.

Sansores, J. C. de, and C. Flores-Covarrubias. 1972. *Foraminíferos bentonicos del Terciario Superior de la Cuenca Salina del Istmo de Tehuantepec, Mex.* Mexico City: Inst. Mex. Petrol.

Saunders, J. B., N. T. Edgar, T. W. Donnelly, and W. W. Hay. 1973. Cruise synthesis. *Init. Repts. DSDP* 15:1077–1111.

Saunders, J. B., P. Jung, and B. Biju-Duval. 1986. Neogene paleontology in the Northern Dominican Republic. Part 1, Field surveys, lithology, environment, and age. *Bulls. Am. Paleontol.* 89:1–79.

Savin, S. M., and R. G. Douglas. 1985. Sea level, climate, and the Central American land bridge. In *The Great American Biotic Interchange,* ed. F. G. Stehli and S. D. Webb, 303–24. New York: Plenum Press.

Savin, S. M., R. G. Douglas, and F. G. Stehli. 1975. Tertiary marine paleotemperatures. *Geol. Soc. Am. Bull.* 86:1499–1510.

Stainforth, R. M., J. L. Lamb, H. Luterbacher, J. H. Beard, and R. M. Jeffords. 1975. *Cenozoic planktonic foraminiferal zonation and characteristics of index forms.* University of Kansas Paleontological Contributions, no. 62. Lawrence: University of Kansas Press.

Stanley, S. M. 1986. Anatomy of a regional mass extinction: Plio-Pleistocene decimation of the western Atlantic bivalve fauna. *Palaios* 1:17–36.

Sykes, L. R., W. R. McCann, and A. L. Kafka. 1982. Motion of the Caribbean Plate during last 7 million years and implications for earlier Cenozoic movements. *J. Geophys. Res.* 87:10656–76.

Tedford, R. H. 1987. Faunal succession and biochronology of the Arikareean through Hemphillian Interval (Late Oligocene through earliest Pliocene epochs) in North America. In *Cenozoic mammals of North America,* ed. M. O.

Woodburne, 153–210. Berkeley and Los Angeles: University of California Press.

Todd, R., and D. Low. 1976. Smaller foraminifera from deep wells on Puerto Rico and St. Croix. *U.S. Geol. Surv. Prof. Pap.* 863:1–57.

Trask, P. D. 1937. Relation of salinity to calcium carbonate content of marine sediments. *U.S. Geol. Surv. Prof. Pap.* 186-N:273–99.

Van Morkhoven, F. P. C. M., W. A. Berggren, and A. S. Edwards. 1986. *Cenozoic cosmopolitan deep-water benthic foraminifera.* Bulletin du Centre de Recherches de Pau, Mémoire, vol. 11. Elf Aquitaine, Pau, France.

Vermeij, G. J. 1978. *Biogeography and adaptation.* Cambridge: Harvard University Press.

Vermeij, G. J., and E. J. Petuch. 1986. Differential extinction in tropical American mollusks: Endemism, architecture, and the Panama land bridge. *Malacologia* 27:29–41.

Wantland, K. F. 1967. Recent benthonic foraminifera of the British Honduras Shelf. Ph.D. diss., Rice University, Houston, Texas.

Weyl, P. L. 1968. The role of the oceans in climatic change: A theory of the ice ages. *Meteorol. Monogr.* 8:37–62.

Whitmore, F. C., and R. H. Stewart. 1965. Miocene mammals and Central American seaways. *Science* 148:180–85.

Whittaker, J. E. 1988. *Benthic Cenozoic foraminifera from Ecuador.* London: British Museum (Nat. Hist.).

Woodring, W. P. 1982. Geology and paleontology of Canal Zone and adjoining parts of Panama: Description of Tertiary mollusks (Pelecypods: Propeamussiidae to Cuspidariidae; additions to families covered in P306-E; additions to gastropods, cephalopods). *U.S. Geol. Surv. Prof. Pap.* 306-F:542–845.

Woodruff, F., S. M. Savin, and R. G. Douglas. 1981. Miocene stable isotope record: A detailed Pacific deep ocean study and its paleoclimatic implications. *Science* 212:665–68.

Wüst, G. 1964. *Stratification and circulation in the Antillean-Caribbean Basins.* Part 1, *Spreading and mixing of the water types with an oceanographic atlas.* New York: Columbia University Press.

Zimmerman, H. B. 1982. Lithologic stratigraphy and clay mineralogy of the western Caribbean and eastern equatorial Pacific, leg 68, Deep Sea Drilling Project. *Init. Repts. DSDP* 68:383–95.

7

Plio-Pleistocene Turnover and Extinctions in the Caribbean Reef-Coral Fauna

Ann F. Budd, Kenneth G. Johnson, and Thomas A. Stemann

INTRODUCTION

Because of physiological constraints imposed by their algal endosymbionts, scleractinian reef corals are highly sensitive to extremes of light, temperature, salinity, and nutrients; and they are restricted to a narrow range of shallow (less than 50 m), clear, nutrient-poor tropical marine habitats (Smith and Buddemeier 1992). Similarly, reef biotas are widely regarded among paleontologists as exceptionally vulnerable to extinction during times of widespread environmental stress or global climate change (Jablonski 1991), and they are commonly used by geologists as indicators of tropical marine conditions. Still, very little is known about the long-term evolutionary patterns of scleractinian reef corals, owing to inadequate systematics, poor chronostratigraphic resolution, and the patchy geographic distribution of reefal deposits.

A case in point is the Caribbean reef-coral fauna over the past 20 million years. Species are notoriously difficult to distinguish, because of subtle morphologic differences among closely related species, high ecophenotypic plasticity, and a shortage of independent diagnostic characters (see discussion in Budd, Johnson, and Potts 1994). Until recently, few fossil taxa had been rigorously studied using multivariate statistical procedures. Similarly, despite recent advances in microfossil and isotope stratigraphy, corals in many Cenozoic Caribbean reef sequences have not been collected and dated since Vaughan (1919). Thus, the only ages that can reasonably be assigned to many specimens span unnecessarily broad time intervals, encompassing the entire range of ages that have more recently been obtained for the geologic unit from which they were collected. Finally, as mentioned earlier, reefs form only under unique environmental conditions, and lagoonal and nearshore patch-reef assemblages tend to be more readily preserved in the rock record than forereef assemblages. Consequently, Cenozoic reef deposits are unevenly scattered through sequences across the Caribbean region, and some reef environments have been more intensively sampled than others.

In this chapter, we reexamine the evolution of Caribbean reef corals over the past 20 million years using a new compilation designed to reduce some of the problems mentioned above. Our most significant finding is that a major episode of accelerated faunal turnover and extinction occurred during Plio-Pleistocene time (4–1 Ma). During the episode, the primary components of Caribbean reef communities changed dramatically. Pliocene reef communities were dominated by *Stylophora*, *Goniopora*, and a suite of agariciid and poritid species that more closely resemble modern Indo-Pacific species than modern Caribbean species (fig. 7.1). A diversity of free-living meandroid corals (*Teleiophyllia*, *Thysanus*, *Trachyphyllia*, *Placocyathus*) inhabited nearby Pliocene seagrass and soft-bottom areas. In contrast, today's Caribbean reefs are dominated by *Acropora*, *Diploria*, and a complex of species resembling *Montastraea annularis* (fig. 7.2). Seagrass and soft-bottom coral taxa are relatively few in number. We estimate that during the period of turnover, 75% of the Early to early Late Pliocene coral species ($n = 82$) became extinct in the Caribbean. Many of the survivors are more common in deeper forereef habitats (Goreau 1959), such as *Agaricia lamarcki* and *Mussa angulosa* (fig. 7.3).

Despite the significance of this episode for understanding reef-coral extinction and assessing the effects of large-scale disturbance on tropical reef biotas, it has been largely overlooked in previous work (e.g., Frost 1977; Stanley 1986). For example, Frost (1977) showed that Plio-Pleistocene extinctions occurred in more than 30% of the forty-six reef-coral species then known from the Early Pliocene of the Caribbean, and that half of the forty-eight modern species surveyed had pre-Pliocene distributions. Our new compilation of stratigraphic ranges of Neogene Caribbean reef corals (Budd, Stemann, and Johnson 1994) is more complete, because it includes additional Pliocene material. It incorporates recently obtained data from sequences in the Dominican Republic (Saunders et al. 1986) and Costa Rica (Coates et al. 1992), as well as published occurrences in Late Pliocene to Early Pleistocene Caribbean reef sites that were previously not considered. Our new compilation also benefits from more refined methods of species recognition and higher resolution dating for some of the included units.

To detect accelerated turnover and extinction during Plio-Pleistocene time, we analyze species occurrence data using quantitative methods wherever possible. We examine the timing, magnitude, and selectivity of extinctions and originations in the overall fauna. Special

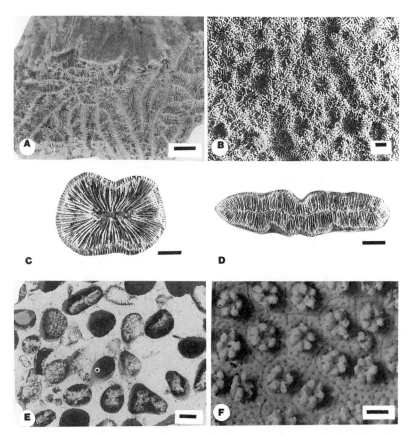

Fig. 7.1 Major reef-building Caribbean corals that became extinct in the Caribbean during the Plio-Pleistocene: (A) *Gardineroseris planulata*, SUI 63664, Late Miocene, TS loc. GG3 (= NMB loc. 15848), Rio Gurabo, Gurabo Formation, Dominican Republic (scale bar = 5 mm); (B) *Goniopora imperatoris*, NMB D5859, Late Miocene, NMB loc. 16853, Rio Cana, Cercado Formation, Dominican Republic (scale bar = 1 mm); (C) *Trachyphyllia bilobata*, NMB D6025, Late Miocene, NMB loc. 15899, Rio Gurabo, Cercado Formation, Dominican Republic (scale bar = 1 cm); (D) *Teleiophyllia* sp. B, NMB D6026, Late Miocene, NMB loc. 16821, Rio Cana, Gurabo Formation, Dominican Republic (scale bar = 1 cm); (E) *Caulastraea portoricensis*, SUI 84537, Late Miocene, AFB loc. 5, Lirio Limestone, Mona (scale bar = 5 mm); (F) *Stylophora granulata*, Late Pliocene, Unda 324, Bahamas cores, Great Bahama Bank (scale bar = 1 mm).

emphasis is given to sampling bias. Our analyses address four major questions: (1) Do Plio-Pleistocene extinctions and originations represent an acceleration over Miocene background rates? (2) Over what time interval did accelerated extinctions and originations occur, and were periods of accelerated extinction and origination synchronous? (3)

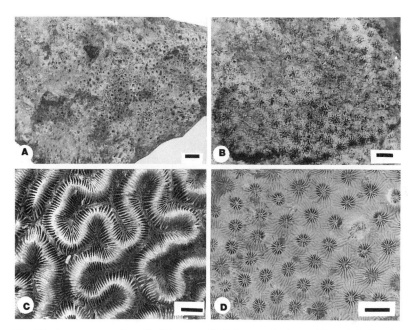

Fig. 7.2 Important modern Caribbean reef-building corals that originated during the Late Pliocene to Pleistocene: (A) *Acropora palmata*, USNM 93889 (AFB 170), Early Pleistocene, KJ-32-1, Lomas del Mar, Moin Formation, Costa Rica (scale bar = 5 mm); (B) *Porites astreoides*, USNM 93890 (AFB 379), Early Pleistocene, TS-CR9-2, Lomas del Mar, Moin Formation, Costa Rica (scale bar = 2 mm); (C) *Diploria strigosa*, BM(NH) 1928-3-1-19, Recent, Matthai-Gardiner collection, St. Thomas, V.I. (scale bar = 5 mm); (D) *Montastraea faveolata*, USNM 93891, Early Pleistocene, KJ-LM-26, Lomas del Mar, Moin Formation, Costa Rica (scale bar = 5 mm).

What proportion of the fauna was affected? and (4) Were different reef habitats affected equally? The role of ecological factors such as reproductive mode, life history, and population dynamics in determining species resistance to extinction are treated elsewhere (Johnson et al. 1995), and investigations of detailed patterns of faunal replacement and their geographic variation are in progress (see Budd, Johnson, and Jackson 1994).

STRATIGRAPHY

The data in our new compilation consist of occurrences of 175 species of reef corals over the past 24 million years (fig. 7.4) and are derived from new collections and all available published information. We have included all recorded hermatypic species except those belonging to fam-

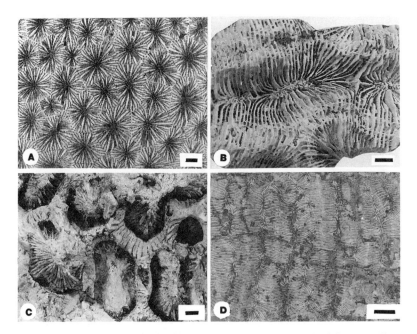

Fig. 7.3 Caribbean reef corals that survived the turnover event: (A) *Siderastrea siderea*, NMB D6027, Late Miocene, NMB loc. 15848, Rio Gurabo, Gurabo Formation, Dominican Republic (scale bar = 1 mm); (B) *Colpophyllia natans*, NMB D6022, Early Pliocene, NMB loc. 16818, Rio Cana, Gurabo Formation, Dominican Republic (scale bar = 5 mm); (C) *Mussa angulosa*, SUI 84538, Late Miocene, AFB loc. 17, Lirio Limestone, Mona (scale bar = 5 mm); (D) *Agaricia lamarcki*, USNM 93892, Early Pleistocene, CJ 92-06-21, Lomas del Mar, Moin Formation, Costa Rica (scale bar = 5 mm).

ilies Oculinidae and Rhizangiidae. These two families are represented by species that only questionably or variably contain zooxanthellae and are therefore not important shallow-water tropical reef builders. We have similarly excluded ahermatypic taxa. The data (fig. 7.5) were obtained from (1) collections made in continuous Neogene sections exposed along four rivers in the Cibao Valley of the northern Dominican Republic (Saunders et al. 1986; Budd, Stemann, and Johnson 1994); (2) occurrences in two Late Miocene to Recent cores (Clino, Unda) from the Great Bahama Bank (Ginsburg 1993); (3) collections made from isolated exposures of Plio-Pleistocene biostromes in the Moin Formation near Limón, Costa Rica (Coates et al. 1992); (4) twenty-one faunal lists for Early Miocene to Pleistocene Caribbean reef corals (app. 7.1); (5) published lists of modern reef corals occurring on the north coast of Jamaica (Wells and Lang 1973) and in the Abrolhos Islands of Brazil (Laborel 1969).

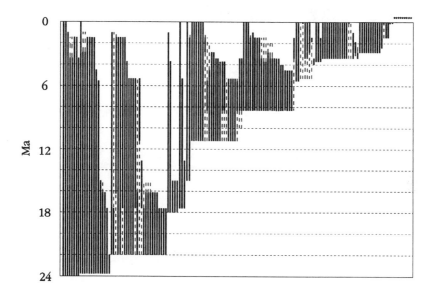

species

Fig. 7.4   Ranges of all Caribbean species of reef corals that lived during the past 24 million years. Ends of range lines are maxima of age estimates for localities in which first and last occurrences appear. Taxa are arranged in order of origination. Asterisks represent living species with no fossil representatives; dashed lines indicate occurrences in deposits with poorly understood chronostratigraphy. Data are in Budd, Stemann, and Johnson 1994.

Our compilation is therefore based on an uneven distribution of localities through geologic time (fig. 7.6), with many localities having age dates somewhere between 8 and 6 Ma and between 3 and 1 Ma. Corals in some localities have also been much more intensively collected than others. Nevertheless, even with these sampling irregularities, the compilation includes more than twice as many Late Miocene to Early Pleistocene sequences as Frost 1977, the best earlier study. Frost's review included more Early to Middle Miocene localities than used here, but many occurrences in these additional Miocene sequences were taken from his still unpublished field notes, and therefore require further study.

We recorded coral occurrences in the northern Dominican Republic using material collected as part of a multidisciplinary project on the paleontology and stratigraphy of the Neogene siliciclastic sequence in the Cibao Valley region (Saunders et al. 1986). Reefs in the sequence consist primarily of patches that formed on a thin shelf and slope along a nar-

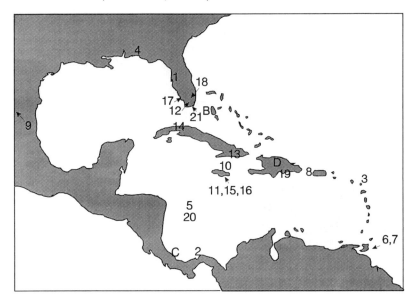

Fig. 7.5   General location of faunas included in the analysis: D = sections in the northern Dominican Republic; B = Bahamas cores; C = exposures in Costa Rica. Numbers refer to faunal lists given in appendix 7.1.

row, eastward-opening trough located north of the rapidly uplifting Cordillera Central. The coral-rich portions of the sequence were part of a thick sedimentary wedge (as much or more than 1000 m thick) that prograded northward during Late Miocene to early Late Pliocene time. Lower portions of the sequence consist of shallow nearshore deposits that gradually deepen (up to 200–400 m deep) upsection with few intervening hiatuses. Corals in the lower portions of the sequence are largely in place, whereas those in the upper portions occur in transported lenses (Evans 1986). Macrofossils were collected in the sequence by haphazardly extracting well-preserved material from the surface of the outcrop. Sampling therefore was roughly representative of species composition, but not of abundance. Age dates for the corals have been determined through the study of nannofossils and planktonic foraminifers (Saunders et al. 1986).

In contrast to the siliciclastics of the northern Dominican Republic, we recorded coral occurrences in the Bahamas using material in two cores (Unda, Clino) drilled through a series of prograding and platform carbonate sequences on the Great Bahama Bank. Plio-Pleistocene and Late Miocene portions of the cores are extremely rich in corals and contain as many as forty-nine species (Budd and Kievman 1993), many of

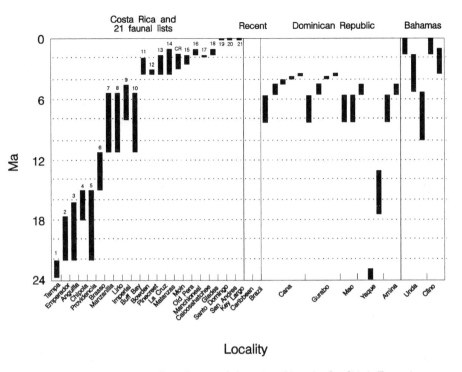

Fig. 7.6 Age ranges of the forty-five sampled stratigraphic units (localities). Formation or informal names are indicated. Numbers refer to faunal lists given in appendix 7.1. Late Miocene to Early Pliocene sections in the Dominican Republic have been divided into calcareous nannoplankton zones. The Bahamas cores have been subdivided as described in the text.

which are first or last occurrences. For the purposes of the present compilation, the cores were subdivided into three stratigraphic intervals in Unda (0–212 m, 212–963 m, > 963 m) and two intervals in Clino (0–420 m, > 420 m). Ages assigned to the intervals (fig. 7.6) were based on preliminary study of microfossils, paleomagnetics, and strontium isotopes (McNeill et al. 1993).

We recorded coral occurrences in Costa Rica using collections made in biostromes near the top of the siliciclastic Moin Formation at the Lomas del Mar housing community near Limón. The Moin Formation consists of a 200 m sequence of blue silty claystones and volcaniclastic litharenites that contain interbedded deeper-water (greater than 30 m) reef lenses, and is interpreted as shallow inner-shelf deposits filling a back-arc basin northeast of the rapidly uplifting Talamanca Cordillera (Coates et al. 1992). We identified forty-four species of reef corals in these collections, some of which form large colonies (more than 0.5 m in diameter). Study of microfossils suggests an approximate age of 1.9–1.7 Ma for the biostromes.

To augment the compilation, we added occurrences taken from twenty-one faunal lists. The faunal lists were based on specimens collected from reefal units that formed in a variety of water depths and in both carbonate and siliciclastic settings. They include six Early to Middle Miocene reef sequences (Florida [2], Panama, Anguilla, Providencia, Trinidad), four Late Miocene to Early Pliocene reefs (Trinidad; Mona, California; Jamaica), eight early Late Pliocene to Early Pleistocene reef sequences (Jamaica [3], Florida [3], Cuba [2]), and three Late Pleistocene terraces (Dominican Republic; San Andrés, Florida). We determined age dates for these assemblages using published reports of microfossils and occasional macrofossils occurring in the same formation (app. 7.1). Because of uncertainties in these age dates, we assigned long time intervals (mean locality duration = 2.8 my) to many of the study units, even though the deposits almost certainly accumulated over shorter periods of time.

The coral composition of modern reefs has been reported to be fairly uniform across the Caribbean region, with a higher diversity center positioned over Jamaica, Panama, and Curaçao (Liddell and Ohlhorst 1988). Since most modern species are widespread, we selected a faunal list from a location within the high-diversity center (the north coast of Jamaica; Wells and Lang 1973) to represent the species composition of the entire modern Caribbean. We modified the list to incorporate recent discoveries of new species (Zlatarski 1990) and recently resurrected species (Knowlton et al. 1992; Weil and Knowlton 1994). Unlike the Caribbean, numerous endemic species are reported from the southeast coast of Brazil, and some are thought to be descendants of a relict Tertiary Caribbean fauna that survived the Plio-Pleistocene environmental perturbations in refugia near Brazil (Leão et al. 1988). Because of the similarity of some coral species (e.g., *Meandrina, Favia, Mussismilia*) in the Mio-Pliocene of the Dominican Republic with modern Brazilian species, we also included a faunal list from Brazil (Laborel 1969) in the compilation.

Ongoing research indicates that several unrecognized endemic species may exist within well-established modern Caribbean "species" in genera such as *Manicina, Porites,* and *Agaricia* (e.g., Johnson 1991). We plan to investigate this problem in more detail in subsequent research on fossil material collected using a more rigorous geographic and stratigraphic sampling scheme. Although the incorporation of more endemic taxa in our analyses might aid in the interpretation of evolutionary patterns and their causes, these additions would only accentuate the finding of a major episode of Plio-Pleistocene extinction and turnover in Caribbean reef corals that we report in this chapter.

SPECIES RECOGNITION

As a first step in identifying material in the Dominican Republic, Bahamas, and Costa Rica collections, we sorted the specimens into genera using a set of diagnostic morphologic criteria derived from Wells (1956) and summarized by Budd and others (1992) and Budd, Stemann, and Johnson (1994). We identified species within each genus using quantitative characters selected on the basis of multivariate statistical analyses. In the Dominican Republic material, we recognized species in most genera by performing cluster and canonical discriminant analyses on ten or more linear distance measurements and counts made on five to ten corallites per colony (Foster 1986, 1987; Budd 1991; Stemann 1991). We compared the results with similar measurements made on type material described by Vaughan (1919) and by Vaughan and Hoffmeister (1925, 1926). In the Bahamas and Costa Rica material, we made identifications by visually examining five or more characters that were most heavily weighted in the Dominican Republic analyses. This use of morphometric methods in species recognition is discussed in detail in Budd and Coates 1992 and has been evaluated by comparison with molecular data (Potts et al. 1993; Budd, Johnson, and Potts 1994). Preliminary results indicate that morphometric methods are capable of distinguishing morphologically similar species in species complexes, such as the complex containing *Montastraea annularis* and its sibling species (Knowlton et al. 1992; Weil and Knowlton 1994; Budd 1993). We provide detailed descriptions of the morphologic criteria used in diagnosing genera and species in the compilation in Budd, Stemann, and Johnson 1994.

We verified species and genera recorded in the twenty-one faunal lists by reexamining specimens in museum collections and reassessing identifications using the same morphologic criteria outlined above. In the few instances where it was not possible to make these examinations (as noted in app. 7.1), we checked original identifications by carefully studying published photographs and systematic descriptions. Wherever necessary, we corrected misidentifications and modified the original faunal lists.

Preliminary comparisons with similar-aged taxa in the Mediterranean and Indo-Pacific regions (c.f. Chevalier 1961, 1968; Veron and Kelley 1988) suggest that no taxa similar to the species in our compilation occurred in the Mediterranean after Middle to Late Miocene time, and in the Indo-Pacific after Early to early Late Pliocene time. Therefore, it appears that the Caribbean fauna became increasingly geographically isolated during the Neogene and that post-Miocene first and last occurrences represent true global originations and extinctions.

ANALYSIS OF TURNOVER

To detect overall patterns of origination and extinction in the fauna, we divided the entire time period into intervals, and calculated species richness and rates of extinction and speciation (evolutionary rates) within each interval. We did not calculate the Lyellian percentages commonly used in this sort of analysis (e.g., Stanley 1986), because Lyellian percentages only show survivorship of living taxa. They cannot be used to estimate rates of extinction, and they assume that species richness has not changed over time. We also did not calculate confidence intervals for our stratigraphic ranges (Marshall 1990), because (1) unlike fossil occurrence data obtained from bulk samples, our sample sizes are not equal, and (2) our samples were not taken from well-defined horizons that are randomly distributed through geologic time. Our data were derived from collections of varying quality with relatively poor age resolution, and the use of more powerful quantitative methods (e.g., Strauss and Sadler 1989) could not be justified.

To minimize bias associated with uneven sampling among localities, we dropped nine rare species (species occurring in only one locality) from the data set, thus providing consistent, although admittedly low, estimates of species richness and evolutionary rates. Five of the nine deleted species were endemic to the Imperial Formation of south-central California, which contains a fauna with strong Caribbean affinities (Vaughan 1917). Two species were from the Pinecrest Sandstone of south Florida (Meeder 1987). One of the remaining two species was found only at one collecting locality in the Dominican Republic; the other was found only as a cast in the Bahamas cores.

We divided the total time period represented in the compilation into intervals of equal amounts of absolute time. Because a variety of methods have been used to assign age dates to the material, we opted against using microfossil zones or other stratigraphic age assignments as subdivisions. We estimated species richness by counting the number of species within each time interval. Even if they were not actually encountered within a particular time interval, species were considered to be present in an interval as long as the interval was between their first and last occurrence. Localities with earliest age dates on the boundary between time intervals were arbitrarily assigned to the lower interval.

We estimated per species speciation and extinction rates following the methods of Lasker (1978) and Wei and Kennett (1986). Because age determinations for many units have poor resolution, first and last occurrences for species in units extending over several time intervals were weighted so that each interval received an equal proportion of the occurrence (methods modified after Barry et al. 1990). For instance, a last occurrence in a locality with a resolution of four time intervals would be treated as one-fourth of an extinction in each of the four intervals. This scheme tends to smooth out the effects of any one poorly resolved fauna by spreading out its first and last occurrences among many time intervals. Therefore, it obscures rather than enhances the true magnitude of accelerated turnover. To compare rates of speciation and extinction among time intervals with differing richness, we divided the sum of the weighted first or last occurrences for each time interval by the total number of species living during the interval. We then determined evolutionary rates by dividing this proportion by the interval duration.

To determine the effects of interval duration on our estimates of species richness and evolutionary rates, we estimated species richness and evolutionary rates using a range of different interval durations. We also ran the analyses using two different sets of stratigraphic ranges for each species: (1) a long range, defined as the range from the lower age estimate of the first occurrence to the upper age estimate of the last occurrence of a species; and (2) a short range, defined as the range between the upper estimate of the age of the first occurrence to the lower estimate of the age of the last occurrence. Our results (fig. 7.7) show that, regardless of interval duration and which set of ranges was used, the resulting overall evolutionary patterns remained the same: low rates of extinction and origination before 4 Ma followed by an accelerated period of turnover. When short durations or ranges were used, values for species richness were decreased, and evolutionary rates were increased. When long durations or ranges were used, trends were dampened.

To determine the effects of poor age resolution and influential observations on our estimates of evolutionary rates, we dropped selected localities from the data set and reran the analyses. Localities were removed singly and in various combinations. First, units with poor-resolution age dates (e.g., Providencia, Isla Mona, and the lower por-

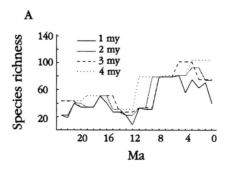

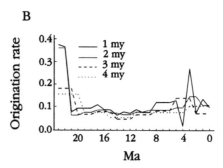

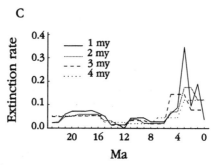

Fig. 7.7 Species richness (A) and evolutionary rates (B, C) of the Caribbean reef-coral fauna during the past 22 million years. Each plot shows patterns calculated with interval durations of 1, 2, 3, and 4 million years (my). Richness increases and variation in evolutionary rates is dampened with decreasing temporal resolution. The pattern suggests an increase in species richness during the Late Miocene and Early Pliocene, followed by a decrease from the Late Pliocene to Recent. Increases in both origination and extinction rates are evident after 4 Ma.

tions of the Bahamas cores) were excluded. The resulting overall patterns were visually similar to that obtained when all data were included. Second, units with high numbers of extinctions and originations (e.g., the Bowden Formation, Pinecrest Sandstone, and Moin Formation) were deleted and subsequently replaced. The results showed that the sizes of the peaks remained roughly the same, although their positions may have sometimes shifted by a million years.

To determine if the number of localities (sampling intensity) affected our estimates of species richness and evolutionary rates, we visually examined bivariate plots of each estimate with sampling intensity, and performed statistical tests for correlation (fig. 7.8). The results suggest a significant correlation between richness and sampling intensity, especially before 8 Ma, but not between evolutionary rate and sampling intensity. Increased richness with more samples is generally expected, because most species tend to be rare (Buzas et al. 1982). Nevertheless, the extremely high correlation before 8 Ma suggests that these richness values may be heavily biased by sampling. This bias appears to be considerably reduced after 8 Ma. Outliers on figure 7.8 also suggest that rates of origination and extinction were high between 4 and 3 Ma, and that richness was low between 1 Ma and the Recent.

To examine sampling biases associated with species richness in further detail, we compared observed distributions of species ranges with expected distributions based solely on sampling intensity (following Koch and Morgan 1988). Our results suggest that the observed patterns of extinction and origination cannot be explained as sampling artifact alone. Examination of residuals calculated by subtracting observed from expected range distributions (fig. 7.9) indicates that higher than expected extinctions and originations occurred at 4 Ma. Higher than expected originations also occurred at 22 Ma, 12 Ma, 8 Ma, and 0 Ma. Although the high value at 8 Ma may indeed reflect increased originations during Late Miocene time, the other three high values may be artifactual. Specifically, the high at 22 Ma may be related to its being the oldest interval in the compilation, that at 12 Ma to limited sampling, and that at 0 Ma to sampling in a greater range of reef habitats in the Recent. With regard to extinctions, other than the highs at up to 4 Ma, high values occurred only in Early Miocene intervals.

MAGNITUDE OF TURNOVER

Our results suggest that species richness was relatively low (30–50) throughout much of the Early to Middle Miocene (22–9 Ma), high (80–

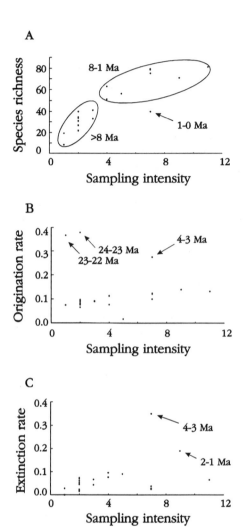

Fig. 7.8 Relationship between sampling intensity (number of localities) and evolutionary rate over time (1 my intervals). (A) For time intervals with ages older than 8 Ma, species richness is significantly correlated with the number of localities (Pearson correlation $r =$ .56, $n = 15$, $p\ (> r) < .029$). However, for time intervals with ages younger than 8 Ma, species richness is not significantly correlated with the number of localities (Pearson correlation $r = .44$, $n = 9$, $p\ (> r) < .239$). Note that species richness in the Recent (1–0 Ma as marked by arrow) is slightly lower than species richness in intervals between 8 and 1 Ma. (B) No relationship exists between origination rate and number of localities ($r = .02$, $p\ (> r) = .933$), although extremely high rates of origination occur at the beginning of the study period (24–23 Ma, 23–22 Ma) and at 4–3 Ma. (C) A marginally significant relationship exists between extinction rates and number of localities ($r = .41$, $p\ (> r) = .05$). The intervals at 4–3 Ma and at 2–1 Ma are characterized by high rates of extinction.

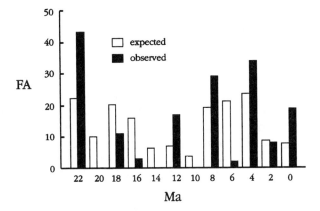

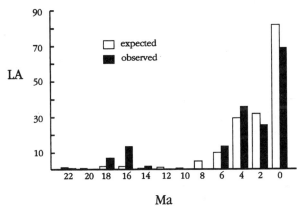

Fig. 7.9 Total number of first appearances (FA) and last appearances (LA) within 2 my intervals. Expected values are determined solely on the basis of sampling intensity (after Koch and Morgan 1988).

100) from the Late Miocene to Early Pleistocene (9–1 Ma), and inter-mediate (40–60) in the Late Pleistocene to Recent (1–0 Ma) (fig. 7.7A). However, owing to the strong correlation between sampling intensity and species richness before 8 Ma, only the period from 1 Ma to Recent is less rich than is expected from sampling intensity alone (fig. 7.8A). Although we only included one Recent Caribbean locality in our data, sampling of Recent material was significantly better than sampling of fossil species. As discussed earlier, even with the addition of currently unrecognized endemic and sibling species, richness would still be rela-tively low from 1 Ma to Recent.

Both extinction and origination rates were accelerated between 4 and 3 Ma (fig. 7.7), and were higher than predicted by sample intensity between 4 and 3 Ma (fig. 8B and C). Although less than at 4–3 Ma, extinction rates were also high at 2–1 Ma. In general, background rates of extinction ranged between 0% and 8% of standing richness per million years. Beginning approximately at 4 Ma, extinction rates doubled, with extinction rates of 10% to possibly more than 35% per million years. Species richness dropped from 93 species to 56 species in the Caribbean region as a result of the turnover and extinction episode. Overall, 16% of the 87 species occurring between 8 and 6 Ma, 22% of the 95 species between 6 and 4 Ma, and 36% of the 106 species between 4 and 2 Ma survived in the Caribbean to the present.

We tested the observed 1 my interval patterns statistically using the simulation methods of Hubbard and Gilinsky (1992). In this test, total numbers of observed extinctions and originations per genus for each time interval are compared with a bootstrapped distribution obtained by repeated random sampling of numbers of extinctions and originations per genus within all of the intervals in the data set. In our case, this was accomplished by performing ten thousand simulations within each time interval. As indicated by the p-values well below .01 in table 7.1, we found significantly higher than expected numbers of originations at 4–3 Ma, and higher than expected numbers of extinctions both at 4–3 Ma and at 2–1 Ma.

To further confirm accelerated turnover, we also performed average linkage cluster analysis on species occurrences for each locality in figure 7.6 (except the Imperial Formation of California) using Jaccard's similarity coefficients. We used Jaccard's coefficients because they use only presences (not absences) to indicate increased similarity. The results reveal two major groups, one consisting of units younger than 2 Ma and the other of localities older than 3.5 Ma (fig. 7.10). Localities with age dates between 2 and 3.8 Ma belonged to either of the two groups. For example, the Bowden Formation of Jamaica (3.8–2 Ma) clustered with the older material; whereas the Pinecrest Sandstone of Florida (3.5–3 Ma) clustered with the younger material. This last result suggests that the reef-coral faunas in the Caribbean region before and after approximately 3.5–3 Ma were distinctly different and separated by a major period of faunal change. This change has been further confirmed in ongoing studies of species occurrences in more precisely dated sites using multidimensional scaling (Budd, Johnson, and Jackson 1994).

To determine if accelerated turnover was taxonomically widespread,

Table 7.1 Bootstrapping Analyses for the Significance of Observed Origination and Extinction Rates

Time	Originations		Extinctions	
(Ma)	observed	p-value	observed	p-value
22–21	3.17	.521	2.13	.601
21–20	3.17	.523	2.13	.596
20–19	3.17	.526	2.13	.605
19–18	3.17	.526	2.13	.609
18–17	5.83	.122	3.53	.330
17–16	3.83	.507	2.73	.516
16–15	2.67	.730	1.40	.822
15–14	1.75	.937	0.40	.997+
14–13	1.75	.937	0.40	.997+
13–12	1.75	.998+	0.00	.999+
12–11	2.89	.995+	1.14	.989+
11–10	2.14	.999+	1.14	.988+
10–9	2.14	.999+	1.14	.988+
9–8	9.39	.203	1.64	.958
8–7	9.39	.207	1.64	.956
7–6	9.39	.204	1.64	.957
6–5	10.39	.143	4.64	.493
5–4	1.00	.999	5.00	.437
4–3	21.00	.000*	29.67	.000*
3–2	6.50	.640	7.17	.073
2–1	10.00	.235	14.17	.000*
1–0	3.50	.973	1.00	.936

Note: p-values are listed for a two-tailed test of the null hypothesis that observed rates are equal to expected rates.
*significantly higher than expected.
+significantly lower than expected.

we calculated extinction rates for genera, and again found accelerated extinction between 4 and 1 Ma (fig. 7.11). Of the thirty-eight genera occurring in the Pliocene of the Caribbean, 68% survived in the region (table 7.2). Nine of the thirteen genera that became extinct in the Caribbean, however, still occur today in the Indo-Pacific. Although our data show that three genera (*Dendrogyra, Mussismilia, Scolymia*) appear for the first time in the Caribbean after the Late Miocene, no new genera may have actually originated in the Caribbean. *Dendrogyra* and *Mussismilia* both occur earlier in the Tertiary of the Mediterranean region (table 7.2; Chevalier 1961), and *Scolymia* occurs in the Early Neogene of the Indo-Pacific (Veron and Kelley 1988). Better sampling is needed in order to determine if, how, and when these genera migrated to the Caribbean.

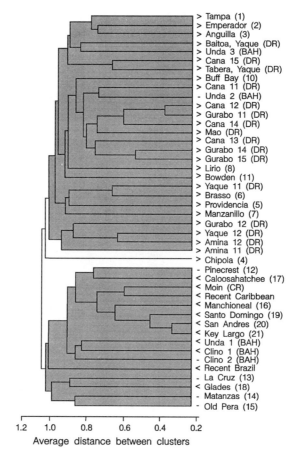

Fig. 7.10 Dendrogram showing the results of average linkage cluster analysis on species occurrences in forty-four localities. Two main clusters of localities are suggested. Localities older than 3.5 Ma are identified by >; those younger than 2 Ma are indicated by <; localities intermediate in age are identified by =. DR = sequences in the northern Dominican Republic; BAH = Bahamas cores; CR = exposures in Costa Rica. Numbers in parentheses refer to faunal lists given in appendix 7.1. Numbers in Dominican Republic river sections refer to calcareous nannoplankton zones. Ages are shown in figure 7.6.

SELECTIVITY OF EXTINCTION

To better understand the turnover process, we performed a preliminary comparison of susceptibility to extinction in different ecological subgroups. Because our best-sampled sequence is that of the Cibao Valley in the Dominican Republic, we made this comparison using assemblages associated with different paleoenvironments interpreted in the

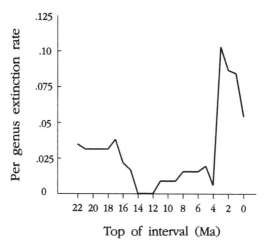

Top of interval (Ma)

Fig. 7.11   Extinction rates determined for coral genera since the Early Miocene.

Dominican Republic sequence. To identify these assemblages, we performed a cluster analysis on species occurrences in different lithostratigraphic units shown on the stratigraphic sections of Saunders and co-workers (Saunders et al. 1986). We selected eighteen intervals that showed little evidence of transport for analysis. All of the eighteen intervals, except two from the Baitoa Formation, are Late Miocene to Early Pliocene in age (calcareous nannoplankton zones NN11–13). The two Baitoa Formation units have Early to Middle Miocene ages (calcareous nannoplankton zones NN4–6).

To perform the analysis, we used average linkage clustering and Jaccard's similarity coefficients. Using an average distance of 1.0 as a cutoff, the resulting dendrogram revealed three distinct clusters (fig. 7.12; table 7.3): (1) a low-diversity assemblage (species richness = 32; Shannon-Wiener diversity = 3.72) consisting primarily of free-living (47%) and branched (28%) corals; (2) a low-diversity assemblage (richness = 28; Shannon-Wiener diversity = 4.09) consisting primarily of mound-shaped (61%) and branched (21%); and (3) a high-diversity assemblage (richness = 71; Shannon-Wiener diversity = 5.12) consisting of a mixture of mound-shaped (32%), free-living (23%), branched (22%), and platy (19%) corals. To test the validity of the three statistically defined assemblages, we ran cluster analyses separately within each of the observed clusters and for every pairwise combination of clusters. The results showed that the three clusters invariably formed discrete and stable entities. A nonparametric discriminant analysis (SAS Inst. 1989) of the three clusters further demonstrated the lithostratigraphic

Table 7.2 Ranges of Caribbean Reef Corals Compared to Mediterranean and Indo-Pacific Occurrences

Genus	Caribbean Origination[1]	Mediterranean[2] (Cenozoic)	Caribbean Extinction	Indo-Pacific[3] (Recent)
Astrocoenia	Eocene	x	Early Miocene	.
Stephanocoenia	Eocene	x	.	.
Stylophora	Eocene	x	?Early Pleistocene	x
Pocillopora	Eocene	x	Late Pleistocene	x
Madracis	Eocene	x	.	x
Acropora	Eocene	x	.	x
Astreopora	Eocene	x	Early Miocene	x
Agaricia	Late Miocene	x	.	.
Undaria	?Oligocene	.	.	.
Gardineroseris	Late Miocene	.	early Late Pliocene	x
Pavona (Pseudo.)	Early Miocene	x	Early Pliocene	x
Helioseris	Late Miocene	.	.	.
Leptoseris	Oligocene	x	.	x
Psammocora	Early Miocene	.	Late Miocene	x
Siderastrea	Eocene	x	.	.
Pironastrea	Eocene	x	Early Miocene	.
Porites II	Eocene	x	.	x
Porites I	Oligocene	x	.	x
Goniopora	Eocene	x	early Late Pliocene	x
Alveopora	Eocene	x	Early Miocene	x
Caulastraea	Oligocene	x	latest Pliocene	x
Cladocora	Eocene	x	.	.
Favia	Eocene	x	.	x
Goniastrea	Eocene	.	Early Miocene	x
Diploria	Late Miocene	x	.	.
Manicina (Man.)	Late Miocene	.	.	.
Manicina (Teleio.)	Late Miocene	.	early Late Pliocene	.
Thysanus	Early Miocene	.	latest Pliocene	.
Colpophyllia	Eocene	x	.	.
Antiguastrea	Eocene	x	Early Miocene	.
Montastraea I	Eocene	x	.	x
Montastraea II	Eocene	x	.	x
Solenastrea	Oligocene	x	.	.
Agathiphyllia	Eocene	x	Early Miocene	.
Trachyphyllia	Early Miocene	.	early Late Pliocene	x
Antillophyllia	Early Miocene	.	Late Miocene	.
Meandrina (Mea.)	Eocene	x	.	.
Meandrina (Placo.)	Late Miocene	.	latest Pliocene	.
Dichocoenia	Eocene	x	.	.
Dendrogyra	Late Pleistocene	x	.	.
Galaxea	Early Miocene	.	Late Miocene	x
Antillia	Early Miocene	x	latest Pliocene	.
Scolymia	Early Pliocene	.	.	x
Mussa	Late Miocene	.	.	.

Table 7.2 *continued*

Genus	Caribbean Origination[1]	Mediterranean[2] (Cenozoic)	Caribbean Extinction	Indo-Pacific[3] (Recent)
Mussismilia	early Late Pliocene	x	Recent-Brazil	.
Isophyllia	Late Miocene	.	.	.
Isophyllastrea	Late Miocene	x	.	.
Mycetophyllia	Late Miocene	x	.	.
Eusmilia	Late Miocene	.	.	.

Sources: [1]Budd et al. 1992; [2]Chevalier 1962; [3]Veron 1986.
Note: Caribbean extinctions have been determined by study of the new compilation. Taxonomic designations conform with the usage in Budd, Stemann, and Johnson 1994.

units to be 100% correctly classified, and that free-living corals best distinguished the first cluster, mound-shaped corals the second cluster, and platy corals the third cluster (table 7.3).

We compared other, more or less independent paleoenvironmental interpretations for each lithostratigraphic unit (Saunders et al. 1986; Bold 1988) among clusters to determine if the three clusters could be generally interpreted to represent different reef environments. Our findings suggested that paleoenvironments interpreted for units in the first cluster were generally silty, soft-bottom areas. At least two of the units that belong to this cluster (C3, C4) contain abundant, unmistakable evidence of seagrass (Cheetham and Jackson, this volume, chap. 8). Lesser amounts of seagrass have been detected at a third unit (C7) and near the three remaining units (G3, C6, M1) in the cluster. We therefore use the term "seagrass" to refer to this assemblage. These grass beds differ from shallow (less than 10 m) nearshore grass flats common today in the Caribbean in that they may have extended to depths of 20–30 m on forereef slopes (Cheetham and Jackson, this volume, chap. 8).

In contrast, we found paleoenvironments interpreted for units in the second cluster to be clear, shallow reefal areas. Units in the third cluster were interpreted as deeper (more than 10–20 m), more turbid reef habitats. In general, the three clusters did not differ in age, but instead appeared to be associated with different river sections. Since the different river sections themselves have been broadly interpreted as having been deposited in slightly different environments (Saunders et al. 1986), we therefore conclude that the three assemblages were associated with different paleoenvironments or reef habitats.

To assess the relative susceptibilities of the three assemblages to ex-

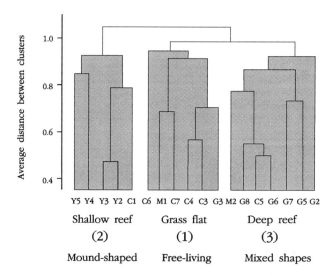

Fig. 7.12 Results of cluster analysis on species occurrences in different lithostratigraphic units in the Dominican Republic. Three clusters are suggested. Eighteen intervals that showed little evidence of transport were included in the analyses. In the Rio Gurabo section, these consisted of one interval in the Cercado Formation (140–150 m [G2]) and five intervals in the Gurabo Formation (150–158 m [G3], 210–250 m [G5], 250–290 m [G6], 310–325 m [G7], 360–386 m [G8]). In the Cana section, they consisted of two intervals in the Cercado Formation (165–185 m [C1], 225–280 m [C3]) and four intervals in the Gurabo Formation (280–330 m [C4], 330–375 m [C5], 375–410 m [C6], 425–440 m [C7]). In the Mao section, two distinctive units (NMB16910 [M1], NMB16911 [M2]) were included. In the Rio Yaque del Norte section, two intervals were included from the Baitoa Formation (one below NMB17286 in the Lopez section [Y2], the other above NMB17286 [Y3]), and localities at Arroyo Lopez (Y4) and Angostura Gorge (Y5) were included as two separate intervals.

tinction, we compared patterns in extinction rates among assemblages. In these analyses, we calculated extinction rates for each assemblage by considering all of the species occurring in the lithostratigraphic units belonging to each cluster. All three assemblages had accelerated extinction rates at 4–3 Ma and 2–1 Ma (fig. 7.13). Using methods described earlier, we tested these patterns by bootstrapping. The seagrass assemblage was found to be most affected by the turnover, with extinction rates of between 30% and 50% per million years. The other assemblages were less affected, and extinction rates were less than 30% of standing diversity per million years. Therefore, whatever caused increased rates of taxonomic turnover on Caribbean reefs, its most potent effects were felt in soft-bottom, seagrass communities characterized by

Table 7.3 Relative Abundance of Coral Species from Dominican Republic Assemblages

Genus	Species	Growth Form	Shallow Reef	Soft Bottom	Deep Reef
Stephanocoenia	*duncani*	M	R	R	C
Stylophora	*affinis*	B	.	R	R
Stylophora	*granulata*	B	R	R	C
Stylophora	*minor*	B	C	C	A
Stylophora	*monticulosa*	B	C	R	C
Stylophora	*canalis*	B	.	.	R
Pocillopora	*baracoaensis*	B	.	.	C
Pocillopora	*crassoramosa*	B	C	R	R
Madracis	*decactis*	B	.	.	C
Madracis	*decaseptata*	B	.	R	C
Madracis	*mirabilis*	B	.	.	R
Acropora	*saludensis*	B	.	R	R
Agaricia	*lamarcki*	P	.	.	C*
Agaricia	*undata*	P	.	.	C
Undaria	*agaricites*	PM	R	.	A
Undaria	*crassa*	PM	.	R	C
Undaria	sp. A	P	.	.	C*
Gardineroseris	*planulata*	M	.	.	C
Pavona (*Pseudo.*)	sp. A	B	.	R	C
Pavona (*Pseudo.*)	sp. B	B	.	.	A
Leptoseris	*cailleti*	B	.	.	R
Leptoseris	*gardineri*	B	.	.	C
Leptoseris	*glabra*	P	.	.	C*
Leptoseris	sp. A	P	.	.	C
Leptoseris	sp. B	B	.	.	C
Psammocora	*trinitatis*	M	A*	.	.
Siderastrea	*hillsboroensis*	M	C	.	.
Siderastrea	*silecensis*	M	C	.	R
Siderastrea	*siderea*	M	C	.	R
Porites-II	*baracoaensis*	B	C	R	A
Porites-II	*convivatoris*	E	.	C	R
Porites-I	*macdonaldi*	PM	.	.	A*
Porites-I	*portoricensis*	B	A	.	A
Porites-I	*waylandi*	M	A*	R	C
Goniopora	*calhounensis*	M	.	.	R
Goniopora	*hilli*	M	C	R	C
Goniopora	*imperatoris*	M	C	.	R
Favia	*dominicensis*	M	C	.	.
Favia	sp. A	M	R	.	.
Diploria	sp. A	M	.	.	R
Diploria	sp. B	M	.	.	C
Manicina (*Man.*)	*puntagordensis*	F	.	R	C
Manicina (*Teleio.*)	*grandis*	F	C	A*	C
Manicina (*Teleio.*)	*navicula*	F	.	R	R
Manicina (*Teleio.*)	sp. A	F	.	C*	R

Table 7.3 *continued*

			Assemblage Type		
Genus	Species	Growth Form	Shallow Reef	Soft Bottom	Deep Reef
Manicina (*Teleio.*)	sp. B	F	.	A*	R
Manicina (*Teleio.*)	sp. C	F	.	R	R
Manicina (*Teleio.*)	sp. D	F	.	.	C
Thysanus	*excentricus*	F	.	R	R
Thysanus	*floridanus*	F	.	R*	.
Colpophyllia	*natans*	M	.	.	R
Montastraea-I	*brevis*	M	.	.	C
Montastraea-II	*canalis*	M	C	.	C
Montastraea-II	*cavernosa*-2	M	R	.	C*
Montastraea-II	*cylindrica*	M	P	.	C
Montastraea-II	*endothecata*	M	.	.	C
Montastraea-I	*limbata*	M	A	R	A*
Montastraea-I	*trinitatis*	M	A*	.	.
Solenastrea	*bournoni*	M	A	R	C
Solenastrea	*hyades*	M	C*	.	R
Trachyphyllia	*bilobata*	F	.	A	A
Trachyphyllia	sp. A	F	.	.	R
Antillophyllia	*sawkinsi*	F	.	.	R
Meandrina (*Mea.*)	*braziliensis*	F	R	A	C
Meandrina (*Mea.*)	*meandrites*	M	R	.	R
Meandrina (*Placo.*)	*alveolus*	F	.	C	R
Meandrina (*Placo.*)	*costatus*	F	.	C	A
Meandrina (*Placo.*)	*trinitatus*	F	.	A	A
Meandrina (*Placo.*)	*variabilis*	F	R	A*	C
Dichocoenis	*tuberosa*	M	.	.	R
Galaxea	*excelsa*	M	.	.	R
Antillia	*dentata*	F	R	C	A
Scolymia	*cubensis*	S	.	.	R
Mussa	*angulosa*	B	.	.	R
Eusmilia	*carrizensis*	B	.	.	R
Eusmilia	sp. A	B	.	R	R

Note: Taxonomic designations conform with the usage in Budd, Stemann, and Johnson 1994. For growth forms, B = branching; E = encrusting; F = free living; M = mound-shaped; S = solitary. For relative abundances in each assemblage, R = rare; C = common; A = abundant.
*Most heavily weighted by discriminant analysis.

small free-living corals with patchily distributed populations (Johnson 1992). Subsequent work comparing the susceptibilities of taxa with different colony sizes to extinction during the Plio-Pleistocene turnover episode (Johnson et al. 1995) confirm these results, and indicate that colony size (not reproductive mode or sex) was the most important ecological characteristic determining extinction rate.

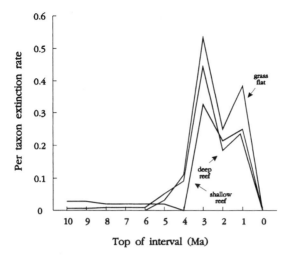

Fig. 7.13 Extinction rates for species characteristic of seagrass, shallow-reef, and deep-reef habitats, using 1 my interval durations. All three communities experienced accelerated rates of extinction, but species that lived in shallow nonreef habitats were more likely to suffer extinction.

POSSIBLE CAUSES OF TURNOVER

Although chronostratigraphic resolution is improved in the present compilation, lack of refined age dates still presents a major problem in determining the cause of the observed patterns of extinction and origination in reef communities. The lack of sampling through continuous sequences is also a significant drawback, since all of the extinctions and originations that we have observed in this study occur between (not within) sequences. At this point, all that we can conclude with certainty is that a major episode of accelerated turnover and extinction occurred in the Caribbean reef-coral fauna sometime between 4 and 1 Ma. Both accelerated extinction and high origination rates were involved, and occurred at approximately the same time. Although the overall pattern of faunal change was generally punctuated, it cannot be ascertained whether extinction and origination events were steady or occurred in pulses. Similarly, it is unclear if origination occurred immediately before or after extinction, or if the processes were somehow linked. Study of patterns of succession in these sorts of events at different Caribbean locations will be essential to assessing the role of biotic factors in turnover. Ongoing studies, however, suggest that faunal change was geographically patchy and followed different pathways in different places (Budd, Johnson, and Jackson 1994).

Nonetheless, we can conclude that approximately 64% of the Early Pliocene coral fauna went extinct, a number comparable to the reported 65% species loss in subtropical bivalve mollusks since the Early Pliocene (Stanley 1986). Species extinctions in both corals and mollusks were similar in that they were accompanied by roughly equal numbers of originations (Allmon et al. 1993; Jackson et al. 1993). The major difference is that molluscan turnover in the Caribbean and western Atlantic appears to have begun at approximately 2.4 Ma. Although temporal resolution is far less refined in our coral compilation, the fact that species common in the younger Caribbean coral fauna are abundant in the Pinecrest Sandstone (3.5–3 Ma) suggests that coral turnover may have preceded molluscan turnover by about one million years. These differences in timing would imply that differences in dispersal capabilities and life history characteristics among taxonomic groups may be important factors controlling the turnover process.

Still, on a larger scale, the approximately synchronous extinction in the Caribbean region of at least three groups of unrelated shallow benthic marine organisms (mollusks, reef corals, planktonic foraminifers [Stanley et al. 1988]) argues in favor of abiotic factors as important causal agents. Owing to low chronostratigraphic resolution and a still incomplete environmental history, the direct relationship between specific abiotic factors and the observed evolutionary patterns cannot be assessed by correlation with regional or global sea level or temperature curves (or other forms of environmental proxy data) using the present compilation. Yet comparisons among the affected taxa do provide preliminary insight into which combinations of abiotic factors may have been more important. Most important, as in mollusks (Stanley 1986), no evidence currently exists for a similar turnover event in the central and western Pacific reef-coral fauna (Potts 1984; Veron and Kelley 1988), implying that the causes are regional in scale, and that global sea level change and associated species-area effects were not responsible. Faunal change was also striking, however, in the eastern Pacific after the Early to early Late Pliocene (Budd 1989; Jackson et al., this volume, chap. 9), indicating that accelerated turnover in the Caribbean may have also occurred prior to or in association with closure of the Isthmus of Panama at approximately 3.5 Ma (Coates et al. 1992).

In general, turnover appears to have been associated with a number of interrelated regional environmental agents that cannot be disentangled using currently available data. These include changes in air or water temperature, siltation, salinity, and nutrients. If temperature alone was a factor, it is puzzling that many of the observed extinctions and originations appear to have occurred before the onset of Northern Hemisphere glaciations at approximately 2.4 Ma. In particular, all of the observed extinctions and originations occurred long before the most extreme, highest amplitude, highest frequency climatic fluctuations, which began around 0.7 Ma (Raymo et al. 1990). In fact, stability persisted in the Caribbean reef-coral fauna throughout the period of the most extreme Pleistocene climatic perturbations (see also Jackson 1992). Nevertheless, ongoing work (Budd, Johnson, and Jackson 1994) suggests that patterns of turnover differed between the northern and southern Caribbean, suggesting that climate change may have played at least an indirect role in the observed faunal change. By the same token, if terrestrial siltation was the only important factor, it is puzzling that extinctions occurred equally both in coastal areas and on offshore carbonate platforms. Similarly, both nutrient-rich and nutrient-poor areas within the Caribbean were affected.

Comparisons within the coral fauna itself show that species characteristic of reef-marginal seagrass communities were more likely to suffer extinction than taxa living in more diverse reef communities. However, as mentioned earlier, unlike their predominance in protected shallow (less than 10 m) areas today, Mio-Pliocene seagrass communities may have extended from shallow-water depths to depths of as great as 20–30 m. Ongoing research indicates that coral taxa with ecological traits (life history, reproductive biology, etc.) suitable for life in patchy, disturbed habitats dominated by seagrass have intrinsically higher evolutionary rates and are more likely to become extinct in response to changing regional environmental conditions (Johnson et al. 1995). Thus, as mentioned earlier, biological factors appear to have played an important role in determining susceptibility to extinction and may explain why different sorts of evolutionary patterns occur in benthic foraminiferal assemblages living in similar habitats (L. S. Collins, personal communication).

CONCLUSIONS

Regardless of its actual cause, our analyses clearly show that an episode of accelerated turnover and extinctions occurred between 4 and 1 Ma, and that it played a major role in shaping the modern reef-coral fauna of the Caribbean region. Extinction and origination rates increased simultaneously, at least initially, and species richness remained relatively constant throughout the turnover episode. Extinction was heaviest in seagrass communities. The most profound consequence of the turnover and extinction episode on modern Caribbean reefs is reduced generic diversity and possibly lowered species diversity, especially in seagrass communities. Nevertheless, it is remarkable that throughout this period of ecological crisis and ecosystem reorganization, reef-coral communities remained common across the Caribbean region. In the future, more detailed study of the transition between these two seemingly stable sets of communities will provide conservation biologists with essential information to assess the effects of addition and removal of species from reef ecosystems.

ACKNOWLEDGMENTS

We thank Alan H. Cheetham, Laurel S. Collins, Jörn Geister, Carl Koch, and Jeremy B. C. Jackson for comments on the manuscript; Jörn Geister, Peter Jung, and John B. Saunders for providing collections in the Dominican Republic; and Jeremy B. C. Jackson, Anthony G. Coates, and Jorge A. Obando for help with fieldwork and collections in Costa Rica. Robert N. Ginsburg and Carrie Kievman assisted with sampling of the Bahamas cores. Alan H. Cheetham, Laurel S. Collins, Thomas M. Cronin, Jörn Geister, Peter Jung, and Steven M. Stanley provided additional collection and locality information. Stephen D. Cairns, René Panchaud, and Julia Golden assisted in specimen curation. Funds were provided by National Science Foundation grant (EAR-9219138) to Ann Budd.

APPENDIX 7.1: DESCRIPTIONS OF LOCALITIES FOR FAUNAL LISTS

1. **Tampa Formation, Florida**
   Environment: argillaceous limestones and siltstones with small patch reefs, interpreted as extremely shallow nearshore deposits (Puri and Vernon 1964)
   Age: Early Miocene, Aquitanian (Puri and Vernon 1964; Carter and Rossbach 1989)
   Faunal list: modified after Weisbord 1973; not all material examined
   Assigned age: 23.7–22 Ma
2. **Emperador Limestone, La Boca Formation, Panama**
   Environment: patchy, thin (less than 15–20 m thick) coralliferous limestones, interspersed among volcanic-rich siliciclastics (Woodring 1957, 1964)
   Age: foraminiferal zone N6 or older (Blacut and Kleinpell 1969)

Faunal list: modified after Vaughan 1919; all material examined
Assigned age: 22–17.6 Ma

3. Anguilla Formation, Anguilla

Environment: 65 m thick sequence of skeletal grainstones and packstones
with coral-rich lenses, interpreted as patch reefs on an isolated shallow
carbonate platform (Budd, Johnson, and Edwards 1995)
Age: late Early Miocene (Drooger 1951; Bold 1970)
Faunal list: Budd, Johnson, and Edwards 1995; all material examined
Assigned age: 22–16.2 Ma

4. Chipola Formation, Florida

Environment: 30 m thick, extremely fossiliferous, sandy limestones interpre-
ted as a nearshore shelf environment (see Bryant et al. 1992 for review)
Age: late Early to early Middle Miocene (Bryant et al. 1992)
Faunal list: modified after Weisbord 1971; not all material examined
Assigned age: 18–15 Ma

5. Providencia Island

Environment: coralliferous calcarenites and tuffs representing shallow la-
goonal and shallow island slope deposits
Age: Early to Middle Miocene (Geister 1992)
Faunal list: Geister 1992; all material examined
Assigned age: 22–15 Ma

6. Brasso and Tamana Formations, Trinidad

Environment: siliciclastics
Age: Middle Miocene (Maurrasse 1990)
Faunal list: modified after Vaughan and Hoffmeister 1926; all material ex-
amined
Assigned age: 15–11.2 Ma

7. Manzanilla Formation, Trinidad

Environment: siliciclastics
Age: Late Miocene (Maurrasse 1990)
Faunal list: modified after Vaughan and Hoffmeister 1926; all material ex-
amined
Assigned age: 11.2–5.3 Ma

8. Lirio Limestone, Mona

Environment: more than 30 m thick reefal limestones, with interspersed car-
bonate grainstones, packstones, and wackestones, interpreted as an iso-
lated platform-margin barrier reef
Age: Late Miocene (González et al. 1992)
Faunal list: González et al. 1992; all material examined
Assigned age: 11.2–5.3 Ma

9. Imperial Formation, California

Environment: up to 1500 m thick accumulations of terrigenous sandstones
and claystones (mostly Colorado River deposits), only 200 m of which is
marine and contains small nearshore patch reefs and oyster beds (Foster
1979)
Age: Late Miocene to Early Pliocene (Kerr and Kidwell 1991)

Faunal list: modified after Vaughan 1917; all material examined
Assigned age: 8–4.5 Ma

10. Buff Bay Formation, Jamaica

Environment: 30–40 m thick sequence of bluish claystones with occasional corals, interpreted as deeper-water talus deposits (Robinson 1969)

Age: Late Miocene (calcareous nannoplankton NN9-NN10) (Aubry 1993)

Faunal list: collections made by P. Jung at NMB; all material examined

Assigned age: 11.2–5.3 Ma

11. Bowden Formation, Jamaica

Environment: calcareous claystones with interbedded bioclastic limestones (possible slump deposits), interpreted as forereef talus deposits (Robinson 1969)

Age: early Late Pliocene (calcareous nannoplankton zone NN16) (Aubry 1993)

Faunal list: collections made by P. Jung at NMB, and modified after Vaughan 1919; all material examined

Assigned age: 3.8–2 Ma

12. Pinecrest Sandstone, Florida

Environment: up to 20 m thick nearshore carbonate deposits comprised of shell beds and patch reefs, the latter being unaffected by coastal sedimentation (Meeder 1987; Allmon 1993)

Age: Early to early Late Pliocene (foraminiferal zone N20) (Jones et al. 1991; Stanley 1991)

Faunal list: modified after Meeder 1987; not all material examined

Assigned age: 3.5–3 Ma

13. La Cruz Marl, Cuba (Guantanamo Basin)

Environment: shallow reefal limestones (Bold 1975)

Age: Late Pliocene (Bold 1975)

Faunal list: modified after Vaughan 1919; all material examined

Assigned age: 3.5–1.6 Ma

14. Matanzas, Cuba (Rio Yumuri)

Environment: 80 m thick sequence of shallow reefal limestones alternating with estuarine deposits (Iturralde-Vinent 1969)

Age: Late Pliocene to earliest Pleistocene (Bold 1975)

Faunal list: modified after Vaughan 1919; all material examined

Assigned age: 3.5–1 Ma

15. Old Pera Beds, Jamaica

Environment: shallow nearshore carbonate sandstones (Robinson 1969)

Age: latest Pliocene (Bold 1971)

Faunal list: collections made by P. Jung at NMB; all material examined

Assigned age: 2.5–1.6 Ma

16. Manchioneal Formation, Jamaica

Environment: shallow rubbly reefal limestone (Robinson 1969)

Age: *Gl. truncatulinoides*, Early Pleistocene (Bolli 1970)

Faunal list: collections made by P. Jung at NMB; all material examined

Assigned age: 1.6–1 Ma

17. Caloosahatchee Formation, Florida
Environment: shallow nearshore carbonate deposits (less than 10 m thick), consisting primarily of shell beds (Weisbord 1974)
Age: latest Pliocene (Bender 1973; Carter and Rossbach 1989)
Faunal list: modified after Weisbord 1974; all material examined
Assigned age: 1.8–1.6 Ma
18. Glades Formation, Florida
Environment: nearshore shell beds (Weisbord 1974)
Age: Early Pleistocene (Weisbord 1974; Carter and Rossbach 1989)
Faunal list: modified after Weisbord 1974; not all material examined
Assigned age: 1.6–1 Ma
19. Pleistocene terraces, Santo Domingo, Dominican Republic
Environment: well-developed fringing reefs and barrier reef (Geister 1982)
Age: 121–235 ky (Geister 1982)
Faunal list: Geister 1982; all material examined
Assigned age: 500–100 Ka
20. Pleistocene terrace, San Andrés
Environment: well-developed lagoonal patch reefs and fringing reefs (Geister 1975)
Age: Late Pleistocene (Geister 1975)
Faunal list: Geister 1975; material not examined
Assigned age: 500–100 Ka
21. Key Largo Limestone, Florida
Environment: well-developed lagoonal patch reefs (Weisbord 1974)
Age: 100–250 ky (Weisbord 1974)
Faunal list: Weisbord 1974; material not examined
Assigned age: 500–100 Ka

REFERENCES

Allmon, W. D. 1993. Age, environment and mode of formation of the densely fossiliferous Pinecrest Sand (Pliocene of Florida): Implications for the role of biological productivity in shell bed formation. *Palaios* 8:183–201.

Allmon, W. D., G. Rosenberg, R. W. Portell, and K. S. Schindler. 1993. Diversity of Atlantic coastal plain mollusks since the Pliocene. *Science* 260:1626–29.

Aubry, M. P. 1993. Calcareous nannofossil stratigraphy of the Neogene formations of eastern Jamaica. In *Biostratigraphy of Jamaica*, ed. E. R. Robinson, J. B. Saunders, and R. M. Wright, 131–78. Geological Society of America Memoir 182. Boulder, Colo.

Barry, J. C., L. J. Flynn, and D. R. Pilbeam. 1990. Faunal diversity and turnover in a Miocene terrestrial sequence. In *Causes of evolution: A paleontological perspective*, ed. R. M. Ross and W. D. Allmon, 381–421. Chicago: University of Chicago Press.

Bender, M. L. 1973. Helium-uranium dating of corals. *Geochim. Cosmochim. Acta* 39:1229–47.

Blacut, G., and R. M. Kleinpell. 1969. A stratigraphic sequence of benthonic smaller Foraminifera from the La Boca Formation, Panama Canal Zone. *Contrib. Cushman Found. Foram. Res.* 20:1–22.

Bold, W. A. van den. 1970. Ostracoda of the Lower and Middle Miocene of St. Croix, St. Martin, and Anguilla. *Carib. J. Sci.* 10:35–61.

———. 1971. Ostracoda of the Coastal Group of formations of Jamaica. *Trans. Gulf Coast Assoc. Geol. Soc.* 21:325–48.

———. 1975. Ostracodes from the Late Neogene of Cuba. *Bulls. Am. Paleontol.* 68:121–67.

———. 1988. Neogene paleontology of the northern Dominican Republic. 7, The subclass Ostracoda (Arthropoda: Crustacea). *Bulls. Am. Paleontol.* 94: 1–105.

Bolli, H. M. 1970. The Foraminifera of sites 23–31, leg 4. *Init. Repts. DSDP* 4:577–643.

Bryant, J. D., B. J. MacFadden, and P. A. Mueller. 1992. Improved chronologic resolution of the Hawthorn and the Alum Bluff Groups in northern Florida: Implications for Miocene chronostratigraphy. *Geol. Soc. Am. Bull.* 104: 208–18.

Budd, A. F. 1989. Biogeography of Neogene Caribbean reef corals and its implications for the ancestry of eastern Pacific reef corals. *Mem. Assoc. Australas. Palaeontols.* 8:219–30.

———. 1991. Neogene paleontology in the northern Dominican Republic. 11, The family Faviidae (Anthozoa: Scleractinia), part 1. *Bulls. Am. Paleontol.* 101:5–83.

———. 1993. Variation within and among morphospecies of *Montastraea*. *Courier Forschungs-Institut Senckenberg* 164:241–54.

Budd, A. F., and A. G. Coates. 1992. Non-progressive evolution in a clade of Cretaceous *Montastraea*-like corals. *Paleobiology* 18:425–46.

Budd, A. F., K. G. Johnson, and J. C. Edwards. 1995. Caribbean coral biofacies during the Early to Middle Miocene: An example from the Anguilla Formation. *Coral Reefs* 14:431–39.

Budd, A. F., K. G. Johnson, and J. B. C. Jackson. 1994. Patterns of replacement in Late Cenozoic Caribbean coral communities. *Geol. Soc. Am. Ann. Meeting; Abstr. Progr.* 25:454.

Budd, A. F., K. G. Johnson, and D. C. Potts. 1994. Recognizing morphospecies of colonial reef corals. Part 1, Landmark based methods. *Paleobiology* 20: 484–505.

Budd, A. F., and C. M. Kievman. 1993. Coral assemblages and reef environments in the Bahamas Drilling Project cores. Part 3 of *Final draft report of the Bahamas Drilling Project*. Coral Gables, Fla.: Rosenstiel School of Marine and Atmospheric Sciences, University of Miami.

Budd, A. F., T. A. Stemann, and K. G. Johnson. 1994. Stratigraphic distributions of genera and species of Neogene to Recent Caribbean reef corals. *J. Paleontol.* 68:951–77.

Budd, A. F., T. A. Stemann, and R. H. Stewart. 1992. Eocene Caribbean reef corals: A unique fauna from the Gatuncillo Formation of Panama. *J. Paleontol.* 66:578–602.

Buzas, M. A., C. F. Koch, S. J. Culver, and N. F. Sohl. 1982. On the distribution of species occurrence. *Paleobiology* 8:142–50.

Carter, J. G., and T. J. Rossbach. 1989. Correlation chart: Gulf and Atlantic Coasts of North America. *Biostrat. Newslett.* (Chapel Hill, N.C.) 3:1–48.

Chevalier, J. P. 1962. *Recherches sur les Madréporaires et les formations récifales Miocènes de la Méditerranée occidentale.* Mémoires de la Société géologique de France, vol. 93. Paris.

Chevalier, J. P. 1968. Les Madréporaires fossiles de Maré. In *Expédition française de la Nouvelle-Calédonie*, 3:91–155. Paris: Éditions de la Fondation Singer-Polignac.

Coates, A. G., J. B. C. Jackson, L. S. Collins, T. M. Cronin, H. J. Dowsett, L. M. Bybell, P. Jung, and J. A. Obando. 1992. Closure of the Isthmus of Panama: The near-shore marine record of Costa Rica and western Panama. *Geol. Soc. Am. Bull.* 104:814–28.

Drooger, C. W. 1951. Foraminifera from the Tertiary of Anguilla, St. Martin, and Tintamarre (Leeward Islands, West Indies). *Proc. Akad. Wetenschap.* (Amsterdam) B-54:54–65.

Evans, C. C. 1986. *A field guide to the mixed reefs and the siliciclastics of the Neogene Yaque Group, Cibao Valley, Dominican Republic.* Coral Gables, Fla.: Comparative Sedimentology Laboratory, University of Miami.

Foster, A. B. 1979. Environmental variation in a fossil scleractinian coral. *Lethaia* 12:245–64.

———. 1986. Neogene paleontology in the northern Dominican Republic. 2, The Family Poritidae (Anthozoa: Scleractinia). *Bulls. Am. Paleontol.* 90:47–123.

———. 1987. Neogene paleontology in the northern Dominican Republic. 4, The genus *Stephanocoenia* (Anthozoa: Scleractinia: Astrocoeniidae). *Bulls. Am. Paleontol.* 93:5–22.

Frost, S. H. 1977. Miocene to Holocene evolution of Caribbean Province reef-building corals. *Proc. Third Int. Coral Reef Symp.* 2:353–60.

Geister, J. 1975. *Riffbau und geologische Entwicklungsgeschichte der Insel San Andrés (westliches Karibisches Meer, Kolumbien).* Stuttgarter Beiträge zur Naturkunde, vol. B-15. Stuttgart.

———. 1982. Pleistocene reef terraces and coral environments at Santo Domingo and near Boca Chica, southern coast of the Dominican Republic. *Trans. Ninth Carib. Geol. Conf.* 2: 689–703.

Geister, J. 1992. Modern reef development and Cenozoic evolution of an oceanic island/reef complex: Isla de Providencia (western Caribbean Sea, Columbia). *Facies* 27:1–70.

González, L. A., H. M. Ruiz, A. F. Budd, and V. Monnell. 1992. A Late Miocene barrier reef in Isla de Mona, Puerto Rico. *Geol. Soc. Am. Ann. Meeting, Abstr. Progr.* 27:A-350.

Goreau, T. F. 1959. The ecology of Jamaican coral reefs. Part 1, Species composition and zonation. *Ecology* 40:67–90.

Hubbard, A. E., and N. L. Gilinsky. 1992. Mass extinctions as statistical phenomena: An examination of the evidence using chi-square tests and bootstrapping. *Paleobiology* 18:148–60.

Iturralde-Vinent, M. A. 1969. Principal characteristics of Cuban Neogene stratigraphy. *Am. Assoc. Petrol. Geol. Bull.* 53:1938–55.

Jablonski, D. 1991. Extinctions: A paleontological perspective. *Science* 253: 754–57.

Jackson, J. B. C. 1992. Pleistocene perspectives on coral reef community structure. *Am. Zool.* 32:719–31.

Jackson, J. B. C., P. Jung, A. G. Coates, and L. S. Collins. 1993. Diversity and extinction of tropical American mollusks and emergence of the Isthmus of Panama. *Science* 260:1624–26.

Johnson, K. G. 1991. Population ecology of a free-living coral: Reproduction, population dynamics, and morphology of *Manicina areolata* (Linneaus). Ph.D. diss., University of Iowa, Iowa City, Iowa.

———. 1992. Population dynamics of a free-living coral: Recruitment, growth, and survivorship of *Manicina areolata* (Linnaeus) on the Caribbean coast of Panama. *J. Exper. Mar. Biol. Ecol.* 164:171–91.

Johnson, K. G., A. F. Budd, and T. S. Stemann. 1995. Extinction selectivity and ecology of Neogene Caribbean reef corals. *Paleobiology* 21:52–73.

Jones, D. S., B. J. MacFadden, S. D. Webb, P. A. Mueller, D. A. Hodell, and T. M. Cronin. 1991. Integrated geochronology of a classic Pliocene fossil site in Florida: Linking marine and terrestrial biochronologies. *J. Geol.* 99: 637–48.

Kerr, D. R., and S. M. Kidwell. 1991. Late Cenozoic sedimentation and tectonics, western Salton Trough, California. In *Geological excursions in southern California and Mexico: Guidebook of the 1991 Annual Meeting of the Geological Society of America*. ed. M. J. Walawender and B. B. Hanan, 397–416. Boulder, Colo.: Geological Society of America.

Knowlton, N., E. Weil, L. A. Weigt, and H. M. Guzman. 1992. Sibling species of *Montastraea annularis*, coral bleaching, and the coral climate record. *Science* 255:330–33.

Koch, C. F., and J. P. Morgan. 1988. On the expected distribution of species ranges. *Paleobiology* 14:126–38.

Laborel, J. 1969. Madréporaires et hydrocoralliaires récifaux des côtes brésiliennes. *Ann. Inst. Océanogr.* 47:171–229.

Lasker, H. R. 1978. The measurement of taxonomic evolution: Preservational consequences. *Paleobiology* 4:135–49.

Leão, Z. M. A. N., T. M. F. Araujo, and M. C. Nolasco. 1988. The coral reefs off the coast of eastern Brazil. *Proc. Sixth Int. Coral Reef Symp.* 3:339–47.

Liddell, W. D., and S. L. Ohlhorst. 1988. Comparison of western Atlantic coral reef communities. *Proc. Sixth Int. Coral Reef Symp.* 3:281–86.

Marshall, C. R. 1990. Confidence intervals on stratigraphic ranges. *Paleobiology* 16:1–10.

Maurrasse, F. 1990. Stratigraphic correlation for the Circum-Caribbean Region. In *The Caribbean Region*, vol. H of *The geology of North America*, ed. G. Dengo and J. E. Case. Boulder, Colo.: Geological Society of America.

McNeill, D. F., B. H. Lidz, T. J. Bralower, J. M. Kenter, G. P. Eberli, and P. K. Swart. 1993. Bahamas Drilling Project chronostratigraphy and correlation: A multidisciplinary approach in a dynamic platform margin and slope setting. Part 12 of *Final draft report of the Bahamas Drilling Project* Coral Gables, Fla.: Rosenstiel School of Marine and Atmospheric Sciences, University of Miami.

Meeder, J. F. 1987. The paleontology, petrology, and depositional model of the Pliocene Tamiami Formation, southwest Florida (with special reference to corals and reef development). Ph.D. diss., University of Miami, Coral Gables, Fla.

Potts, D. C. 1984. Generation times and the Quaternary evolution of reef-building corals. *Paleobiology* 10:48–58.

Potts, D. C., A. F. Budd, and R. L. Garthwaite. 1993. Soft tissue vs. skeletal approaches to species recognition and phylogeny reconstruction in corals. *Courier Forschungs-Institut Senckenberg* 164:221–31.

Puri, H. S., and R. O. Vernon. 1964. *A summary of the geology of Florida and a guidebook to classic exposures.* Florida Geological Survey Publication 5. Tallahassee, Fla.

Raymo, M. E., W. F. Ruddiman, N. J. Shackleton, and D. W. Oppo. 1990. Evolution of Atlantic-Pacific $\delta^{13}C$ gradients over the last 2.5 m.y. *Earth Planet. Sci. Lett.* 97:353–68.

Robinson, E. 1969. *Geological field guide to Neogene sections in Jamaica, West Indies: Field Guidebook, Nineteenth Annual Meeting, Gulf Coast Association of Geological Societies, SEPM Field Trip.* New Orleans, La.

Saunders, J. B., P. Jung, and B. Biju-Duval. 1986. Neogene paleontology in the northern Dominican Republic. Part 1, Field Surveys, lithology, environment, and age. *Bulls. Am. Paleontol.* 89:1–79.

SAS Institute. 1989. *SAS/STAT user's guide: Version 6.* 4th ed. Vol. 1. Cary, N.C.: SAS Institute.

Smith, S. V., and R. W. Buddemeier. 1992. Global change and coral reef ecosystems. *Ann. Rev. Ecol. System.* 23:89–118.

Stanley, S. M. 1986. Anatomy of a regional mass extinction: Plio-Pleistocene decimation of the western Atlantic bivalve fauna. *Palaois* 1:17–36.

———. 1991. Evidence from marine deposits in Florida that the Great American Interchange of mammals began prior to 3 Ma. *Geol. Soc. Am. Ann. Meeting, Abstr. Progr.* 23:A-405.

Stanley, S. M., K. L. Wetmore, and J. P. Kennett. 1988. Macroevolutionary differences between two clades of Neogene planktonic Foraminifera. *Paleobiology* 14:235–49.

Stemann, T. A. 1991. Evolution of the reef-coral family Agariciidae (Anthozoa: Scleractinia) in the Neogene through Recent of the Caribbean. Ph.D. diss., University of Iowa, Iowa City, Iowa.

Strauss, D., and P. M. Sadler. 1989. Classical confidence intervals and Bayesian probability estimates for ends of local taxon ranges. *Math. Geol.* 21:411–27.

Vaughan, T. W. 1917. The reef-coral fauna of Carrizo Creek, Imperial County, California, and its significance. *U.S. Geol. Surv. Prof. Pap.* 98-T:355–86.

———. 1919. Fossil corals from Central America, Cuba, and Porto Rico with an account of the American Tertiary, Pleistocene, and Recent coral reefs. *U.S. Nat. Mus. Bull.* 103:189–524.

Vaughan, T. W., and J. E. Hoffmeister. 1925. New species of fossil corals from the Dominican Republic. *Bull. Mus. Comp. Zool.* (Harvard) 67:315–26.

———. 1926. Miocene corals from Trinidad. *Pap. Dept. Mar. Biol., Carnegie Inst. Wash.* 23:107–32.

Veron, J. E. N. 1986. *Corals of Australia and the Indo-Pacific.* North Ryde, NSW, Austral.: Angus and Robertson.

Veron, J. E. N., and R. Kelley. 1988. Species stability in reef corals of Papua New Guinea and the Indo Pacific. *Mem. Assoc. Australas. Palaeontols.* 6:1–69.

Wei, K. Y., and J. P. Kennett. 1986. Taxonomic evolution of Neogene planktonic Foraminifera and paleoceanographic relations. *Paleoceanography* 1:67–84.

Weil, E., and N. Knowlton. 1994. A multi-character analysis of the Caribbean coral *Montastraea annularis* (Ellis & Solander, 1786) and its two sibling species *M. faveolata* (Ellis & Solander, 1786) and *M. franksi* (Gregory, 1895). *Bull. Mar. Sci.* 55:151–75.

Weisbord, N. E. 1971. Corals from the Chipola and Jackson Bluff Formations of Florida. *Fla. Bur. Geol., Geol. Bull.* 53:1–100.

———. 1973. New and little-known corals from the Tampa Formation of Florida. *Fla. Bur. Geol., Geol. Bull.* 56:1–147.

———. 1974. Late Cenozoic corals of south Florida. *Bulls. Am. Paleontol.* 66:255–544.

Wells, J. W. 1956. Scleractinia. In *Treatise on invertebrate paleontology*, ed. R. C. Moore, part F, Coelenterata, 328–444. Lawrence: Geological Society of America and University of Kansas Press.

Wells, J. W., and J. C. Lang. 1973. Systematic list of Jamaican shallow-water Scleractinia. *Bull. Mar. Sci.* 23:55–58.

Woodring, W. P. 1957. Geology and paleontology of Canal Zone and adjoining parts of Panama. *U.S. Geol. Surv. Prof. Pap.* 306-A:1–145.

———. 1964. Geology and paleontology of Canal Zone and adjoining parts of Panama. *U.S. Geol. Surv. Prof. Pap.* 306-C:241–97.

Zlatarski, V. N. 1990. *Porites colonensis*, new species of stony coral (Anthozoa: Scleractinia) off the Caribbean coast of Panama. *Proc. Biol. Soc. Wash.* 103: 257–64.

8

Speciation, Extinction, and the Decline of Arborescent Growth in Neogene and Quaternary Cheilostome Bryozoa of Tropical America

Alan H. Cheetham and Jeremy B. C. Jackson

Cheilostome bryozoans are a major component of the tropical American marine benthos, in both abundance and number of species, with a rich fossil record beginning in mid-Cretaceous time (McKinney and Jackson 1989 and references therein). A great variety of encrusting, free-living, and erect growth forms have characterized cheilostomes from the latest Cretaceous throughout the Cenozoic. However, beginning in the Neogene, faunas underwent a striking shift in morphology, especially in the tropics (Jackson and McKinney 1990). Assemblages of the Paleogene and much of the Neogene are dominated volumetrically by genera, such as *Adeonellopsis* and *Metrarabdotos*, that grow rigidly erect in the shape of regularly branching, miniature trees (arborescent colonies, fig. 8.1A; Cheetham 1986a). In contrast, arborescent taxa are much diminished in abundance in living cheilostome faunas, and nonarborescent taxa show a distinct dichotomy in their modern geographic distribution. Sandy substrata, especially on the North and South American continental margins, are characterized by a great abundance of free-living colonies, typically belonging to the genera *Cupuladria* and *Discoporella* (Marcus and Marcus 1962; Maturo 1968; Cadée 1975; Winston and Håkansson 1986), whereas on hard substrata, encrusting species of genera such as *Parasmittina*, *Reptadeonella*, *Steginoporella*, and *Stylopoma* are volumetrically dominant (fig. 8.1B). Species of these predominantly encrusting genera are conspicuous on coral reefs, but not restricted to them (Jackson et al. 1985; Winston 1986; Hughes and Jackson 1992).

The major decline in abundance of arborescent cheilostomes was accompanied by only slightly less striking changes in taxonomic composition. Arborescent species have always been outnumbered by encrusting ones, even in such bryozoan-rich Paleogene deposits as the Danian of Scandinavia, where they far exceed encrusting species in abundance (Cheetham 1971). Even so, the number of species of predominantly arborescent genera decreased by approximately 60% in tropical America

A

B

Fig. 8.1 Neogene and living cheilostome bryozoan assemblages. Scale bar = 1 cm. (A) branch fragments of arborescent colonies predominantly of *Metrarabdotos*, with some *Adeonellopsis*, and few free-living colonies of *Discoporella* and other genera, Cercado Formation (NN11), Bluff 3, Rio Mao, Dominican Republic (locality TU 1294, Saunders et al. 1986); (B) fragments of foliaceous expansions of encrusting colonies of *Steginoporella*, encrusted by *Parasmittina*, *Celleporaria*, and other genera, and few arborescent colonies of *Bracebridgia* and other genera, dredge haul (Albatross station D.2414, Canu and Bassler 1928), shell and sand bottom, west coast of Florida (150 m).

from the Late Paleogene to the present, while the number of genera has decreased by 26% (tables 8.1 and 8.2). Especially noteworthy is the loss of arborescent species in genera such as *Steginoporella, Schizoporella,* and *Hippopleurifera,* now represented exclusively by encrusting species, and the concomitant rise of encrusting species in genera such as *Metrarabdotos* that formerly were exclusively arborescent. Some encrusting species in these genera can produce irregularly branching, foliaceous or tubular erect expansions superficially similar to arborescent growth, and perhaps reflecting arborescent ancestry (*Steginoporella* in fig. 8.1B; Cheetham 1968).

Table 8.1 Numbers of Late Paleogene to Quaternary Aborescent Cheilostome Species Reported from Caribbean and Adjacent Regions

Genus	Late Paleogene	Neogene	Living
Adeonellopsis	3	2	1
Bracebridgia	0	0	1
Cellaria	1	2	2
Cigclisula	0	1	1
Cystisella	0	0	1
Enoplostomella	7	0	0
Euginoma	0	0	1
Gemelliporella	0	2	1
Hippopleurifera	5 (+encrusting)	(encrusting)	(encrusting)
Kleidionella	2	0	0
Leiosella	3	0	0
Margaretta	1	2	1
Membraniporella	1 (+encrusting)	0	(encrusting)
Membraniporidra	2 (+encrusting)	0	0
Meniscopora	1	0	0
Metrarabdotos	2	14	1 (+encrusting)
Nellia	1	2	1
Phoceana	0	0	1
Poricellaria	1	2	0
Porina	1	0	0
Schizomavella	2 (+encrusting)	0	(encrusting)
Schizoporella	(encrusting)	1 (+encrusting)	(encrusting)
Semihaswellia	1	1	1
Skylonia	0	1	0
Stamenocella	2 (+encrusting)	0	0
Steginoporella	1 (+encrusting)	(encrusting)	(encrusting)
Tetraplaria	0	1	1
Umbonula	0	0	1 (+encrusting)
Vincularia	1	1	0

Sources: McKinney and Jackson 1989; Canu and Bassler 1920, 1928; Logaaij 1963; Cheetham 1963, 1973, unpublished data.

Note: Some genera include encrusting species, as indicated.

Table 8.2 Percentage of Change in Numbers of Arborescent Cheilostome Species and Genera from Late Paleogene to Present

Taxa	Late Paleogene (n)	Neogene (n)	Living (n)	Change (%)		
				P–N	N–L	Total
Species	38	32	15	−16	−53	−61
Genera	19	13	14	−32	+8	−26

Although the broad outlines of these changes are well known, their timing within the Neogene and Quaternary and the underlying patterns of speciation, extinction, and shifts in geographic range are not. We have studied new material from extensive collections in the Dominican Republic (Saunders et al. 1982; Saunders et al. 1986) and Panama and Costa Rica (Coates et al. 1992) that are more tightly controlled stratigraphically than those previously available. These collections allow refinement of species ranges in both space and time, and examination of their possible relationship to major environmental changes associated with the shoaling and final closure of the Isthmus of Panama (Duque-Caro 1990; Coates et al. 1992; Coates and Obando, this volume, chap. 2; L. Collins, this volume, chap. 6).

Here we describe patterns of speciation, distribution, and extinction for two genera, *Stylopoma*, which is exclusively encrusting, and *Metrarabdotos*, which is predominantly arborescent. Both are ascophoran cheilostomes with similar degrees of zooidal morphologic complexity and polymorphism (Cheetham 1986b; Jackson and Cheetham 1990, 1994). Both occur outside the Americas (Harmer 1957; Cheetham 1967, 1968; Cook 1986) but apparently originated in tropical America, where much of their evolution has taken place. Based on a consistent approach to species discrimination and the best evidence for phylogenetic relationships among species, the tropical American Neogene and Quaternary records of the two genera show similar episodes of radiation resulting in nearly identical numbers of species over the same time interval. However, rates of morphologic evolution are quite different in the two genera (Cheetham and Jackson 1995), and differences in species durations, geographic ranges, and rates of origination and extinction suggest disparity in the responses of these two genera to environmental changes that may help to explain the shift in dominance from arborescent to encrusting growth.

NATURE OF THE CHEILOSTOME RECORD

Geographic distributions of cheilostomes are influenced by the fact that most species, regardless of colony form, brood larvae that are competent to settle more-or-less immediately after release by the parent (McKinney and Jackson 1989 and references therein). Dispersal of all such species over distances of more than a few kilometers, and especially among islands separated by deep ocean, must depend therefore upon "rafting" of sessile colonies attached to floating objects rather than larval dispersal (Jackson et al. 1985; Jackson 1986). Encrusting species typically occur in a wider range of depths and on more kinds of substrata than arborescent species (Lagaaij and Gautier 1965; Harmelin 1976), and are thus more likely to disperse as adults by rafting. Even with the likelihood of rafting, however, many of the species that are reported to range circumtropically or from the Carolinas to Brazil, and to persist from early in the Miocene to the present, may comprise suites of cryptic species with much more restricted geographic and temporal ranges (Lagaaij and Cook 1973; Jackson et al. 1985; Banta 1991).

Major revisions of geographic and temporal ranges are likely to result from more detailed, quantitative morphologic analysis of cheilostome species. Genetic tests of tropical American species of *Stylopoma*, *Steginoporella*, and *Parasmittina* have greatly increased confidence in cheilostome morphospecies, but only if species are split to the limits of statistical significance; the number of species in these genera was increased by four- to fivefold (Jackson and Cheetham 1990, 1994). Thus, morphologically variable, widespread, and long-lived species that have not been studied in morphometric detail almost certainly comprise suites of cryptic species.

Reliable estimates of changes in the numbers of cheilostome species through the Cenozoic in tropical America, and of underlying changes in rates of origination and extinction, depend strongly on resolution of such hidden taxonomic variation. The effects of our new stratigraphic information and taxonomic perspective on the numbers of tropical species and genera (table 8.1) are likely to be quite large, as exemplified by the increase from six Caribbean Neogene and Quaternary species (or subspecies) included in *Metrarabdotos* by Cheetham (1968) to seventeen species known today (Cheetham 1986b). If the changes in *Metrarabdotos* are representative, estimates of species numbers will increase more strongly for fossil than for living faunas and thus accentuate or leave unchanged the pattern of decline of arborescent species.

PHYLOGENY, RADIATION, AND EXTINCTION

Comparison of patterns of speciation, distribution, and extinction in *Stylopoma* and *Metrarabdotos* requires a uniform approach to species discrimination and well-supported hypotheses of phylogenetic relationships among species. We therefore employed the same morphometric and statistical procedures to discriminate morphospecies on characters of skeletal morphology in each genus (Cheetham 1986b, 1987; Jackson and Cheetham 1990, 1994). Electrophoretic studies of *Stylopoma* have established not only that morphospecies so discriminated are genetically distinct, but also that the genetic distances between species are strongly correlated with morphologic differences, so that phylogenies can be confidently reconstructed using morphology (Jackson and Cheetham 1990, 1994). Consistency between genetic and cladistic distances (number of branching points separating species in cladistically constructed trees) in *Stylopoma* increased greatly with the increased fineness of morphospecies discrimination and the use of fossil species in the construction and rooting of the phylogenetic trees. Rooting on the earliest-included species, rather than on species belonging to the most closely related genera (outgroups), also increased the consistency of trees with the stratigraphic record for both *Stylopoma* and *Metrarabdotos* (Jackson and Cheetham 1994).

Phylogenies constructed for both *Stylopoma* and *Metrarabdotos* (fig. 8.2) are not without uncertainties due to uneven stratigraphic and geographic sampling. The entire first half (16.5 to 8 Ma) of the Neogene and Quaternary record of these genera in tropical America is represented by only 5% of the 48 *Stylopoma*-bearing horizons and 10% of the 153 *Metrarabdotos*-bearing horizons (Cheetham and Jackson 1995). Between 8 and 6 Ma, ten species of *Stylopoma* and eleven of *Metrarabdotos* appear in the record, and the first occurrences of all of these species are in the Dominican Republic (tables 8.3 and 8.4). Confidence intervals on stratigraphic ranges make the order of first occurrences of many of the species of *Stylopoma* and some of *Metrarabdotos* uncertain (fig. 8.2; Marshall 1990, 1991, 1995).

PATTERN OF SPECIATION IN STYLOPOMA

For *Stylopoma*, stratigraphic data are insufficient to support reconstructing phylogeny stratophenetically, by connecting species to their morphologically and stratigraphically nearest neighbors. Confidence limits on first occurrences average more than 1 my, and almost all spe-

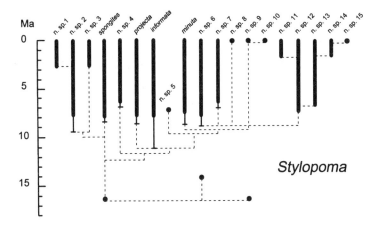

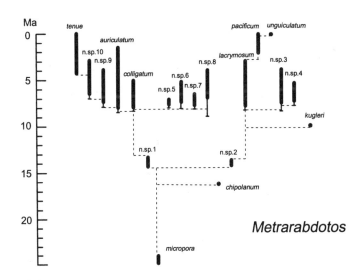

Fig. 8.2 Phylogenetic trees of *Stylopoma* and *Metrarabdotos*, tropical American Neogene to Holocene (from Jackson and Cheetham 1995, figs. 6.3 and 6.4. Reprinted with permission). *Stylopoma* tree is modified cladogram rooted on *S. spongites*. *Metrarabdotos* tree is stratophenetic. Scale bars are 95% confidence intervals on first occurrences for species originating in detailed sampling interval from 8 to 4 Ma (Cheetham and Jackson 1995).

Table 8.3 Neogene and Quaternary *Stylopoma* Species Distributions

Species	Oldest	Youngest	Duration (my)	Regions						
				A	B	C	D	E	F	G
New species 1	Plio.?, Panama	Living	1.8	L/F					L/F	
New species 2*	NN11, Dom. Rep.	Living	7.8		L/F	L	L/F	L		
New species 3*	NN17/20, C. Rica	Living	2.4	L/F	L/F	L	L/F	L	L/F	
*Spongites**	NN4/5, Florida	Living	16.5	L/F	L/F	L	F		L/F	
New species 4	NN11, Dom. Rep.	Living	6.4	L			F			
*Projecta**	NN11, Dom. Rep.	Living	8.0	L/F	L/F	L	L/F	L	L/F	
Informata	NN11, Dom. Rep.	Living	8.0		L/F		F			
New species 5	NN11, Dom. Rep.	Same	0.1				F			
Minuta	NN11, Dom. Rep.	Living	7.6	F	L	F	F	L	F	L
New species 6	NN11, Dom. Rep.	Living	8.0	L/F			F			
New species 7	NN11, Dom. Rep.	Living	6.6				L/F			
New species 8*	Living	Same	0.0					L		
New species 9	Living	Same	0.0	L						
New species 10	Living	Same	0.0	L						
New species 11	NN17/18, C. Rica	Living	1.8	L			L		F	
New species 12	NN11, Dom. Rep.	Living	6.8				L/F			
New species 13	NN11, Dom. Rep.	Living	6.0				L/F		F	
New species 14*	NN17/18, C. Rica	Living	1.8						L/F	
New species 15*	Living	Same	0.0					L	L	

Note: L = living; F = fossil. Regions refer to figure 8.3.
*Analyzed electrophoretically (Jackson and Cheetham 1994).

Table 8.4 Neogene and Quaternary *Metrarabdotos* Species Distributions

Species	Oldest	Youngest	Duration (my)	Regions						
				A	B	C	D	E	F	G
New species 1	NN5/6, Haiti	Same	0.1				F			
Tenue	NN13, Dom. Rep.	Living	4.5				L/F	L		
Colligatum	NN11, Dom. Rep.	NN16, Panama	3.0				F		F	
Auriculatum	NN11, Dom. Rep.	NN19, Florida	6.2	F	F		F			
New species 5	NN11, Dom. Rep.	Same	1.8				F			
New species 6	NN11, Dom. Rep.	NN13, Dom. Rep.	2.2				F			
New species 7	NN11, Dom. Rep.	NN12, Dom. Rep.	1.4				F			
New species 8	NN11, Dom. Rep.	NN13, Dom. Rep.	3.2				F			
New species 9	NN11, Dom. Rep.	NN16, Panama	3.4				F		F	
New species 10	NN11, Dom. Rep.	NN16, Panama	3.5				F		F	
Chipolanum	NN4/5, Florida	Same	0.1	F						
New species 2	NN5/6, Haiti	Same	0.1			F	F			
Lacrymosum	NN11, Dom. Rep.	NN16, Jamaica	5.0				F	F	F	
Kugleri	NN10, Trinidad	Same	0.1			F				
New species 3	NN11, Dom. Rep.	Same	3.8				F			
New species 4	NN11, Dom. Rep.	Same	1.6				F			
Pacificum	NN16, Panama	Living	2.9	L	L			L		
Unguiculatum	Living	Same	0.0							L/F

Note: Regions refer to figure 8.3.

cies range to the present (fig. 8.2). Instead, we used stratigraphic occur-rences only to root and modify a cladistically derived tree based on the morphologic characters alone (Jackson and Cheetham 1994; Cheetham and Jackson 1995). Ancestor-descendant relationships determined from the cladistic analysis (cladogram) were used to construct the modified tree (fig. 8.2), but two Middle Miocene occurrences morphologically indistinguishable from new species 6 and 9 are assumed to be conver-gent, rather than conspecific, with later populations (Jackson and Cheetham 1994). Without this adjustment, half the species would be required to extend back more than 17 my (Jackson and Cheetham 1994; Cheetham and Jackson 1995), despite the absence of *Stylopoma* in older bryozoan-rich deposits (Canu and Bassler 1920, 1923).

Based on ranges implied by the modified cladistic tree, the major pe-riod of radiation involving thirteen of the nineteen species of *Stylopoma* took place between 12 and 7 Ma. Twelve of these thirteen species, along with one that originated earlier (*spongites*), continued through the Neo-gene and Quaternary. A second wave of originations followed during the past 3 my, with at least five new species distributed among all major subclades (eight species based on actual occurrences). Only one species (new species 5) has become extinct in the last 12 my during which the two waves of originations occurred.

PATTERN OF SPECIATION IN METRARABDOTOS

The denser fossil record of *Metrarabdotos* provides greater support for a stratophenetic reconstruction (fig. 8.2; Jackson and Cheetham 1994; Cheetham and Jackson 1995), despite uncertainties in the order of ap-pearance of some species (Marshall 1990, 1991, 1995). Confidence lim-its on first occurrences average only half those of *Stylopoma* (0.5 my), far less than the 8 to 10 my extensions that would be required of ten species for adherence to the cladogram (Jackson and Cheetham 1994; Cheet-ham and Jackson 1995). Moreover, cladograms ignore the morphologic and geographic proximity of the *Metrarabdotos* species that first occur at 8 to 6 Ma in the Dominican Republic to either new species 1 or 2, which occur in the same area at about 14 Ma (Jackson and Cheetham 1994; Cheetham and Jackson 1995).

Stratophenetic reconstruction based on confidence limits for first oc-currences suggests that, like *Stylopoma*, *Metrarabdotos* underwent major radiation in the Late Miocene (9–7 Ma), when eleven of the seventeen Neogene species originated (fig. 8.2). Unlike *Stylopoma*, however, all of these species became extinct later in the Neogene or Quaternary, and only three additional species (*tenue, pacificum,* and *unguiculatum*) evolved after the Late Miocene radiation. Significantly, two of the three later species (*pacificum* and *unguiculatum*), belonging to the same subclade (fig. 8.2), differ from all other American *Metrarabdotos* in exhibiting encrusting growth.

GEOGRAPHIC DISTRIBUTION OF STYLOPOMA AND METRARABDOTOS SPECIES

For biogeographic analysis, we divided the Caribbean and adjacent areas into seven regions, including the Pacific coast of Panama, each 1000 km or more in major dimension (fig. 8.3). Unifying features were used wherever possible to delineate regions (e.g., A = Gulf of Mexico, C = predominantly carbonate areas connected by Nicaragua Rise), but boundaries are necessarily arbitrary. Despite obvious differences in the numbers of occurrences of species of the two genera in given regions, the distributions of *Stylopoma* and *Metrarabdotos* are similar in including all seven regions (fig. 8.3).

Sampling of living *Stylopoma* populations is moderately dense and evenly distributed in all seven regions, except along the Pacific coast of Panama, represented by a single occurrence of a single species, and the Gulf of Mexico, with no known occurrences in the western half (fig. 8.3). Material includes collections made by dredging and scuba diving (Canu and Bassler 1928; Jackson 1984; Winston and Jackson 1984; Hughes and Jackson 1992). Fossil collections are much less extensive but lacking only in regions E and G. *Stylopoma* reported from the Pliocene of Veracruz, Mexico (Herrera Anduaga 1985), was not available for this study.

Relative frequencies of fossil and living collections of *Metrarabdotos* are opposite to those of *Stylopoma* (fig. 8.3), reflecting the greater incidence of extinction in *Metrarabdotos* in the Late Neogene (fig. 8.2). Collections from living populations all represent dredging, because *Metrarabdotos* only rarely occurs at depths as shallow as many of the reef-associated occurrences of *Stylopoma* (Canu and Bassler 1928; Cheetham 1967). Fossils occur in all seven regions, whereas living *Metrarabdotos* populations are absent from regions C and F. This may partly reflect the paucity of dredging in these regions, although *Metrarabdotos* is generally uncommon wherever it occurs throughout the Caribbean region, except off the coast of Puerto Rico.

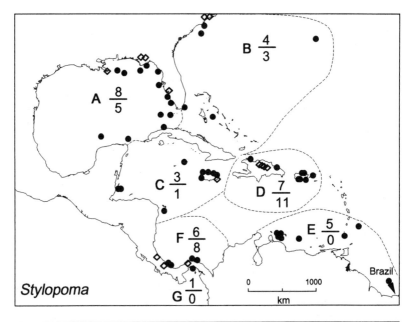

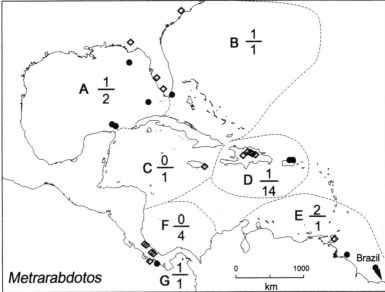

Fig. 8.3 Distribution of collections of *Stylopoma* and *Metrarabdotos*. Living populations are indicated by filled circles; fossil populations by unfilled diamonds. Number of morphospecies, living (above line) and fossil (below line), is given for each of seven regions (A–G).

Stylopoma and *Metrarabdotos* differ considerably in numbers of species and numbers of occurrences in different regions (fig. 8.3; tables 8.3 and 8.4). Both living and fossil, *Stylopoma* includes more species with more occurrences than *Metrarabdotos* in regions peripheral to the tropics, including the Gulf of Mexico (region A) and the northwest Atlantic (region B), perhaps reflecting broader thermal tolerance. In the Caribbean, living species of *Metrarabdotos* are found only at depths of 20 m to 220 m, where temperatures range from 16°C to 28°C (Cheetham 1967). *Stylopoma* occurs at both shallower depths (less than 1 m) and greater ones (more than 300 m) in both the Caribbean and the Gulf of Mexico, and thus is most likely exposed to a greater range of temperatures.

Nevertheless, the largest numbers of species of both genera, both living and fossil, occur in the more strictly tropical regions (C to G). Taken at face value, the stratigraphic records of both genera suggest that the major Neogene radiations were centered on this tropical belt (initially region D, later region F). Such an interpretation is supported for *Metrarabdotos* by the close morphologic similarity between the Late Miocene species first occurring in the Dominican Republic and their earlier Miocene congeners (new species 1 and 2) from the same area. It is more conjectural for *Stylopoma*.

Geographic ranges of *Stylopoma* and *Metrarabdotos* species are summarized in tables 8.3 and 8.4, and examples of species occurrences are shown in figures 8.4 and 8.5. *Stylopoma* includes a greater number of widely distributed species (21% in five or more regions) than *Metrarabdotos* (none in more than three regions; fig. 8.6), but the statistical significance of the difference is uncertain. Assumptions underlying a parametric test ($t = 2.48$, one-tailed $p < .001$) are almost certainly violated, and a nonparametric test (Mann-Whitney $U = 146.5$, one-tailed $p = .10$) is compromised by the large number of ties.

Half the species in each genus are known from a single region. Eight of eleven such *Metrarabdotos* species are from Hispaniola (region D). Only four of ten similarly endemic *Stylopoma* species occur in a single region (F); the others occur in three different regions (fig. 8.4). Eight of the ten species of *Stylopoma* known from only one region occur no earlier than 3 Ma; these species might be expected to expand their ranges over time. However, mean durations of endemics (based on actual occurrences) are similar in the two genera (*Stylopoma* 2.0 my, *Metrarabdotos* 1.6 my). Thus, the truncated ranges appear not to bias in favor of endemism in *Stylopoma* any more than in *Metrarabdotos*.

The maximum number of regions for any species of *Metrarabdotos* is three, but distributions of four species include noncontiguous regions

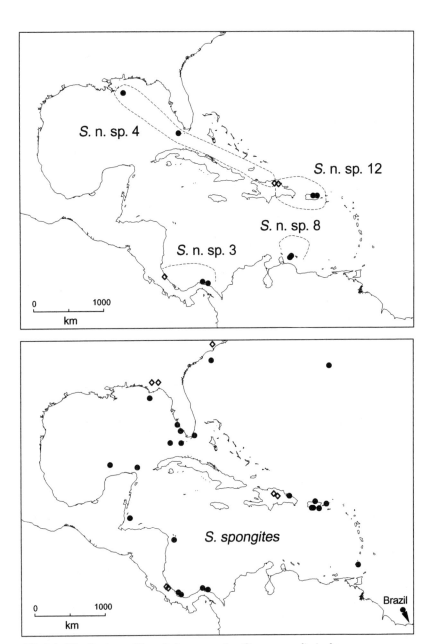

Fig. 8.4 Distributions of four *Stylopoma* species. Occurrences limited to one or two regions are shown above, and one species distributed widely through the Caribbean and adjacent areas below. Filled circles indicate living populations; unfilled diamonds, fossils.

(fig. 8.5). Thus the true maximum is likely to be at least four, most probably for the living species *M. unguiculatum* (known from regions A, B, and E, but not from connecting region D). In contrast, *Stylopoma* includes at least four species known from five or more regions (fig. 8.6), and two of these species (*spongites*, fig. 8.4, and new species 2) are nearly coextensive with the American distribution of the genus.

On balance, broad geographic distributions appear to be more characteristic of *Stylopoma* than *Metrarabdotos*, even though most species in both genera have restricted distributions, and there is no hard statistical support for the difference between genera.

SPECIES DURATION, GEOGRAPHIC RANGE, AND FREQUENCY OF SPECIATION

Durations of *Stylopoma* species based on actual occurrences are mostly underestimates. Most species are not yet extinct, confidence intervals on first occurrences are long, and cladograms suggest still-earlier origins for some species (fig. 8.3). Nevertheless, *Stylopoma* includes a much larger proportion (52%) of long-lived species based on actual occurrences (durations of 6 my or more, fig. 8.6) than *Metrarabdotos* (5%), but again the statistical significance of the difference is uncertain (parametric test, $t = 2.34$, one-tailed $p = .001$; nonparametric test, Mann-Whitney $U = 127.0$, one-tailed $p = .094$). Results of both parametric and nonparametric tests remain the same if 95% confidence limits are substituted for observed stratigraphic ranges, and extensions of range suggested by the cladogram would only accentuate the difference between genera.

Species durations and geographic ranges are significantly positively correlated in both genera (fig. 8.7; table 8.5). The correlation in *Stylopoma* is higher and more significant, if based on actual stratigraphic occurrences, but lower and less significant than that for *Metrarabdotos* if based on ranges implied by the cladogram. Thus, the degree of correlation may not differ between genera, implying that the apparent differences in duration and geographic range are proportional.

All four of the most widely distributed *Stylopoma* species (*spongites, projecta, minuta*, new species 2) have durations based on observed stratigraphic occurrences of 7.6 to 16.5 my. These values comfortably exceed the durations of all of the *Metrarabdotos* species, despite the much more extensive fossil record of *Metrarabdotos* compared to *Stylopoma*. Two of the three most widely distributed species of *Metrarabdotos* (*lacrymosum, auriculatum*) have the longest durations in that genus (5.0 and 6.2 my).

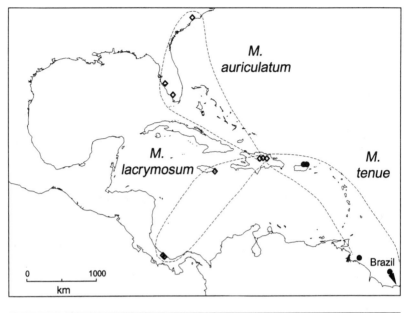

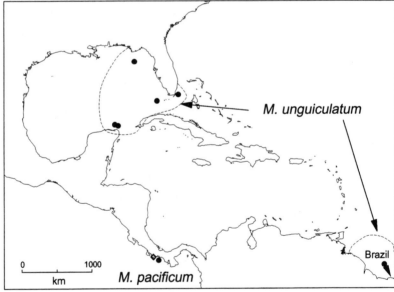

Fig. 8.5 Distributions of five *Metrarabdotos* species. Filled circles indicate living populations; unfilled diamonds, fossils.

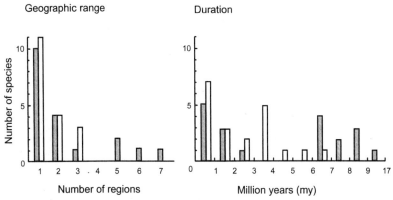

Fig. 8.6 Geographic ranges and durations for species of *Stylopoma* (filled bars) and *Metrarabdotos* (unfilled bars).

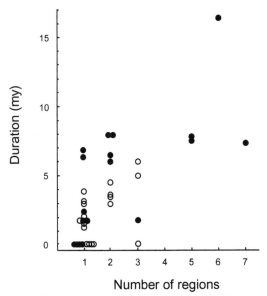

Fig. 8.7 Relationship between geographic range and duration for species of *Stylopoma* (filled circles) and *Metrarabdotos* (unfilled circles).

Table 8.5   Spearman Rank-Order Correlations between Species Durations, Geographic Ranges, and Numbers of descendant species for *Stylopoma* and *Metrarabdotos*

Correlation	*Stylopoma*	*Metrarabdotos*
Duration/Range	.655***	.483**
	(.414*)	
Duration/Descendants	.659***	.116
	(.546**)	
Range/Descendants	.587***	.139

Note: Values in parentheses are based on durations implied in cladogram.
*p < .10 ** p < .05 *** p < .01

Most species with restricted distributions have short durations in both genera. There are exceptions to the general pattern, however: one species in each genus has ranged into two or three regions in less than 2 my (*Stylopoma* new species 11, *Metrarabdotos unguiculatum*), and two species of *Stylopoma* have been restricted to a single region for 6.6 to 6.8 my (new species 7 and 12).

In contrast to their apparent similarity in the correlation of duration with geographic range, *Stylopoma* and *Metrarabdotos* differ markedly in the correlations of the frequency of speciation, as measured by numbers of inferred descendants, with species durations and with geographic ranges (table 8.5). In *Stylopoma*, these correlations are approximately as high as that between duration and geographic range, but are low and clearly nonsignificant in *Metrarabdotos*. The most widespread and long-lived species of *Stylopoma* tend to lie near the base of the Neogene radiation, and vice versa (fig. 8.2). Moreover, the youngest descendent species (with first occurrences after 2.5 Ma) are restricted to peripheral regions, rather than occurring in the central Caribbean; peripheral regions include Panama (new species 15), Costa Rica (new species 3, 11, and 14), Curaçao (new species 8), and the Gulf of Mexico (new species 9 and 10). All these patterns are consistent with speciation by geographic isolation on the periphery of the generally large ranges of ancestral species, as proposed by Mayr (1954, 1963).

In contrast, *Metrarabdotos* shows no such relationships. Two of the most widespread species (*tenue* and *unguiculatum*, fig. 8.5) appeared relatively recently (after 5 Ma, fig. 8.2), and only three (*colligatum, auriculatum,* and *lacrymosum*) of the seven to eleven species that appeared at 8 Ma or earlier occur in more than one region (table 8.4, figs. 8.2 and 8.5). The apparent restriction of many species to the same region (D) as their ancestors may reflect the bias of sampling concentrated in the interval from 8 to 4 Ma in the Dominican Republic, but the few known earlier Neogene occurrences show the same degree of geographic restriction of species even though sampling extends from the Gulf of Mexico to Trinidad (table 8.4). These patterns suggest that speciation in *Metrarabdotos* may involve differentiation of populations on more localized geographic scales than for *Stylopoma*. Such a difference in the geographic scale of speciation is consistent with the greater likelihood of dispersal by rafting in the exclusively encrusting *Stylopoma* than in the predominantly erect *Metrarabdotos*.

TIMING AND RATES OF SPECIATION AND EXTINCTION IN STYLOPOMA AND METRARABDOTOS

We compared the timing and rates of origination and extinction of species of *Stylopoma* and *Metrarabdotos* using 1 to 2 my intervals, except for the earliest interval represented by the least detailed sampling (7.4 my; table 8.6). Species origins were counted at the beginning of 95% confidence intervals. For *Stylopoma*, we also counted origins at the levels implied by the modified cladistic tree (fig. 8.2). We followed Gilinsky and Bambach (1987) for calculation of rates of origination and extinction.

The major radiations of both *Stylopoma* and *Metrarabdotos* seem to have been concentrated in the Late Miocene interval from 10 to 7 Ma, when rates of origination peaked with the appearance of nearly half the species in each genus (table 8.6, figs. 8.8B and 8.9B). Sampling bias may make this concentration more apparent than real for *Stylopoma*, but in *Metrarabdotos* all species extant before 8 Ma are different from the ones produced by the Late Miocene radiation.

Epitomizing the ascendancy of encrusting over arborescent taxa in the Neogene, *Stylopoma* species have increased in numbers without decline, while those of *Metrarabdotos* have declined steadily since the Late Miocene peak (figs. 8.8A and 8.9A). At least by the beginning of the Pliocene, *Stylopoma* appears to have overtaken *Metrarabdotos* in number of species. However, *Metrarabdotos* continued to predominate in abundance into the Pliocene, and rates of speciation in the two genera were comparable (figs. 8.8B and 8.9B). Indeed, the number of species in *Stylopoma* may have exceeded that in *Metrarabdotos* even earlier in the Miocene, if species ranges are extended as implied by the cladogram (fig. 8.8A). However, the rate of speciation based on cladistic ranges shows an earlier decline in speciation rate than that based on stratigraphic ranges (fig. 8.8B).

Before about 8 Ma, the speciation rate in *Metrarabdotos* apparently exceeded that in *Stylopoma*, perhaps reflecting the distinctly higher rates

Table 8.6 Numbers of Species and Rates of Origination and Extinction in Tropical American Neogene and Quaternary *Stylopoma* and *Metrarabdotos*

Interval (Ma)	Stylopoma					Metrarabdotos				
	N_{pi}	N_{oi}	N_{ei}	R_{oi}	R_{ei}	N_{pi}	N_{oi}	N_{ei}	R_{oi}	R_{ei}
2.0–0.0	18 (18)	6 (4)	0 (0)	0.17 (0.11)	0 (0)	4	1	1	0.12	0.12
3.4–2.0	12 (14)	2 (1)	0 (0)	0.12 (0.05)	0 (0)	6	1	2	0.12	0.24
4.4–3.4	10 (13)	0 (0)	0 (0)	0 (0)	0 (0)	7	1	2	0.14	0.29
5.6–4.4	10 (13)	0 (0)	0 (0)	0 (0)	0 (0)	9	0	3	0	0.28
6.8–5.6	10 (13)	3 (1)	0 (0)	0.08 (0.06)	0 (0)	10	1	1	0.08	0.08
8.0–6.8	10 (13)	4 (0)	1 (1)	0.33 (0)	0.08 (0.06)	10	3	1	0.28	0.08
10.0–8.0	6 (13)	4 (9)	0 (0)	0.33 (0.35)	0 (0)	8	8	1	0.50	0.06
17.4–10.0	4 (6)	4 (6)	2 (2)	0.14 (0.14)	0.07 (0.05)	3	3	3	0.23	0.23

Note: N_{pi} = number of species present; R_{oi} = rate of origination; R_{ei} = rate of extinction.
$R_{oi} = N_{oi}/(N_{pi}D_i)$ and $R_{ei} = N_{ei}/(N_{pi}D_i)$, where N_{oi} and N_{ei} are the numbers of species with first and last occurrences, respectively, in interval i; and D_i is the duration of the interval (Gilinsky and Bambach 1987). Values in parentheses for *Stylopoma* are based on ranges implied by the cladogram.

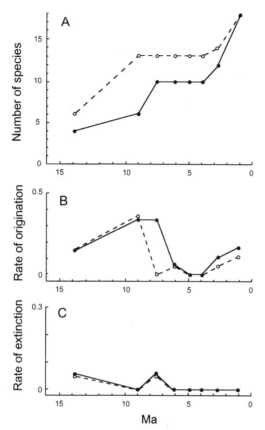

Fig. 8.8 Neogene and Quaternary patterns of species numbers (A), originations (B), and extinctions (C) in *Stylopoma* based on stratigraphic ranges (filled circles, solid lines) and ranges implied by cladogram (unfilled circles, dashed lines). Points are plotted at medians of time intervals in table 8.6

of morphologic divergence and greater overall morphologic distances between *Metrarabdotos* species (Cheetham and Jackson 1995). Nevertheless, patterns of change in speciation rate are quite similar in the two genera, with the Late Miocene major rise followed by an equally profound decline, and then a less profound rebound from mid-Pliocene time onward (figs. 8.8B and 8.9B). This second, minor wave of radiation appears to have started in *Metrarabdotos* (5 to 4 Ma) slightly earlier than in *Stylopoma* (4 to 3 Ma). The similarity of speciation patterns makes it very likely that differences in species numbers are the result of differences in rates of extinction.

Patterns of extinction are extremely different in the two genera. In

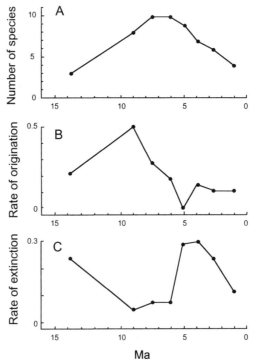

Fig. 8.9 Neogene and Quaternary patterns of species numbers (A), originations (B), and extinctions (C) in *Metrarabdotos*. Points are plotted at medians of time intervals in table 8.6.

Stylopoma extinction rates are low, with only a slight rise between 8 and 7 Ma; since the latest Miocene (6 Ma) extinction has dropped to zero (fig. 8.8C). In contrast, extinction rates in *Metrarabdotos* (fig. 8.9C) were initially more than twice those of *Stylopoma*, fell to a level about equal to the highest *Stylopoma* rates in the Late Miocene interval from 10 to 6 Ma, then increased more than threefold between 6 and 5 Ma, peaking just as extinction in *Stylopoma* dropped to zero. Through the Pliocene and Quaternary, *Metrarabdotos* extinction again declined, although still exceeding the highest rates in *Stylopoma*. Thus, extinction rates differ between genera both in amount and timing.

TIMING OF CHEILOSTOME EVOLUTION AND CLOSURE OF THE PANAMANIAN PORTAL

The patterns of speciation and extinction in *Stylopoma* and *Metrarabdotos* are not simply the outcome of closure of the Isthmus of Panama, but they do appear to be related to changes in oceanic and benthic circulation associated with tectonic events in the isthmian region (Keigwin 1982; Duque-Caro 1990; Coates et al. 1992; Coates and Obando, this volume, chap. 2; Cronin and Dowsett, this volume, chap. 4; L. Collins, this volume, chap. 6). The Late Miocene peaks in speciation rates occurred at least 4.5 my before final closure of the isthmus at about 3.5 Ma (Coates et al. 1992) but lie within the interval between initial disruption of circulation by uplift of the Panamanian sill at 13 to 12 Ma and a return to common circulation between the Caribbean and Pacific at 7 to 6 Ma (Duque-Caro 1990). Whatever these changes may have done to stimulate speciation, the effects were the same for both *Stylopoma* and *Metrarabdotos* and thus could have little to do directly with the decline in either the numbers or abundance of arborescent cheilostome species relative to encrusting ones.

In contrast, the striking difference in extinction rates of these genera at the close of the Miocene does appear to bear directly on the decline in numbers of arborescent species. The rapid increase in *Metrarabdotos* extinction between 6 and 5 Ma also occurred well before final closure of the isthmus, but it does coincide with renewed disruption of circulation and changes in Caribbean oceanographic conditions as the Panamanian sill underwent final uplift leading to closure (Duque-Caro 1990; L. Collins, this volume, chap. 6; Teranes et al., this volume, chap. 5). Onset of high *Metrarabdotos* extinction also preceded that of Caribbean corals, mollusks, and planktonic foraminifers by 1 to 3 my (Stanley 1986; Allmon et al. 1993; Jackson et al. 1993; Budd et al., this volume, chap. 7).

The second, minor wave of radiation in both genera occurred at or near the culmination of these disruptive changes (beginning at 5 to 4 Ma in *Metrarabdotos* and 4 to 3 Ma in *Stylopoma*). Unlike the earlier radiation, this one produced only encrusting species in both genera. One of the new *Metrarabdotos* species (*unguiculatum*) achieved the widest distribution in the genus, which is consistent with the hypothesis that rafting is more likely among encrusting taxa. The failure of any more than one of the three major *Metrarabdotos* subclades extant at that time to produce encrusting species almost surely contributed to the rapid decline of the genus. The contrastingly flourishing history of *Stylopoma* at that time may be related as much to its ability to produce new species across several subclades as to its low extinction rate.

A factor possibly bearing on the decline in abundance of arborescent species of *Metrarabdotos* and other cheilostomes, and thus to their increasing extinction rate, is their association with seagrasses (fig. 8.10). Identifiable patterns of seagrass cells are preserved on 75% of all attach-

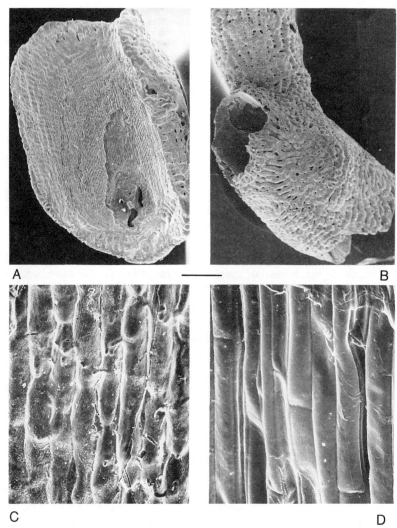

Fig. 8.10 Association of Neogene *Metrarabdotos* with seagrass. Scale bar = 0.5 mm (A and B) or 0.03 mm (C and D). (A, B) attachment bases of two colonies of *Metrarabdotos colligatum* (?), in which pattern (A) and shape (B) of seagrass rhizome are preserved, Cercado Formation (NN11), Arroyo Bajon Bluff, Rio Mao, Dominican Republic (locality NMB16918, Saunders et al. 1986); (C) detail of A to show pattern of seagrass cells; (D) pattern of cells on dried rhizome of living *Halodule wrightii*, Andros Island, Bahamas, showing similarity in size and shape to C.

ment bases of colonies (n = 125) of seven *Metrarabdotos* species (*colligatum, auriculatum,* and new species 5, 6, 7, 9, and 10), ranging through the Dominican Republic Miocene-Pliocene sections from 8 to 3 Ma. Such patterns are also present on bases of three other arborescent genera (*Adeonellopsis, Cigclisula,* and *Gemelliporella*) in the same sections. Direct and indirect evidence reviewed by Ivany and others (1990) suggests that seagrass communities were more extensive in the Caribbean and adjacent regions before closure of the isthmus.

Seagrasses are known to be sensitive to changes in temperature, water clarity, nutrients, and other factors related to oceanographic patterns (Dennison 1987; Larkum and den Hartog 1989; Ivany et al. 1990). At water depths greater than 20 m to which living species of *Metrarabdotos* are restricted in the Caribbean, seagrass communities would be at their lower limits and thus the most sensitive (Dennison 1987). The impressions preserved on the Neogene cheilostomes are similar to modern *Halodule* (fig. 10D), species of which are known from greater depths than typically stenobiont tropical seagrasses (Phillips and Meñez 1988). The decline of seagrass communities thus might have had a direct effect on reducing both the abundance of arborescent species of *Metrarabdotos* and other cheilostomes, and their numbers as well. A linkage with seagrass decline is also possible for coral extinction. Pliocene coral genera that have living species associated with seagrasses suffered higher rates of extinction than those not associated with seagrasses (Budd et al., this volume, chap. 7). Many of these corals are from the same sections in the Dominican Republic as arborescent species of *Metrarabdotos.*

SUMMARY

The Late Neogene decline in abundance and number of species of arborescent cheilostome bryozoans in tropical America and their replacement by encrusting species is well illustrated by the contrasting histories of *Metrarabdotos* and *Stylopoma.* Both clades underwent a strong pulse of speciation in mid–Late Miocene time and a second wave of more modest origination beginning in the Early Pliocene. The later radiation produced only encrusting species in both genera, thus contributing to the relative decline in importance of arborescent growth. However, the major impetus for the change in morphology was the high extinction rate in predominantly arborescent *Metrarabdotos* compared to exclusively encrusting *Stylopoma.* Moreover, *Metrarabdotos* extinction increased dramatically at the close of the Miocene, just as that of *Stylopoma* dropped to zero. Thus, many species of *Stylopoma* have long durations, and some are very widely distributed, apparently with broad thermal tolerance. *Metrarabdotos* species have shorter durations and smaller geographic ranges confined to the tropics. Reduction in abundance and increasing extinction in *Metrarabdotos* may be linked to the Pliocene-Pleistocene decline of Caribbean seagrasses on which many of the Miocene-Pliocene arborescent species grew. Compared with other Caribbean marine taxa, the pattern of evolution in *Metrarabdotos* most closely resembles those of corals inhabiting seagrass beds (high extinction through most of the Pliocene) and strombinid gastropods (striking shifts in morphology and ecology of the few Quaternary survivors). In contrast, *Stylopoma* resembles benthic foraminifers in its virtual lack of Late Miocene–Pliocene extinction. *Stylopoma* and encrusting species of *Metrarabdotos* share an increased Late Pliocene–Quaternary speciation rate with planktonic foraminifers, reef corals, and mollusks.

ACKNOWLEDGMENTS

We thank JoAnn Sanner for measuring specimens and providing much other help with the bryozoans and seagrasses; Yira Ventocilla for processing samples; Peter Jung, John Saunders, Emily Vokes, Kevin Schindler, Roger Cuffey, and Laurel Collins for material and stratigraphic documentation; Amalia Herrera, Javier Jara, David West, Steve Miller, Julio Calderon, and Marcos Sorriano for help with Recent specimens; Francis Hueber for help with seagrass identification; Lee-Ann Hayek for statistical advice; Ann Budd, Dana Geary, and Tony Coates for helpful comments on the manuscript; and the Kuna nation and Government of Panama for permitting work in the San Blas. This work was supported by the Smithsonian Institution Scholarly Studies Program, the Marie Bohrn Abbott Fund of the National Museum of Natural History, and the Smithsonian Tropical Research Institute. Figure 8.2

REFERENCES

Allmon, W. D., G. Rosenberg, R. W. Portell, and K. S. Schindler. 1993. Diversity of Atlantic Coastal Plain mollusks since the Pliocene. *Science* 260: 1626–29.

Banta, W. C. 1991. The Bryozoa of the Galápagos. In *Galápagos marine invertebrates*, ed. M. J. James, 371–89. New York: Plenum Press.

Cadée, G. C. 1975. Lunulitiform Bryozoa from the Guyana Shelf. *Neth. J. Sea Res.* 9:320–43.

Canu, F., and R. S. Bassler. 1920. North American Early Tertiary Bryozoa. *U.S. Nat. Mus. Bull.* 106:1–879.

————. 1923. North American later Tertiary and Quaternary Bryozoa. *U.S. Nat. Mus. Bull.* 125:1–302.

————. 1928. Fossil and Recent Bryozoa of the Gulf of Mexico region. *Proc. U.S. Nat. Mus.* 72(14):1–199.

Cheetham, A. H. 1963. *Late Eocene zoogeography of the eastern Gulf Coast region.* Geological Society of America Memoir 91. Boulder, Colo.

————. 1967. Paleoclimatic significance of the bryozoan *Metrarabdotos. Trans. Gulf Coast Assoc. Geol. Soc.* 17:400–407.

————. 1968. *Morphology and systematics of the bryozoan genus Metrarabdotos.* Smithsonian Miscellaneous Collections, vol. 153, no. 1. Washington, D.C.

————. 1971. *Functional morphology and biofacies distribution of cheilostome Bryozoa in the Danian Stage (Paleocene) of southern Scandinavia.* Smithsonian Contributions to Paleobiology, no. 6. Washington, D.C.

————. 1973. Study of cheilostome polymorphism using principal components analysis. In *Living and fossil Bryozoa,* ed. G. P. Larwood, 385–409. London: Academic Press.

————. 1986a. Branching, biomechanics, and bryozoan evolution. *Proc. Roy. Soc., Lond.* B-228:151–71.

————. 1986b. Tempo of evolution in a Neogene bryozoan: Rates of morphologic change within and across species boundaries. *Paleobiology* 12:190–202.

————. 1987. Tempo of evolution in a Neogene bryozoan: Are trends in single morphologic characters misleading? *Paleobiology* 13:286–96.

Cheetham, A. H., and J. B. C. Jackson. 1995. Process from pattern: Tests for selection versus random change in punctuated bryozoan speciation. In *New approaches to speciation in the fossil record,* ed. D. H. Erwin and R. L. Anstey, 184–207. New York: Columbia University Press.

Coates, A. G., J. B. C. Jackson, L. S. Collins, T. M. Cronin, H. J. Dowsett, L. M. Bybell, P. Jung, and J. A. Obando. 1992. Closure of the Isthmus of Panama: The near-shore marine record of Costa Rica and western Panama. *Geol. Soc. Am. Bull.* 104:814–28.

Cook, P. L. 1986. *Bryozoa from Ghana—a preliminary survey.* Koninklijk Museum voor Midden-Afrika Zoologische Wetenschappen, Annals 238. Teuvren, Belgium.

Dennison, W. C. 1987. Effects of light on seagrass photosynthesis, growth, and depth distribution. *Aquat. Bot.* 27:15–26.

Duque-Caro, H. 1990. Neogene stratigraphy, paleoceanography, and paleobiogeography in northwest South America and the evolution of the Panama seaway. *Palaeogeogr., Palaeoclimatol., Palaeoecol.* 77:203–34.

Gilinsky, N., and R. Bambach. 1987. Asymmetrical patterns of origination and extinction in higher taxa. *Paleobiology* 13:427–45.

Harmelin, J. -G. 1976. *Le sous-ordre des Tubuliporina (Bryozoaires, Cyclostomes) en Méditerranée: Écologie et systématique.* Mémoires de l'Institute Océanographique Monaco 10. Monaco.

Harmer, S. F. 1957. The Polyzoa of the Siboga Expedition. Part 4, Cheilostomata Ascophora. *Siboga-Exped.* 28:641–1147.

Herrera Anduaga, Y. 1985. Pliocene Cheilostomata in the Isthmus of Tehuantepec region (Mexico). In *Bryozoa: Ordovician to Recent*, ed. C. Nielsen and G. P. Larwood, 145–51. Fredensborg, Den.: Olsen and Olsen.

Hughes, D., and J. B. C. Jackson. 1992. Distribution and abundance of cheilostome bryozoans on the Caribbean reefs of central Panama. *Bull. Mar. Sci.* 51:443–65.

Ivany, L. C., R. W. Portell, and D. S. Jones. 1990. Animal-plant relationships and paleobiogeography of an Eocene seagrass community from Florida. *Palaios* 5:244–58.

Jackson, J. B. C. 1984. Ecology of cryptic coral reef communities. Part 3, Abundance and aggregation of encrusting organisms with particular reference to cheilostome Bryozoa. *J. Exper. Mar. Biol. Ecol.* 75:37–57.

———. 1986. Models of dispersal of clonal benthic invertebrates: Consequences for species' distributions and genetic structure of local populations. *Bull. Mar. Sci.* 39:588–606.

Jackson, J. B. C., and A. H. Cheetham. 1990. Evolutionary significance of morphospecies: A test with cheilostome Bryozoa. *Science* 248:521–636.

———. 1994. Phylogeny reconstruction and the tempo of speciation in cheilostome Bryozoa. *Paleobiology* 20:407–23.

Jackson, J. B. C., P. Jung, A. G. Coates, and L. S. Collins. 1993. Diversity and extinction of tropical American mollusks and emergence of the Isthmus of Panama. *Science* 260:1624–26.

Jackson, J. B. C., and F. K. McKinney. 1990. Ecological processes and progressive macroevolution of marine clonal benthos. In *Causes of evolution: A paleontological perspective*, ed. R. M. Ross and W. D. Allmon, 173–209. Chicago: University of Chicago Press.

Jackson, J. B. C., J. E. Winston, and A. G. Coates. 1985. Niche breadth, geographic range, and extinction of Caribbean reef-associated cheilostome Bryozoa and Scleractinia. *Proc. Fifth Int. Coral Reef Cong.* 4:151–58.

Keigwin, L. D., Jr. 1982. Isotopic paleoceanography of the Caribbean and east Pacific: Role of Panama uplift in Late Neogene time. *Science* 217:350–52.

Lagaaij, R. 1963. New additions to the bryozoan fauna of the Gulf of Mexico. *Inst. Mar. Sci., Tex.* 9:162–36.

Lagaaij, R., and P. L. Cook. 1973. Some Tertiary to Recent Bryozoa. In *Atlas of palaeobiogeography*, ed. A. Hallam, 489–98. Amsterdam: Elsevier.

Lagaaij, R., and Y. V. Gautier. 1965. Bryozoan assemblages from marine sediments of the Rhone Delta, France. *Micropaleontology* 11:39–58.

Larkum, A. W. D., and C. den Hartog. 1989. Evolution and biogeography of seagrasses. In *Biology of seagrasses*, ed. A. W. D. Larkum, A. J. McComb, and S. A. Shepherd, 112–56. Amsterdam: Elsevier.

Marcus, E., and E. Marcus. 1962. On some lunulitiform Bryozoa. *Boll. Fac. Fil., Cien., Let. Univ. São Paulo* 261, *Zool.* 24:281–24.

Marshall, C. R. 1990. Confidence intervals on stratigraphic ranges. *Paleobiology* 16:1–10.

———. 1991. Estimation of taxonomic ranges from the fossil record. In *Analytical paleobiology*, ed. N. L. Gilinsky and P. L. Signor, 19–38. Knoxville, Tenn.: Paleontological Society.

————. 1995. Stratigraphy, the true order of species' originations and extinctions, and testing ancestor-descendant hypotheses among Caribbean Neogene bryozoans. In *New approaches to speciation in the fossil record*, ed. D. H. Erwin and R. L. Anstey, 207–35. New York: Columbia University Press.

Maturo, F. J. S., Jr. 1968. The distributional pattern of the Bryozoa on the East Coast of the United States exclusive of New England. *Atti Soc. Ital. Sci. Nat. Mus. Civ. Stor. Nat. Milan.* 108:261–84.

Mayr, E. 1954. Geographic speciation in tropical echinoids. *Evolution* 8:1–18.

Mayr, E. 1963. *Animal species and evolution.* Cambridge: Harvard University Press.

McKinney, F. K., and J. B. C. Jackson. 1989. *Bryozoan evolution.* Boston: Unwin Hyman.

Phillips, R. C., and E. G. Meñez. 1988. *Seagrasses.* Smithsonian Contributions to Marine Science, no. 34. Washington, D.C.

Saunders, J. B., P. Jung, and B. Biju-Duval. 1986. Neogene paleontology in the northern Dominican Republic. Part 1, Field surveys, lithology, and age. *Bulls. Am. Paleontol.* 89:1–79.

Saunders, J. B., P. Jung, J. Geister, and B. Biju-Duval. 1982. The Neogene of the south flank of the Cibao Valley, Dominican Republic: a stratigraphic study. *Trans. Ninth Carib. Geol. Conf.* 1:151–60.

Stanley, S. M. 1986. Anatomy of a regional mass extinction: Plio-Pleistocene decimation of the western Atlantic bivalve fauna. *Palaios* 1:17–30.

Winston, J. E. 1986. An annotated checklist of coral-associated bryozoans. *Am. Mus. Nov.*, no. 2859:1–39.

Winston, J. E., and E. Håkansson. 1986. The interstitial bryozoan fauna from Capron Shoal, Florida. *Am. Mus. Nov.*, no. 2865:1–50.

Winston, J. E., and J. B. C. Jackson. 1984. Ecology of cryptic coral reef communities. Part 4, Community development and life histories of encrusting cheilostome Bryozoa. *J. Exper. Mar. Biol. Ecol.* 76:1–21.

9

Paciphilia Revisited: Transisthmian Evolution of the *Strombina* Group (Gastropoda: Columbellidae)

Jeremy B. C. Jackson, Peter Jung, and Helena Fortunato

The Pliocene emergence of the Isthmus of Panama separated the tropical American ocean into two now strikingly different realms (Birkeland 1977, 1987; Vermeij 1978). Coastal regions of the tropical eastern Pacific have much stronger tides, seasonality, local upwelling, and El Niño/Southern Oscillation (ENSO) climate anomalies than the Caribbean (Rubinoff 1968; Glynn 1988; Cubit et al. 1989; D'Croz et al. 1991). Moreover, planktonic biomass and primary production are higher, predation more intense, and coral reef development depauperate in the eastern Pacific compared to the Caribbean (Bayer et al. 1970; Birkeland 1977, 1987; Vermeij 1978; Glynn and Colgan 1991). Isolation and environmental change also resulted in increased evolutionary divergence of species across the developing isthmus (Rubinoff 1968; Lessios 1981; Knowlton et al. 1993; T. Collins, this volume, chap. 11) and shifts in transisthmian diversity due to differential extinction and radiation (Woodring 1966; Vermeij 1978; Jones and Hasson 1985; Vermeij and Petuch 1986). However, the magnitude and chronology of these events are poorly known (Jackson et al. 1993; Vermeij 1993).

Woodring (1966) coined the term paciphile, or "Pacific loving," to describe fossil marine taxa that occurred abundantly throughout tropical American during the Miocene but were subsequently largely or entirely restricted to the eastern Pacific. Gastropods of the *Strombina* group (hereafter referred to as strombinids) are among the most common and diverse of these paciphile taxa (Vermeij 1978). Strombinids comprise more than one hundred species entirely restricted to tropical and subtropical America (Jung 1989). They are abundant fossils in Neogene and Pleistocene deposits, and are common today in the tropical eastern Pacific but rare in the Caribbean.

Fossil and living strombinids have been sampled extensively throughout tropical America and their taxonomy recently revised (Jung 1989). Thus, they provide a model system to document geographic and temporal changes in species composition, morphology, and diversity of a large paciphile clade, as well as a baseline for evaluation of the possible influence of changing oceanographic conditions versus such factors as predation on that history (Vermeij 1978; Jackson et al. 1993). In this paper we describe patterns of strombinid evolution at the species level and apparent adaptive trends. The phylogeny of strombinids is still under investigation, and will be treated elsewhere.

THE STROMBINA GROUP

The Columbellidae are a large, cosmopolitan family of mostly small buccinacean gastropods dating from the Eocene (Keen 1971; Radwin 1977). The five exclusively tropical American genera of the strombinids, sensu Jung 1989, include most taxa classified as *Strombina* by earlier workers. The relationship of these genera to other American columbellids and to each other is poorly known (Radwin 1977; Jung 1989), and the entire group requires extensive systematic revision (de Maintenon 1994). Nevertheless, cladistic analysis based on shell morphology and anatomy strongly support the classification of the strombinids as a separate columbellid clade, as well as the validity of Jung's (1989) major genera and subgenera (Fortunato and Jung 1995).

The three principal strombinid genera are *Sincola* Olsson and Harbison (1953), *Strombina* Morch (1852), and *Cotonopsis* Olsson (1942) (fig. 9.1, table 9.1). Species of *Sincola* are generally small and squat, with strong apertural armor and moderate sculpture; the four subgenera of *Sincola* were most diverse in the Miocene and Early Pliocene and declined thereafter. The six subgenera of *Strombina* are generally larger and more elongate with similar armor and sculpture. *Strombina (Strombina)* was most diverse from the Early Miocene through Early Pliocene, but was largely replaced by the other five subgenera from the Late Pliocene to the present. Species of *Cotonopsis* are large and elongate, but weakly armored and sculptured; the two subgenera are exclusively Pliocene to Recent with the exception of one species of questionable affinity from the Middle Miocene. The remaining two strombinid genera (fig. 9.1, table 9.1), *Clavistrombina* Jung (1989) and *Bifurcium* Fischer (1884), most closely resemble *Sincola* and *Cotonopsis* respectively.

Living strombinids occur principally on sandy silts and muds from the shoreline to less than 200 m deep, but have been dredged as deep as 430 m (Jung 1989). There are few observations of life habits or behavior. Two Caribbean species, *Strombina (Strombina) pumilio* and *S. (Lirastrombina) francesae*, were observed buried or crawling in sand (Cipriani and Penchaszadeh 1993). Both of these species lay their large eggs in a

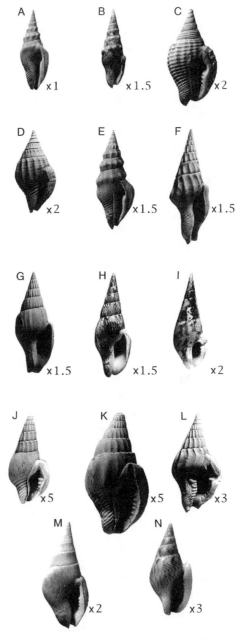

Fig. 9.1 Portraits of the *Strombina* group taken from Jung 1989: (A) *Strombina (Strombina) lanceolata*, (B) *Strombina (Spiralta) maculosa*, (C) *Strombina (Lirastrombina) pulcherrima*, (D) *Strombina (Arayina) arayana*, (E) *Strombina (Recurvina) recurva*, (F) *Strombina (Costangula) angularis*, (G) *Cotonopsis (Cotonopsis) panacostaricensis*, (H) *Cotonopsis (Turrina) turrita*, (I) *Clavistrombina clavulus*, (J) *Sincola (Sincola) sincola*, (K) *Sincola (Sinaxila) bassi*, (L) *Sincola (Sinuina) sinuata*, (M) *Sincola (Dorsina) dorsata*, (N) *Bifurcium bicanaliferum*.

Table 9.1 Strombinid Genera and Subgenera

Strombina	*Strombina*	Morch 1852
Strombina	*Spiralta*	Jung 1986
Strombina	*Lirastrombina*	Jung 1989
Strombina	*Arayina*	Jung 1989
Strombina	*Recurvina*	Jung 1989
Strombina	*Costangula*	Jung 1989
Cotonopsis	*Cotonopsis*	Olsson 1942
Cotonopsis	*Turrina*	Jung 1989
Clavistrombina		Jung 1989
Sincola	*Sincola*	Olsson and Harbison 1953
Sincola	*Sinaxila*	Jung 1989
Sincola	*Sinuina*	Jung 1889
Sincola	*Dorsina*	Jung 1989
Bifurcium		Fischer 1884

Source: Jung 1989

single layer on the body whorl of the shell, where they develop directly into crawling juveniles without a free-living larval stage. Egg masses were also found on shells of *Sincola (Dorsina) gibberula, Strombina (Spiralta) elegans* and *Bifurcium canae* from the Bay of Panama and Gulf of Chiriquí, but these species have smaller eggs and planktonic development (Fortunato 1995, unpublished data; Fortunato et al. 1995). Drill holes and shell repair marks are common on fossil and living species from both oceans (Houbrick 1983; illustrations in Jung 1989).

DIVERSITY IN TIME AND SPACE

Our data include all strombinids reported by Jung (1989) plus 1626 fossils of 33 species from extensive new collections in Panama and Costa Rica (Coates et al. 1991; Jackson et al. 1993) that were also identified by Jung. There are 110 species. The stratigraphic and geographic occurrence of each species is listed in appendix 9.1. Stratigraphic resolution varies greatly among regions, so we recorded occurrences to epoch and stage as follows: Early Miocene (EM); Middle Miocene (MM); Late Miocene (LM); Early Pliocene (EP); Late Pliocene (LP); late Late Pliocene (after 1.9 Ma) and Pleistocene combined, hereafter referred to as Plio-Pleistocene (PP); and Recent (R). In cases of uncertain age assignment (e.g., Early or Middle Miocene), we recorded a species as present in each interval. This conservative procedure, along with including fossil taxa from poorly dated sites and our own new collections, is the rea-

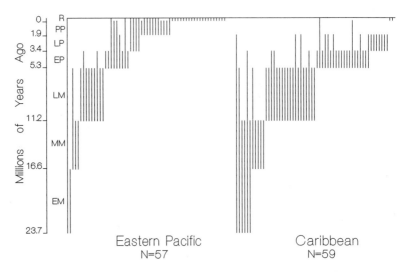

Fig. 9.2 Stratigraphic ranges for eastern Pacific and Caribbean strombinids: R = Recent; PP = Plio-Pleistocene; LP = Late Pliocene; EP = Early Pliocene; LM = Late Miocene; MM = Middle Miocene; and EM = Early Miocene.

son for the different numbers of taxa than reported previously (Jackson et al. 1993). In addition, collections from the eastern Pacific Armuelles and Montezuma Formations and the Lomas del Mar "member" of the Caribbean Moin Formation, which straddle the Pliocene-Pleistocene boundary (Coates et al. 1991, unpublished data), were combined with the Pleistocene because of their strong faunal affinity and striking differences from older Late Pliocene material.

Because of these stratigraphic uncertainties, we did not attempt to calculate ranges of species in millions of years, but instead counted the numbers of stratigraphic intervals occupied as a rough measure of species duration. This has the disadvantage of counting longer (EM, MM, and LM) and shorter (EP, LP, PP) intervals equally, but is an honest reflection of the uncertainties in the data, especially for the older intervals.

The resulting stratigraphic ranges of each species from the eastern Pacific and Caribbean are illustrated in fig. 9.2. The simultaneous origination and extinction of taxa at stage boundaries are artifacts of the coarse scale of analysis which extends species durations to the full length of each sampling interval. Nevertheless, strombinid species were clearly short-lived, with 76 of 110 species recorded from only a single stratigraphic interval (mean and standard error 1.5 ± 0.1 intervals per species).

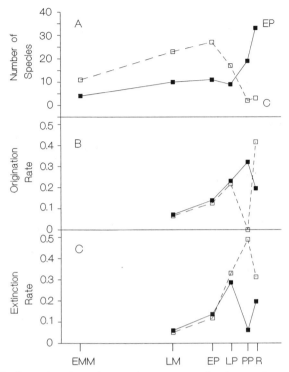

Fig. 9.3 Numbers of species and rates of origination and extinction for strombinids in the eastern Pacific (EP: solid squares) and Caribbean (C: open squares). Origination rate calculated as number of species that first appear at time *t*, divided by the total species at *t*, divided by the time between *t* and *t* − 1 in millions of years. The extinction rate equals the number of species present at *t* − 1 that are absent at *t* and thereafter, divided by the total species at *t* − 1, divided by the time between *t* and *t* − 1 in millions of years. Ages as in figure 9.2 except EMM is Early and Middle Miocene combined because of the small numbers of species in each interval.

Diversity rose slowly in the Caribbean to an Early Pliocene peak of 27 species and then declined to only 3 Plio-Pleistocene and Recent species (fig. 9.3A). In contrast, numbers of eastern Pacific species hardly changed until they nearly doubled in the Plio-Pleistocene. The difference in diversity between the oceans is due to the simultaneous decline in origination rate and rise in extinction rate of Caribbean strombinids in the Plio-Pleistocene, and the opposite pattern in the eastern Pacific (fig. 9.3B,C). This is part of the Plio-Pleistocene turnover event described for Caribbean mollusks in Panama and Costa Rica (Jackson et al. 1993; Robinson 1993).

The Plio-Pleistocene reversal in eastern Pacific and Caribbean

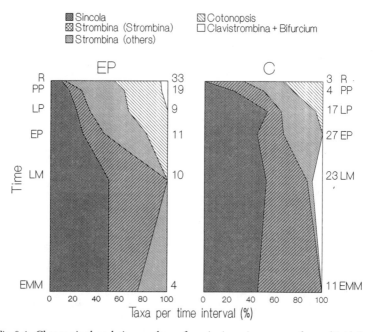

Fig. 9.4 Changes in the relative numbers of species in major groups of strombinids in the eastern Pacific (EP) and Caribbean (C). Numbers of species for each interval indicated at right of boxes for each ocean. Ages as in figure 9.3. The assignment of four species to the Pleistocene of the Caribbean is based on the assumption that *Cotonopsis* and *Strombina* (*Lirastrombina*) must have been represented by at least one species, since each was present in the Late Pliocene and the Recent of the Caribbean (app. 9.1), and communication with the eastern Pacific was by then impossible.

strombinid diversity was preceded by a much more gradual shift in generic and subgeneric composition of the fauna (fig. 9.4). Beginning in the Early Pliocene in the eastern Pacific and the Late Pliocene in the Caribbean, *Sincola* and *Strombina* (*Strombina*) were progressively replaced by *Cotonopsis* and the other five subgenera of *Strombina*.

The geographic density of sampling is uneven. We therefore divided tropical America into eleven regions (fig. 9.5), based wherever possible on unifying geographic features (e.g., A = Gulf of Mexico, C = Nicaragua Rise; see Cheetham and Jackson, this volume, chap. 8), but boundaries are necessarily arbitrary and the length of coastline per region varies about twofold. Nevertheless, strombinid species are narrowly distributed, with 75 species known from only a single geographic region (mean and standard error 1.6 ± 0.1 regions per species).

Somewhat surprisingly, given the generally narrow distributions in

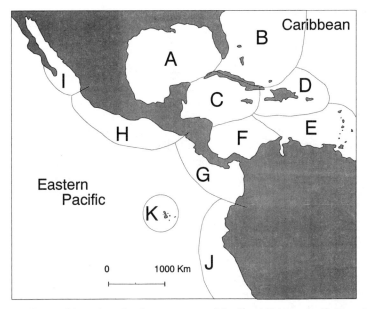

Fig. 9.5 Geographic regions for the occurrence of fossil and living strombinids, arbitrarily defined on the basis of unifying geographic features and approximate size.

time and space and unequal sampling units, there is still a significant positive correlation between species longevity measured in intervals and geographic range measured in regions ($r^2 = 0.37$, $p < .001$), as has been observed previously for many taxa (Jackson 1974; Jablonski et al. 1985). Moreover, this pattern is apparently unbiased by the relative abundance of species. Some of the most abundant species are restricted to a single interval or region (Jung 1989), and both stratigraphic and geographic range are virtually uncorrelated with abundance of specimens in our extensive Panamanian and Costa Rican collections ($r^2 = .06$ and $-.01$, $p = .70$ and $.95$ respectively). Thus, the short durations, high endemism, and general patterns of strombinid distributions and diversity are probably real.

LARVAL ECOLOGY AND DISTRIBUTIONS

The mode of larval development is correlated with dispersal ability, geographic range, and longevity of a wide variety of marine benthic invertebrates (Thorson 1950; Hansen 1978, 1980; Jablonski and Lutz 1983; Jablonski 1986; Bhaud 1989, 1993; Scheltema 1989). In general,

Table 9.2   Characters of the Larval Shell (Protoconch)

1. Width perpendicular to axis of coiling = D
2. Maximum number of whorls (volutions) = V
3. Ratio = D/V

Note: See appendix 9.2 for data.

planktotrophic species (broadcast feeding larvae that develop in the plankton) are more widely distributed and survive longer before extinction than nonplanktotrophic species. The latter either release nonfeeding lecithotrophic larvae, which drift for a much shorter period in the plankton than do planktotrophic species, or exhibit direct development and lack free-living larvae entirely.

The morphology of the larval shell can be used to infer mode of development for fossil species and the great majority of living species whose larval ecology is unknown (Jablonski and Lutz 1983). Two important characters for this purpose are the maximum number of volutions (V) and maximum diameter (D) of the larval shell, and the ratio of D in millimeters divided by V (D/V). Species with low values of V and high values of D/V are mostly nonplanktotrophic, whereas species with high V and low D/V are generally planktotrophic (Jablonski and Lutz 1983).

We measured D and V as described by Jablonski and Lutz (1983) using the best SEM photographic illustrations with similar orientation of specimens published in Houbrick 1983 and Jung 1989 (table 9.2; data in app. 9.2). Values of D versus V for 54 species including all five genera are plotted in fig. 9.6. The 5 species whose development is known from direct observations (Cipriani and Penchaszadeh 1993; Fortunato 1995, unpublished data; Fortunato et al. 1995) are indicated by larger filled symbols. Both species with direct development have values of $V = 1.5$, whereas they differ in diameter by nearly half the entire range of D for all 54 species. We therefore assigned all species with $V \leq 1.5$ as having direct development (category D in fig. 9.6), regardless of the diameter of the protoconch. In addition, values of $V \geq 3.0$ and $D/V \geq 0.3$ strongly suggest planktotrophic development (Jablonski and Lutz 1983), designated by category P in fig. 9.6. Intermediate values (between the two vertical dashed lines) may be planktotrophic or lecithotrophic (category P/L in fig. 9.6), but we need more data on living strombinids to refine the analysis.

As defined above, the three categories D, P/L, and P in figure 9.6 are

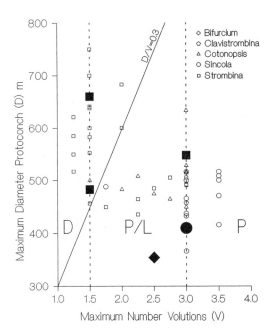

Fig. 9.6   Relation between the maximum diameter (*D*) and maximum number of volutions (*V*) of the larval shell (protoconch). The five species whose development is known independently from direct observations are indicated by large filled symbols. The vertical dashed lines separate the graph into three fields of inferred development based on the maximum number of whorls: D indicates direct development (*V* < 1.5); P/L indicates planktotrophic or lecithotrophic development (1.5 < *V* < 3.0); P indicates planktotrophic development (*V* ≥ 3.0). Solid oblique line designates all values of *D* and *V* for *D/V* = 0.30.

distinguished primarily by the maximum number of volutions *V*, whereas the ratio *D/V* contributes little additional information. Values of *V* alone are available for 19 additional species in appendix 9.2 whose maximum diameters could not be reliably measured from photographs. We added these species to those in fig. 9.6 for a total of 73 species. We then used these data on inferred development to examine the frequency of the three developmental modes among strombinids since the Early Miocene (fig. 9.7), and the possible association of development with stratigraphic or geographic range.

The first strombinids were apparently entirely planktotrophic, but the proportion of nonplanktotrophic species increased thereafter. However, the nature of this increase was very different in the two oceans (fig. 9.7), with possibly lecithotrophic species (the P/L category in fig. 9.6)

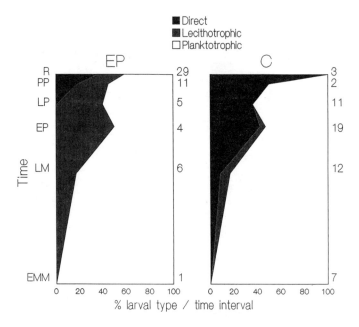

■ Direct
■ Lecithotrophic
□ Planktotrophic

Fig. 9.7 Temporal changes in the relative numbers of species inferred to possess plankto-trophic (P), either planktotrophic or lecithotrophic (P/L), and direct development (D). Data from appendix 9.2 include the fifty-four species in figure 9.6 plus nineteen additional species with data for *V* only that permits their classification as D, P/L, or P. Numbers of species for each age indicated to right of boxes for each ocean.

more frequent in the eastern Pacific but direct development taking over in the Caribbean. Moreover, direct development was common among Caribbean strombinids by the Early Pliocene but did not appear among eastern Pacific species until the Pleistocene.

There was no significant association between inferred mode of de-velopment and geographic ranges for the 73 species in fig. 9.7 (3 × 4 contingency table analysis for inferred developmental mode versus number of geographic regions per species, chi square = 10.3, p = .11). However, the geographic ranges of Recent species can be measured more precisely in kilometers of coastline rather than numbers of re-gions. Geographic ranges for the three inferred developmental modes of these 30 recent species are very different (means and standard errors, D = 1182 ± 574 km, P/L = 2860 ± 577 km, P = 3416 ± 613 km; 1-way ANOVA, F = 3.382, p = .049; multiple range tests, D different from P at p < 0.05).

There was not even a hint of an association, however, between devel-opmental mode and species longevity measured in stratigraphic intervals (3×3 contingency table, chi square $= 3.54$, $p = .47$). This was also true for 31 species from Panama and Costa Rica whose stratigraphic ranges are sufficiently well known to justify using durations in millions of years instead of stratigraphic intervals. Durations of the 22 of these species with a fossil record do not differ among inferred developmental modes (mean durations with standard errors in millions of years, D $= 3.1 \pm 0.8$, L/P $= 2.1 \pm 1.2$, P $= 3.1 \pm 0.7$; 1-way ANOVA, F $= 0.28$, $p = .76$). The absence of association between development and longevity of strombinid species is very different than reported for other gastropods (Jablonski 1986; Scheltema 1989). Thus, other factors such as eurytopy and stenotopy may also have an important influence on their durations and distributions (Jackson 1974; Bhaud 1989, 1993).

SHELL MORPHOLOGY

The size, shape, armor, and sculpture of the postlarval shell (teleoconch) of gastropods may reflect differences in life habit, oceanographic conditions, or the intensity of predation (Vermeij 1978, 1987; Stanley 1988). Among mollusks, for example, large size is believed to be associated with high primary productivity, and strong shell armor or sculpture with the presence of crushing predators.

We measured or coded twenty characters of the teleoconch from photographic illustrations in Jung 1989 (fig. 9.8, table 9.3). Measurements were for the largest illustrated specimen of each species because maximum size is a good measure of the potential growth of a species, and nearly half of the specimens originally examined by Jung (1989) are no longer available for measurements of the entire suite of shell characters, which precludes a population approach (e.g., Budd and Johnson 1991). Use of the largest specimen of a species regardless of age or location should cause little bias since two-thirds of the species are restricted to a single time interval and half to a single region. Moreover, only six Late Miocene to Early Pliocene species occurred in both the Caribbean and eastern Pacific. Because these did not differ obviously in size, we used the same data for these species for both oceans.

The twenty characters were used to define seven composite characters of overall size, shape, spire development, sculpture and apertural size, shape, and armor (table 9.4). Data for the ninety-seven species measured and scored are listed in appendix 9.2. Temporal trends in all seven composite morphological characters were evaluated by 2-way ANOVA with age (stratigraphic intervals) and oceans (EP, C) as factors (table 9.5), and four of these are plotted in fig. 9.9. The most striking

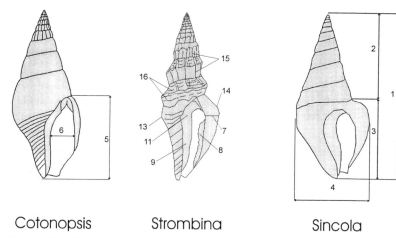

Cotonopsis   Strombina   Sincola

Fig. 9.8 Morphological characters of the shell illustrated for the three most common strombinid genera. For definition of numbered characters, see table 9.3.

Table 9.3 Measurements and Ranked Characters of the Postlarval Shell (Teleoconch)

Size
  1. Height parallel to main axis of coiling
  2. Height of spire whorls
  3. Height of body whorl
  4. Width perpendicular to main axis of coiling
  5. Height shell aperture parallel to axis of coiling
  6. Width shell aperture perpendicular to axis of aperture between two most distant points
Aperture
  7. Thickness outer lip: (0) unthickened, (1) weakly thickened, (2) moderately thickened, (3) prominent
  8. Teeth outer lip: (0) absent, (1) few, (2) numerous
  9. Columelar callus: (0) absent, (1) weak, (2) well developed, (3) prominent
  10. Teeth columelar callus: (0) absent, (1) present
  11. Parietal callus: (0) absent, (1) weak, (2) well developed, (3) prominent
Sculpture
  12. Parietal ridge: (0) absent, (1) weak, (2) well developed, (3) prominent
  13. Hump to left of aperture: (0) absent, (1) weak, (2) well developed, (3) prominent
  14. Dorsal hump: (0) absent, (1) weak, (2) well developed, (3) prominent, (4) several prominent knobs
  15. Vertical sculpture spire: (0) absent, (1) weak, (2) prominent
  16. Horizontal sculpture spire: (0) absent, (1) weak, (2) prominent
  17. Vertical sculpture central part of body whorl: (0) absent, (1) weak, (2) prominent
  18. Horizontal sculpture central part of body whorl: (0) absent, (1) weak, (2) prominent
  19. Vertical sculpture base of body whorl: (0) absent, (1) weak, (2) prominent
  20. Horizontal sculpture base of body whorl: (0) absent, (1) weak, (2) prominent

Table 9.4 Composite Characters of the Teleoconch

1. Size[1] (cm) = $\sqrt{\text{char 1} \times \text{char 4}}$
2. Shape = char 1/char 4
3. Spire = char 2/char 3
4. Size Aperture[1] (cm) = $\sqrt{\text{char 5} \times \text{char 6}}$
5. Shape Aperture = char 5/char 6
6. Aperture Armor[2] = $\dfrac{\Sigma \text{ character numbers 7–11}}{12}$
7. Sculpture[3] = $\dfrac{\Sigma \text{ character numbers 12–20}}{19}$

Note: Character numbers are from table 9.3; data in appendix 9.2.
[1]Geometric mean size
[2]The sum of the ranks is divided by 12 (the maximum possible score from table 9.3) so that values of aperture armor range from 0 to 1.
[3]The sum of the ranks is divided by 19 (the maximum possible score from table 9.3) so that values of sculpture range from 0 to 1.

Table 9.5 Significance of Temporal Trends for Changes in Strombinid Morphology

	Size	Shape	Spire	Size Aperture	Shape Aperture	Aperture Armor	Sculpture
Main Effects							
F	10.56	5.76	3.84	10.29	1.08	2.87	3.49
p	.000	.000	.001	.000	.375	.011	.003
Age							
F	3.09	2.05	3.12	3.35	0.57	1.92	3.84
p	.011	.075	.011	.007	.720	.095	.003
Ocean (EP, C)							
F	13.86	5.18	0.01	12.26	0.81	0.32	0.056
p	.000	.024	.923	.001	.017	.570	.813
Age × Ocean							
F	1.22	1.06	0.50	1.32	0.31	0.39	0.41
p	.305	.385	.775	.258	.906	.855	.844

Note: Morphological characters defined in table 9.4. Boldface indicates significant differences.

differences between the oceans are in total shell and aperture size. However, aperture size scales perfectly to shell size ($r^2 = .96$) with a ratio of aperture size/shell size = .35–.39 for all time intervals in both oceans, and is therefore not considered further. Differences in shell size with time are not significant for Caribbean strombinids (1-way ANOVA, F = 1.47, p = .21), despite the large size of the three surviving species after the Plio-Pleistocene turnover. In contrast, size of eastern Pacific strombinids increased by 61%, mostly during the Pliocene (1-way ANOVA, F = 2.84, p = .02). Thus, the difference between the two

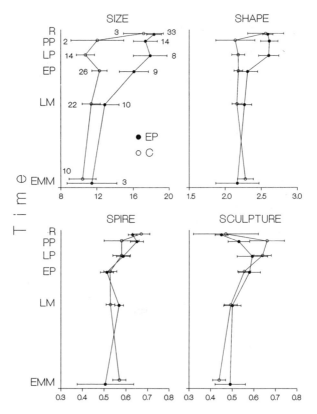

Fig. 9.9 Temporal changes in morphological characters for strombinids in the eastern Pacific (EP) and Caribbean (C). Numbers of species for each interval indicated along side of standard error bars in graph for size. Characters defined in table 9.2; ages as in figure 9.3.

oceans is due to much greater size increase in the Pacific. This geographic difference also explains the weak pattern of strombinid size increase reported by Budd and Johnson (1991), who combined species from both oceans in their analyses.

Shell shape is also different between the oceans owing to the marked increase in height to width of eastern Pacific species between the Early and Late Pliocene. However, this temporal increase is not statistically significant (table 9.5). In contrast, the ratio of spire height to body whorl height increased significantly with time in both oceans after the Early Pliocene. Total sculpture also increased in both oceans to a Late Pliocene–Pleistocene maximum and then declined abruptly. Apertural armor did not change significantly with time.

ENVIRONMENTAL CORRELATES OF STROMBINID EVOLUTION

Strombinid evolution exhibits two general temporal and geographical patterns. The first includes gradual divergence between Caribbean and eastern Pacific faunas that can be somehow attributed to environmental changes across the developing isthmus since the Late Miocene. The second includes the relatively sudden changes that occurred in both oceans at the very end of the Pliocene and therefore must require some more general explanation.

Transisthmian Divergence

Deep-water exchange down to bathyal depths between the Caribbean and eastern Pacific continued until the end of the Miocene or even the Early Pliocene (Keigwin 1982; Hasson and Fischer 1986; Whittaker 1988; Collins and Coates 1993). Shallow seaways connected the Caribbean and eastern Pacific through much of the Early Pliocene (Coates and Obando, this volume, chap. 2), but emergence coupled with declining sea levels (Haq et al. 1987; Matthews 1988) had almost certainly reduced flow to a trickle by 3.5 Ma (Keigwin 1982).

We still know very little about the environmental changes in coastal waters due to these events (L. Collins, this volume, chap. 6). Isotopic evidence from sixteen mollusk shells suggests moderate seasonality and possible nutrient upwelling on both sides of the developing isthmus through the Late Miocene and Early Pliocene, after which seasonality declined in the Caribbean and increased in the eastern Pacific (Teranes et al., this volume, chap. 5). This is consistent with isotopic evidence from eight snail shells showing a sharp decline in upwelling off the west coast of Florida in the Late Pliocene, and with limited evidence from planktonic and benthic foraminifers from the isthmian region (Keller et al. 1989; Duque-Caro 1990).

The decline of *Sincola* and *Strombina* (*Strombina*), the rise of *Cotonopsis* and the new subgenera of *Strombina*, and the general increase in strombinid size all began in the eastern Pacific during the Early Pliocene. Similar changes did not begin in the Caribbean until after the Plio-Pleistocene turnover. However, strikingly divergent increase in nonplanktotrophic larval development began in both oceans during the Early Pliocene. All of these changes precede any evidence for environmental divergence, although they are consistent with what little is known about the geographic correlation of larval development and body size with primary productivity.

Patterns of larval development of recent species reflect regional differences in productivity for reasons that are still poorly understood (Strathmann 1985, 1986; Olson and Olson 1989). Thus, regardless of the mechanism, the inferred Pliocene shift toward direct development in the Caribbean but lecithotropic development in the eastern Pacific agrees well with the apparent shift in oceanographic conditions. Egg size of closely related species of echinoderms is also larger in the Caribbean than in the eastern Pacific (Lessios 1990), and the incidence of direct development is greater among Caribbean reef corals than anywhere else (Richmond and Hunter 1990).

Size of mollusks has been suggested to increase with primary production (Vermeij 1978). Present data are inadequate to test this hypothesis although it is consistent with strombinid size increase in the eastern Pacific. In addition, the intensity of scraping and crushing predation is more intense in the eastern Pacific than the Caribbean, and molluscan shell architecture commonly exhibits enemy-related adaptations to these differences (Birkeland 1977, 1987; Vermeij 1978, 1989). In general, large, squat, low-spired, heavily sculptured or armored shells are more resistant to shell breakage than smaller, slender, high-spired, smooth or unarmored shells (Vermeij 1987). Thus, strombinid size increase in the eastern Pacific and constancy in the Caribbean could reflect defensive adaptations against predators. However, eastern Pacific strombinids also became more slender and higher spired during the Pliocene, while sculpture and armor were unchanged or decreased (fig. 9.9, table 9.5), suggesting that differing intensity of predation was not a major factor.

Plio-Pleistocene Faunal Turnover

The relatively sudden shift in strombinid diversity between the two oceans, and turnover of species during the Plio-Pleistocene occurred 1–2 my *after* the above interoceanic differences in strombinid taxonomic composition, morphology, and larval development were well established. Turnover probably reflects environmental changes correlated with the abrupt intensification of Northern Hemisphere glaciation at the end of the Pliocene (Shackleton et al. 1984; Raymo et al. 1989; Joyce et al. 1990; Jansen and Sjoholm 1991; Sikes et al. 1991). Assuming that climate change was somehow responsible, however, the magnitude in the tropics was probably not very great. Fluctuations in oxygen isotopic ratios, and therefore of sea level and temperature, were less than during the last 1 my (Webb and Bartlein 1992). Isotopic data from reef corals and terrestrial pollen records at the end of the Pleistocene indicate a drop in tropical sea surface temperatures (SSTs) of 5–6°C (Beck et al. 1992; Bush et al. 1992; Guilderson et al. 1994; Colinvaux, this volume, chap. 13). Comparison of these data against isotopic fluctuations 2.4 Ma suggests a drop in tropical SSTs no greater than 3–5°C below the preceeding several my, when tropical SSTs were as warm as today (Dowsett and Poore 1991; Cronin and Dowsett, this volume, chap. 4).

The different fates of strombinids on opposite sides of the isthmus suggest that declining temperatures were at least a partial factor. By analogy to the present, Late Pliocene strombinids in the eastern Pacific were regularly exposed to fluctuations in SSTs of more than 10°C due to seasonal upwelling and more extreme ENSO events (Glynn 1988; D'Croz et al. 1991; Diaz and Markgraf 1992). Such fluctuations exceed estimates for the drop in tropical SSTs due to the intensification in glaciation (see above). Thus, eastern Pacific strombinids should have been preadapted to glacial cooling, whereas Late Pliocene Caribbean strombinids would not have been (Jackson 1994). This is consistent with Stanley's (1986) refrigeration hypothesis, albeit at least 1 my later than was originally hypothesized (Stanley and Campbell 1981; Allmon 1993; Allmon et al. 1993, this volume, chap. 10; Jackson et al. 1993).

On the other hand, it is difficult to understand how a drop of only a few degrees centigrade could be directly responsible for strombinid extinction and turnover (Clarke 1993), as well as that of mollusks and corals in general (Jackson et al. 1993; Budd et al. 1994, Budd et al.; this volume, chap. 7). Thus, other factors, including loss of habitat due to sea level fluctuations, declining productivity, or changes in metapopulation dynamics, may also have been involved (Jackson et al., 1996). For example, occurrence of the few surviving Caribbean paciphiles in areas of moderate upwelling and productivity, such as northern Venezuela and Colombia (Jung 1989), may reflect isolated instances of pre-adaptation to glacial cooling as hypothesized for the eastern Pacific. Alternatively, these same occurrences may signal a more pervasive decline in Caribbean upwelling as an important factor in extinction (Vermeij 1978; Vermeij and Petuch 1986). The upwelling hypothesis is supported by isotopic and micropaleontological evidence from west Florida (Allmon et al. 1995, this volume, chap. 10) and changes in ostracode faunas from Costa Rica (Borne et al. 1994). However, the limited isotopic data available from the isthmus (Collins et al. 1995; Teranes et al., this volume, chap. 5) argue against any substantial decline of upwelling as a cause of extinction, at least in the southwestern Caribbean. Data are unfortunately inadequate to distinguish between these conflicting alternatives.

PHYLOGENETIC CONSTRAINTS

The temporal trends could simply be due to unrelated changes in numbers of species among the different strombinid genera, rather than adaptation to changing environments (Felsenstein 1985; Harvey and Pagel 1991). For example, the inferred development of all but one species of *Sincola* is planktotrophic, and *Sincola* are also much smaller, squatter, and more heavily armored and sculptured, on average, than the other strombinid genera (app. 9.2). Thus, the trends summarized in figures 9.7 and 9.9 reflect, at least in part, the demise of *Sincola* and concomitant rise of *Cotonopsis* and the five new subgenera of *Strombina* after the Miocene.

On the other hand, shell size increased for species in all genera in the eastern Pacific after the Miocene, but for no genera in the Caribbean until the Recent. Moreover, species of *Strombina* shifted from almost exclusively planktotrophic development in the Miocene to lecithotrophic/planktotrophic development in the eastern Pacific, versus direct development in the Caribbean. Likewise, *Cotonopsis* are predominantly planktotrophic in the eastern Pacific, but the one living Caribbean species is strongly inferred to have direct development. Thus, the repeated nature of these patterns within and across taxa strongly suggests that they are adaptive.

Strombinids are only one group of paciphiles, so that their evolution may not be representative in some respects. It would therefore be interesting to analyze evolution of size and larval development of turitelline gastropods, which also inhabit sediments, and suffered comparable Caribbean extinction at the end of the Pliocene (Allmon 1988, 1992). For example, developmental data for the only two tropical American species studied suggest a similar pattern to the strombinids: Caribbean *Turritella variegata* have direct development, whereas eastern Pacific *T. gonostoma* are lecithotrophic (Allmon 1988). Similar trends in these two groups would lend strong support to Vermeij's (1993) assertion of the importance of ecological characteristics for understanding extinction, and the particular vulnerability of molluscan clades characteristic of shallow-water sediments.

CONCLUSIONS

Throughout their history, most strombinid species have survived only a few million years and were narrowly restricted to a small portion of the

Caribbean or tropical eastern Pacific. Diversity peaked in the Caribbean during the Early Pliocene and declined to only a few species thereafter, but continued to increase to the present in the eastern Pacific. Reversal of diversity between the two oceans in the Plio-Pliocene was due to a decline in origination and increase in extinction rates in the Caribbean, and the opposite in the eastern Pacific. These patterns are similar in part to the history of reef corals and erect bryozoans, which somewhat earlier also suffered a marked decline in Caribbean diversity during the Late Pliocene, but not an eastern Pacific radiation (Budd et al., this volume, chap. 7; Cheetham and Jackson, this volume, chap. 8).

Changes in distribution and diversity were accompanied by profound shifts in the subgeneric and generic composition of the fauna. There was also a general decline in planktotrophic larval development that was balanced by increased lecithotrophy in the eastern Pacific and direct development in the Caribbean. Average size increased by nearly two-thirds in the eastern Pacific, mostly during the Pliocene, but changed very little in the Caribbean until the Recent. Other morphological trends were generally similar in the two oceans.

Size increase and shifts in development across higher taxa demand an ecological explanation (Vermeij 1987; Jackson 1988). Changing oceanographic conditions were obviously important, but how and why remains to be demonstrated pending better paleoenvironmental data. This will require extensive, systematic, stratigraphically precise and replicated sampling for estimating paleotemperatures, seasonality, and primary productivity, as well as more extensive collections of strombinids for better stratigraphic and geographic control. In contrast, shell sculpture and armor, which commonly differ between oceans in apparent response to differing levels of predation (Vermeij 1978, 1989), do not seem to have been as important as oceanography in this case.

Whatever their explanation, the trends documented here are the result of many generations of speciation and extinction. Thus, like other trends, they appear to reflect differential origination or survival by species with particular suites of characteristics, rather than evolutionary responses within species (Jackson and McKinney 1990; Cheetham and Jackson, this volume, chap. 8). Testing this hypothesis, however, will require detailed phylogenies of the entire strombinid clade. We need more studies of taxa with contrasting ecological and evolutionary histories, along with better comparative data on life histories and environments, before we can explain the interacting biological effects of tectonics and climate change.

ACKNOWLEDGMENTS

Magnolia Calderon, Karl Müller, and Yira Ventocilla prepared the samples, and Anthony Coates, Laurel Collins, René Panchaud, David West and innumerable others assisted in the field and in many other ways. Betzabeth Ríos prepared figure 9.1 and Héctor Barrios prepared figure 9.8. Warren Allmon, Ann Budd, Anthony Coates, Laurel Collins, David Jablonski, Nancy Knowlton, Paul Morris, Jay Schneider, and Geerat Vermeij criticized the manuscript and gave much useful advice. This work was supported by the National Geographic Society, National Science Foundation, Kugler Fund of the Basel Naturhistorisches Museum, Walcott Fund of the Smithsonian Institution, and the Smithsonian Tropical Research Institute.

APPENDIX 9.1: STRATIGRAPHIC AND GEOGRAPHIC OCCURRENCE OF STROMBINID SPECIES IN TROPICAL AMERICA

| Age | Genus | Subgenus | Species | Geographic Regions | | | | | | | | | | |
| | | | | Caribbean | | | | | | Eastern Pacific | | | | |
				A	B	C	D	E	F	G	H	I	J	K
R	Bifurcium		bicanaliferum	0	0	0	0	0	0	1	1	1	1	1
R	Clavistrombina		clavulus	0	0	0	0	0	0	1	1	1	1	0
R	Cotonopsis	Cotonopsis	aff. deroyae	0	0	0	0	0	0	0	0	0	1	0
R	Cotonopsis	Cotonopsis	aff. suteri	0	0	0	0	0	0	1	1	1	0	0
R	Cotonopsis	Cotonopsis	argentea	0	0	0	1	0	0	0	0	0	0	0
R	Cotonopsis	Cotonopsis	crassiparva	0	0	0	0	0	0	0	0	0	0	1
R	Cotonopsis	Cotonopsis	deroyae	0	0	0	0	0	0	0	0	0	0	1
R	Cotonopsis	Cotonopsis	edentula	0	0	0	0	0	0	1	1	1	0	0
R	Cotonopsis	Cotonopsis	jaliscana	0	0	0	0	0	0	0	1	1	0	0
R	Cotonopsis	Cotonopsis	mendozana	0	0	0	0	0	0	0	1	1	0	0
R	Cotonopsis	Cotonopsis	panacostaricensis	0	0	0	0	0	0	1	0	0	0	0
R	Cotonopsis	Cotonopsis	skoglundae	0	0	0	0	0	0	0	0	1	0	0
R	Cotonopsis	Cotonopsis	suteri	0	0	0	0	0	0	0	1	1	0	0
R	Cotonopsis	Turrina	hirundo	0	0	0	0	0	0	1	1	1	1	0
R	Cotonopsis	Turrina	radwini	0	0	0	0	0	0	0	1	0	0	0
R	Cotonopsis	Turrina	turrita	0	0	0	0	0	0	1	1	0	0	0
R	Sincola	Dorsina	dorsata	0	0	0	0	0	0	1	0	1	0	0
R	Sincola	Dorsina	gibberula	0	0	0	0	0	0	1	1	1	1	0
R	Sincola	Sinuina	sinuata	0	0	0	0	0	0	0	1	1	0	0
R	Strombina	Costangula	aff. angularis	0	0	0	0	0	0	0	0	1	0	0
R	Strombina	Costangula	angularis	0	0	0	0	0	0	1	1	1	0	0
R	Strombina	Lirastrombina	carmencita	0	0	0	0	0	0	1	1	1	0	0
R	Strombina	Lirastrombina	colpoica	0	0	0	0	0	0	0	1	1	0	0

Age	Genus	Subgenus	Species	Geographic Regions										
				Caribbean						Eastern Pacific				
				A	B	C	D	E	F	G	H	I	J	K
R	Strombina	Lirastrombina	francesae	0	0	0	0	1	0	0	0	0	0	0
R	Strombina	Lirastrombina	lilacina	0	0	0	0	0	0	0	0	1	0	0
R	Strombina	Lirastrombina	marksi	0	0	0	0	0	0	0	0	1	0	0
R	Strombina	Lirastrombina	pavonina	0	0	0	0	0	0	1	1	1	0	0
R	Strombina	Lirastrombina	pulcherrima	0	0	0	0	0	0	1	1	1	0	0
R	Strombina	Lirastrombina	solidula	0	0	0	0	0	0	0	1	1	0	0
R	Strombina	Recurvina	fusinoidea	0	0	0	0	0	0	1	1	1	1	0
R	Strombina	Recurvina	paenoblita	0	0	0	0	0	0	1	0	1	1	1
R	Strombina	Recurvina	recurva	0	0	0	0	0	0	1	1	0	0	0
R	Strombina	Spiralta	elegans	0	0	0	0	0	0	1	1	1	1	0
R	Strombina	Spiralta	maculosa	0	0	0	0	0	0	1	1	1	1	1
R	Strombina	Strombina	lanceolata	0	0	0	0	1	0	0	0	0	0	1
R	Strombina	Strombina	pumilio	0	0	0	0	1	1	0	0	0	0	0
PP	Bifurcium		bicanaliferum	0	0	0	0	0	0	1	0	0	0	0
PP	Cotonopsis	Cotonopsis	n. sp.	0	0	0	0	0	0	1	0	0	0	0
PP	Cotonopsis	Cotonopsis	sp. 1	0	0	0	0	0	0	1	0	0	0	0
PP	Cotonopsis	Tarrina	hirundo	0	0	0	0	0	0	1	0	0	0	0
PP	Cotonopsis	Tarrina	sp. 1	0	0	0	0	0	0	1	1	1	1	0
PP	Sincola	Dorsina	dorsata	0	0	0	0	0	0	0	1	1	0	0
PP	Sincola	Dorsina	gibberula	0	0	0	0	0	0	1	0	1	0	0
PP	Sincola	Sincola	crassilabrum	0	0	0	0	1	0	0	0	0	0	0
PP	Sincola	Simuina	sinuata	0	0	0	0	1	0	1	0	0	0	0

	Genus	Subgenus	Species													
PP	Strombina	Lirastrombina	carmencita	0	0	0	1	0	0	0	0	0	0	0	0	0
PP	Strombina	Lirastrombina	pulcherrima	0	0	1	1	0	0	0	0	0	0	1	0	0
PP	Strombina	Recurvina	n. sp,	0	0	0	1	0	0	0	0	0	0	0	0	0
PP	Strombina	Recurvina	penita	0	0	0	1	0	0	0	0	0	0	0	0	0
PP	Strombina	Recurvina	recurva	0	0	0	1	0	0	0	0	0	0	0	0	0
PP	Strombina	Spiralta	elegans	0	0	1	1	0	0	0	0	0	0	1	0	0
PP	Strombina	Spiralta	maculosa	0	0	0	1	0	0	0	0	0	0	0	0	0
PP	Strombina	Strombina	aff. pumilio	0	0	0	0	0	0	0	0	0	0	0	0	0
PP	Strombina	Strombina	lanceolata	1	1	0	1	0	0	0	0	0	0	0	0	0
PP	Strombina	Strombina	pumilio	0	0	0	0	0	1	0	0	0	0	0	0	0
LP	Cotonopsis	Cotonopsis	cf. mendozana	0	0	0	0	1	0	0	0	0	1	0	0	0
LP	Cotonopsis	Cotonopsis	panacostaricensis	0	0	0	0	0	0	0	0	0	0	0	0	0
LP	Cotonopsis	Cotonopsis	suteri	0	0	0	0	1	0	0	0	0	0	0	0	0
LP	Cotonopsis	Turrina	hirundo	0	0	0	0	1	0	0	0	0	1	0	0	0
LP	Sincola	Sincola	mexicana	0	0	0	0	0	0	0	0	0	0	0	0	0
LP	Sincola	Dorsina	gibberula	0	0	0	0	0	0	0	0	0	0	0	0	0
LP	Sincola	Sinaxila	cf. lloydsmithi	0	0	0	0	1	0	0	0	0	1	0	0	0
LP	Sincola	Sinaxila	gunteri	0	0	0	0	0	0	0	0	0	0	0	0	0
LP	Sincola	Sinaxila	lloydsmithi	0	1	0	0	1	0	0	0	0	0	0	0	0
LP	Sincola	Sinaxila	matima	0	0	0	0	1	0	0	0	0	0	0	0	0
LP	Sincola	Sincola	aff. sincola	0	0	0	0	1	0	0	0	0	0	0	0	0
LP	Sincola	Sincola	chiriquiensis	0	0	0	0	1	0	0	0	0	0	0	0	0
LP	Sincola	Sincola	crassilabrum	0	0	0	0	1	0	0	0	0	0	0	0	0
LP	Sincola	Sincola	sincola	0	0	0	0	0	0	0	0	0	0	0	0	0
LP	Sincola	Simuina	ecuadoriana	0	0	0	1	1	0	0	0	0	0	0	0	0
LP	Simuina	Simuina	costaricensis	0	0	0	0	0	1	0	0	0	0	0	0	0
LP	Strombina	Lirastrombina	mareana	0	0	0	0	1	0	0	0	0	0	0	0	0
LP	Strombina	Lirastrombina	musanica	0	0	0	0	0	0	0	0	0	0	0	0	0
LP	Strombina	Recurvina	penita	0	0	0	1	0	0	0	0	0	0	0	0	0

APPENDIX 9.1: *continued*

Age	Genus	Subgenus	Species	A	B	C	D	E	F	G	H	I	J	K
						Caribbean						Eastern Pacific		
LP	*Strombina*	*Recurvina*	*radicensis*	0	0	0	0	1	0	0	0	0	0	0
LP	*Strombina*	*Recurvina*	*recurva*	0	0	0	0	0	0	1	0	0	0	0
LP	*Strombina*	*Spiralta*	*guppyi*	0	0	0	1	0	1	0	0	0	0	0
LP	*Strombina*	*Strombina*	*lanceolata*	0	0	0	0	0	0	0	0	0	1	0
LP	*Strombina*	*Strombina*	*pumilio*	0	0	0	0	1	0	0	0	0	0	0
LP	*Strombina*	*Strombina?*	sp. B	0	0	0	0	1	0	0	0	0	0	0
EP	*Cotonopsis*	*Cotonopsis*	*esmeraldensis*	0	0	0	0	0	0	0	0	0	1	0
EP	*Cotonopsis*	*Cotonopsis*	*panacostaricensis*	0	0	0	0	0	0	1	0	0	0	0
EP	*Sincola*	*Dorsina*	*mexicana*	1	0	0	0	0	0	0	0	0	0	0
EP	*Sincola*	*Dorsina*	*colombiana*	0	0	0	0	1	1	0	0	0	0	0
EP	*Sincola*	*Dorsina*	*melajoensis*	0	0	0	0	1	0	0	0	0	0	0
EP	*Sincola*	*Dorsina?*	*pigea?*	0	0	0	0	0	1	0	0	0	0	0
EP	*Sincola*	*Sinaxila*	sp. 1	0	0	0	0	0	0	1	0	0	0	0
EP	*Sincola*	*Sinaxila*	*cunninghamcraigi*	0	0	0	0	1	0	0	0	0	0	0
EP	*Sincola*	*Sinaxila*	*lloydsmithi*	0	0	0	0	0	1	1	0	0	0	0
EP	*Sincola*	*Sinaxila*	*matima*	0	0	0	0	0	1	0	0	0	0	0
EP	*Sincola*	*Sinaxila*	*naufraga*	0	0	1	0	0	0	0	0	0	0	0
EP	*Sincola*	*Sinaxila*	sp. A	0	0	0	0	0	1	0	0	0	0	0
EP	*Sincola*	*Sincola*	*chiriquiensis*	0	0	0	0	0	1	1	0	0	0	0
EP	*Sincola*	*Sincola*	*crassilabrum*	0	0	0	1	1	0	0	0	0	0	0
EP	*Sincola*	*Simuina*	*ecuadoriana*	0	0	0	0	0	0	1	0	0	1	0
EP	*Sincola?*		*anomala*	0	1	0	0	0	0	0	0	0	0	0
EP	*Strombina*	*Araxina*	*araxana*	0	0	0	0	1	0	0	0	0	0	0

				C1	C2	C3	C4	C5	C6	C7	C8	C9	C10	C11	C12	C13	C14	C15
EP	*Strombina*	*Arcyina*	*humboldti*	0	0	0	0	1	0	0	0	0	0	0	0	0	0	0
EP	*Strombina*	*Lirastrombina*	cf. *cricamola*	0	0	1	1	0	0	0	0	0	0	0	0	0	0	0
EP	*Strombina*	*Lirastrombina*	*cricamola*	0	0	1	1	1	0	0	0	0	0	0	0	0	0	0
EP	*Strombina*	*Lirastrombina*	*daddeleyi*	0	0	0	0	0	0	0	0	0	0	0	0	0	0	0
EP	*Strombina*	*Lirastrombina*	*musanica*	0	0	1	1	1	0	0	0	0	0	0	0	0	0	0
EP	*Strombina*	*Lirastrombina*	*rutschi*	0	0	1	1	1	0	0	0	0	0	0	0	0	0	0
EP	*Strombina*	*Recurvina*	n. sp.	0	0	0	0	0	0	0	0	0	0	0	0	0	0	0
EP	*Strombina*	*Recurvina*	*penita*	1	0	1	0	1	0	0	0	0	0	0	0	0	0	0
EP	*Strombina*	*Recurvina*	*radicensis*	1	0	1	1	1	0	0	0	0	0	0	0	0	0	0
EP	*Strombina*	*Recurvina*	*recurva*	0	1	0	0	0	0	0	0	0	0	0	0	0	0	0
EP	*Strombina*	*Recurvina?*	sp. A	1	1	1	1	1	0	0	0	0	0	0	0	0	0	0
EP	*Strombina*	*Spiralta*	*guppyi*	0	0	0	0	0	0	1	0	0	0	0	0	0	0	0
EP	*Strombina*	*Strombina*	cf. *zorritosensis*	0	1	0	0	0	0	0	0	0	0	0	0	0	0	0
EP	*Strombina*	*Strombina*	*colimensis*	0	0	0	1	0	0	0	0	0	0	0	0	0	0	0
EP	*Strombina*	*Strombina*	*gradata*	1	0	1	1	0	0	1	0	0	0	0	0	0	0	0
EP	*Strombina*	*Strombina*	*lessepsiana*	0	0	0	0	0	0	0	0	0	0	0	0	0	0	0
EP	*Strombina*	*Strombina*	*pumilio*	1	0	1	1	1	0	0	0	0	0	0	0	0	0	0
EP	*Strombina*	*Strombina?*	sp. A	1	1	0	1	1	0	0	0	0	0	0	0	0	0	0
EP	*Strombina*	*Strombina?*	sp. B	1	0	0	0	1	0	0	0	0	0	0	0	0	0	0
LM	*Bifarcium*		*canae*	0	0	0	0	0	1	0	0	0	0	0	0	0	0	0
LM	*Bifarcium*		*nuestrasenorae*	0	0	0	0	0	1	0	0	0	0	0	0	0	0	0
LM	*Sincola*	*Dorsina*	*colombiana*	0	1	1	0	1	0	0	0	0	0	0	0	0	0	0
LM	*Sincola*	*Dorsina*	*pigea*	0	1	1	0	1	0	0	0	0	0	0	0	0	0	0
LM	*Sincola*	*Dorsina*	*pigea?*	0	1	1	0	1	0	0	0	0	0	0	0	0	0	0
LM	*Sincola*	*Sinaxila*	*bassi*	0	0	1	0	0	1	0	0	0	0	0	0	0	0	0
LM	*Sincola*	*Sinaxila*	*lloydsmithi*	0	1	0	0	1	0	0	0	0	0	0	0	0	0	0
LM	*Sincola*	*Sincola*	*amphidyma*	0	1	1	0	1	0	0	0	0	0	0	0	0	0	0
LM	*Sincola*	*Sincola*	*chiriquiensis*	0	1	1	0	1	0	0	0	0	0	0	0	0	0	0

APPENDIX 9.1: *continued*

Age	Genus	Subgenus	Species	Geographic Regions										
				Caribbean								Eastern Pacific		
				A	B	C	D	E	F	G	H	I	J	K
LM	*Sincola*	*Sincola*	*daulechia*	0	0	0	0	0	0	0	0	0	1	0
LM	*Sincola*	*Sincola*	*galvestonensis*	1	0	0	0	0	0	0	0	0	0	0
LM	*Sincola*	*Sincola*	*gurabensis*	0	0	0	1	0	0	0	0	0	0	0
LM	*Sincola*	*Sincola*	*barbisonae*	0	0	0	0	0	0	0	0	0	1	0
LM	*Sincola*	*Sincola*	*namniebellae*	0	0	0	1	0	0	0	0	0	0	0
LM	*Sincola*	*Sincola*	*pseudobaitensis*	0	0	0	1	0	0	0	0	0	0	0
LM	*Sincola*	*Sincola*	sp. 1	0	0	0	0	0	1	0	0	0	0	0
LM	*Strombina*	*Spiralta*	*falconensis*	0	0	0	0	1	0	0	0	0	0	0
LM	*Strombina*	*Strombina*	*carrizalensis*	0	0	0	0	1	0	0	0	0	0	0
LM	*Strombina*	*Strombina*	cf. *colinensis*	0	0	0	0	0	0	0	1	0	0	0
LM	*Strombina*	*Strombina*	*colinensis*	0	0	0	0	1	1	0	0	0	0	0
LM	*Strombina*	*Strombina*	*cyphonotus*	0	0	0	0	0	1	1	1	0	0	0
LM	*Strombina*	*Strombina*	*lessepsiana*	0	0	0	0	0	0	1	0	0	1	0
LM	*Strombina*	*Strombina*	*ocbyra*	0	0	0	0	0	1	0	0	0	0	0
LM	*Strombina*	*Strombina*	*zorritosensis*	0	0	0	0	0	0	0	0	0	0	0
LM	*Strombina*	*Strombina?*	*cartagenensis*	0	0	0	0	0	1	0	0	0	1	0
LM	*Strombina*	*Strombina?*	*parca*	0	0	0	0	0	1	0	0	0	0	0
LM	*Strombina*	*Strombina?*	sp. B	0	0	0	0	1	0	0	0	0	0	0

	Genus	Subgenus	Species	A	B	C	D	E	F	G	H	I	J	K
MM	*Cotonopsis?*	*Cotonopsis?*	sp. 1	0	0	0	0	0	0	0	0	0	0	0
MM	*Sincola*	*Sincola*	n. sp. A	0	0	0	0	0	0	0	1	0	0	0
MM	*Sincola*	*Dorsina*	*caribaea*	0	0	1	0	0	0	0	1	0	0	0
MM	*Sincola*	*Dorsina*	*quirosana*	0	0	1	1	0	0	0	0	0	0	0
MM	*Sincola*	*Sincola*	*walli*	1	0	0	1	0	0	0	0	0	0	0
MM	*Sincola?*	*Dorsina?*	*lampra*	1	0	0	0	0	0	0	0	0	0	0
MM	*Strombina*	*Strombina*	*aldrichi*	1	0	0	0	0	0	0	1	0	0	0
MM	*Strombina*	*Strombina*	*colimensis*	0	0	0	0	0	0	0	0	0	0	0
MM	*Strombina*	*Strombina*	sp. C	1	0	0	1	0	1	0	1	0	0	0
MM	*Strombina*	*Strombina*	*waltonia*	0	0	0	0	0	0	0	0	0	0	0
MM	*Strombina*	*Strombina?*	*cartagenensis*	0	0	0	1	1	0	0	0	0	0	0
MM	*Strombina*	*Strombina?*	sp. B	0	0	0	0	0	0	0	0	0	0	0
EM	*Sincola*	*Dorsina*	*caribaea*	0	0	1	0	0	0	0	0	0	0	0
EM	*Sincola*	*Dorsina*	*pigea?*	0	0	0	1	0	0	0	0	0	0	0
EM	*Sincola*	*Dorsina*	*quirosana*	0	0	0	1	0	0	0	0	0	0	0
EM	*Sincola*	*Sinaxila*	*tumbezia*	0	0	0	0	0	0	0	0	0	0	0
EM	*Strombina*	*Strombina*	*aldrichi*	1	0	0	0	0	0	0	0	1	0	0
EM	*Strombina*	*Strombina*	*cyphonotus*	0	0	0	1	0	1	0	0	0	0	0
EM	*Strombina*	*Strombina?*	sp. B	0	0	0	1	0	0	0	0	0	0	0

Note: n = 110; 1 = presence; 0 = absence; R = Recent; PP = Plio-Pleistocene; LP = Late Pliocene; EP = Early Pliocene; LM = Late Miocene; MM = Middle Miocene; EM = Early Miocene. Regions A–K as in figure 9.5.

APPENDIX 9.2: MORPHOLOGY OF LARVAL AND ADULT STROMBIND SHELLS

Genus	Subgenus	Species	DEV	Protoconch DIAM (mm)	WHL	RATIO	Teloconch SIZE (cm)	SHAPE	SPIRE	SZAP (cm)	SHPAP	APARM	SCLPT
Caribbean													
Bifurcium		camae	L/P		1.75		5.537	2.015	0.424	2.522	4.417	0.833	0.579
Bifurcium		nuestrasenorae	L/P		2.00		6.505	2.000	0.333	3.163	4.447	1.000	0.632
Cotonopsis	Cotonopsis	argentea	D		1.50		21.552	3.055	0.738	8.246	4.250	0.500	0.211
Sincola		mexicana	P		3.00		8.195	2.220	0.654	2.967	4.977	0.750	0.789
Sincola	Dorsina	caribaea	P		3.00		6.910	1.910	0.540	2.665	4.933	0.917	0.579
Sincola	Dorsina	colombiana	P	500	3.00	0.20	16.972	1.945	0.552	6.430	4.247	0.500	0.368
Sincola	Dorsina	melajoensis	P		3.00		11.603	2.393	0.760	4.183	4.375	0.750	0.526
Sincola	Dorsina	pigea					10.609	2.297	0.616	3.857	4.857	0.750	0.421
Sincola	Dorsina	pigea?	P		3.00		8.107	2.313	0.682	2.902	4.759	0.667	0.421
Sincola	Dorsina	qurosana	P		3.50		5.200	2.086	0.640	1.921	4.556	0.583	0.421
Sincola	Sinaxila	bassi	L/P		2.50		8.106	1.875	0.461	3.568	3.526	1.000	0.579
Sincola	Sinaxila	cunninghamcraigi	P	508	3.50	0.14	7.016	1.720	0.415	3.340	3.089	0.917	0.737
Sincola	Sinaxila	gunteri	P	438	3.00	0.14	7.318	2.019	0.651	3.209	2.575	0.833	0.632
Sincola	Sinaxila	lloydsmithi	P	433	3.00	0.14	7.982	1.831	0.500	3.182	4.500	1.000	0.842
Sincola	Sinaxila	matima	P	433	3.00	0.14	7.891	2.259	0.596	3.156	3.530	0.750	0.474
Sincola	Sinaxila	naufraga	L/P	488	1.75	0.30	8.725	1.737	0.484	3.664	3.840	0.833	0.474
Sincola	Sinaxila	sp. A	P	516	3.50	0.14	11.811	2.323	0.385	4.421	6.383	0.667	0.526
Sincola	Sincola	amphidyma	P	500	3.50	0.14	5.547	1.972	0.583	1.943	5.900	0.917	0.526
Sincola	Sincola	chiriquiensis	P	416	3.50	0.11	5.553	1.834	0.563	1.876	5.500	0.917	0.632
Sincola	Sincola	crassilabrum	P	366	3.00	0.12	9.120	1.875	0.499	4.281	2.932	0.583	0.579
Sincola	Sincola	galvestonensis	P		3.00		6.314	1.969	0.520	2.785	4.383	0.833	0.474
Sincola	Sincola	gurabensis	P	466	3.00	0.20	7.531	1.981	0.497	3.487	3.368	0.500	0.368
Sincola	Sincola	nanniebellae	P		3.00		9.477	1.833	0.453	4.424	3.132	0.333	0.579
Sincola	Sincola	pseudobaitensis	P		3.00		6.660	1.746	0.443	2.811	3.421	0.833	0.474

Genus	species											
Sincola	*sincola*	P	433	3.00	0.14	5.624	2.168	0.732	1.833	5.250	0.917	0.737
Sincola	*walli*	P	471	3.50	0.13	5.056	1.972	0.543	2.210	3.392	0.917	0.474
Sincola?	*anomala*	L/P		2.50		3.447	2.063	0.394	1.325	4.154	0.750	0.368
Dorsina?	*lampra*	P	450	3.00	0.20	6.243	2.210	0.606	2.590	3.681	0.917	0.474
Arayina	*arayana*	D	641	1.50	0.42	16.054	2.130	0.453	7.181	3.667	0.667	0.632
Arayina	*humboldti*	D	583	1.50	0.40	18.640	2.137	0.453	5.943	6.978	0.667	0.263
Lirastrombina	*costaricensis*	D	550	1.25	0.44	11.910	2.044	0.548	3.935	5.620	0.667	0.737
Lirastrombina	*cricamola*					11.238	1.820	0.318	4.423	4.890	0.583	0.579
Lirastrombina	*duddeleyi*					14.457	2.090	0.353	5.511	6.000	0.667	0.579
Lirastrombina	*francesae*	D	483	1.50	0.32	14.916	2.225	0.618	4.923	6.060	0.667	0.474
Lirastrombina	*mareana*					13.795	1.847	0.293	5.293	5.533	0.667	0.579
Lirastrombina	*musanica*	D	457	1.50	0.30	12.139	2.404	0.614	4.074	6.024	0.667	0.684
Lirastrombina	*rutschi*					17.678	2.000	0.429	6.708	5.000	0.667	0.632
Recurvina	*radicensis*	D	553	1.50	0.40	14.465	2.583	0.632	4.509	6.640	0.667	0.789
Spiralta	*falconensis*					16.255	3.240	0.837	5.892	4.869	0.583	0.368
Spiralta	*guppyi*					16.812	3.023	0.801	6.269	4.367	0.500	0.368
Strombina	*aldrichi*	P	492	3.00	0.20	10.400	2.926	0.722	3.981	4.732	0.667	0.263
Strombina	*carrizalensis*					17.434	2.310	0.497	6.215	6.180	0.583	0.263
Strombina	*colinensis*	P		3.00		16.196	2.623	0.614	5.385	7.250	0.667	0.632
Strombina	*cybbonotus*					15.807	2.254	0.462	5.752	6.094	0.667	0.368
Strombina	*gradata*	D	700	1.50	0.50	15.891	2.290	0.631	6.852	4.792	0.583	0.421
Strombina	*lessepsiana*	L/P	485	2.50	0.20	18.738	2.438	0.530	7.036	5.500	0.750	0.474
Strombina	*ochyra*					14.507	2.178	0.494	4.770	7.429	0.667	0.684
Strombina	*pumilio*	D	660	1.50	0.44	14.979	2.409	0.661	5.060	5.643	0.667	0.737
Strombina	*waltonia*	P	491	3.00	0.20	14.187	2.786	0.613	5.356	5.667	0.583	0.579
Strombina?	*cartagenensis*					17.862	2.331	0.472	6.728	5.691	0.750	0.421
Strombina?	*parca*	D	585	1.25	0.50	14.115	2.602	0.716	4.743	5.625	0.583	0.368
Strombina?	sp. A					11.811	2.323	0.385	4.421	6.383	0.667	0.526
Strombina?	sp. B					14.368	2.012	0.382	5.464	5.898	0.583	0.368

APPENDIX 9.2: *continued*

Genus	Subgenus	Species	DEV	DIAM (mm)	WHL	RATIO	SIZE (cm)	SHAPE	SPIRE	SZAP (cm)	SHPAP	APARM	SCLPT
				Protoconch			**Teloconch**						
Eastern Pacific													
Bifurcium		bicanaliferum	L/P	354	2.50	0.14	9.994	2.039	0.558	4.242	3.856	0.417	0.316
Clavistrombina		clavulus	P	458	3.00	0.20	13.389	2.766	0.976	4.437	3.889	0.750	0.158
Cotonopsis	Cotonopsis	aff. deroyae					25.913	2.324	0.491	11.091	3.417	0.583	0.263
Cotonopsis	Cotonopsis	aff. suteri	L/P	483	2.00	0.24	19.779	2.717	0.642	8.352	3.444	0.250	0.421
Cotonopsis	Cotonopsis	crassiparva	P	514	3.00	0.20	13.831	2.012	0.495	5.374	3.818	0.750	0.474
Cotonopsis	Cotonopsis	deroyae	L/P	475	2.50	0.20	29.618	3.035	0.659	12.962	3.977	0.500	0.368
Cotonopsis	Cotonopsis	edentula	P	633	3.00	0.21	22.555	2.477	0.614	9.375	3.288	0.583	0.368
Cotonopsis	Cotonopsis	esmeraldensis	D		1.50		17.058	2.279	0.471	6.626	4.658	0.583	0.368
Cotonopsis	Cotonopsis	jaliscana	D	500	1.50	0.33	17.140	2.289	0.532	7.209	3.138	0.500	0.368
Cotonopsis	Cotonopsis	mendozana	L/P	508	2.25	0.22	15.335	2.474	0.622	6.092	3.514	0.500	0.316
Cotonopsis	Cotonopsis	panacostaricensis	P	528	3.00	0.20	20.454	2.514	0.567	8.537	3.887	0.583	0.474
Cotonopsis	Cotonopsis	skoglundae					19.424	3.314	0.739	6.182	7.039	0.583	0.526
Cotonopsis	Cotonopsis	suteri	L/P		2.00		24.799	2.733	0.673	9.367	4.333	0.500	0.526
Cotonopsis	Turrina	hirundo	P	450	3.00	0.20	10.828	2.708	0.569	4.830	3.732	0.333	0.211
Cotonopsis	Turrina	radwini	L/P	465	2.75	0.20	13.636	2.429	0.667	5.379	3.825	0.583	0.316
Cotonopsis	Turrina	turrita	P	516	3.00	0.20	19.239	2.883	0.719	7.420	4.087	0.333	0.263
Sincola		n. sp. A					11.486	1.600	0.242	4.578	5.240	0.750	0.421
Sincola	Dorsina	colombiana	P	500	3.00	0.20	16.972	1.945	0.552	6.430	4.247	0.500	0.368
Sincola	Dorsina	dorsata	P	438	3.00	0.14	16.475	2.349	0.741	6.093	4.909	0.667	0.421
Sincola	Dorsina	gibberula	P	410	3.00	0.13	10.200	2.105	0.650	3.887	4.099	0.750	0.632
Sincola	Dorsina	pigea					10.609	2.297	0.616	3.857	4.857	0.750	0.421
Sincola	Sinaxila	tumbezia					6.650	2.284	0.661	2.168	4.700	0.667	0.421
Sincola	Sincola	chiriquiensis	P	416	3.50	0.11	5.553	1.834	0.563	1.876	5.500	0.917	0.632
Sincola	Sincola	daulechica	P		3.50		6.725	2.233	0.689	2.335	5.450	0.917	0.474

Genus	species	DEV	DIAM	WHL	RATIO	SIZE	SHAPE	SPIRE	APARM	SHPAP	SZAP	SCLPT
Sincola	*barbisonae*	P		3.00		7.773	1.860	0.493	2.835	5.583	0.750	0.421
Sincola	*ecuadoriana*					15.379	1.955	0.387	6.874	5.250	0.833	0.684
Sincola	*sinuata*	P	516	3.00	0.20	11.947	2.230	0.712	4.445	4.567	0.667	0.632
Costangula	*aff. angularis*					18.720	2.730	0.643	7.252	3.905	0.500	0.737
Costangula	*angularis*	L/P	450	1.75	0.30	20.874	2.866	0.631	8.546	4.979	0.667	0.632
Lirastrombina	*carmencita*	D	750	1.50	0.50	21.563	2.685	0.643	7.695	5.675	0.667	0.579
Lirastrombina	*colpoica*	L/P	600	2.00	0.30	18.344	2.781	0.631	6.528	5.322	0.833	0.211
Lirastrombina	*lilacina*	D	621	1.25	0.50	15.458	2.119	0.429	5.701	5.200	0.750	0.368
Lirastrombina	*marksi*	D	600	1.50	0.40	16.173	2.067	0.661	4.661	6.214	0.667	0.474
Lirastrombina	*pavonina*	L/P	683	2.00	0.34	18.000	2.250	0.479	6.309	6.368	0.500	0.211
Lirastrombina	*pulcherrima*	D	638	1.50	0.42	16.274	1.918	0.443	6.538	4.750	0.917	0.526
Lirastrombina	*solidula*	D	517	1.25	0.41	15.889	2.086	0.412	6.423	4.583	0.583	0.211
Recurvina	*fusinoidea*	L/P	436	2.25	0.20	21.688	3.557	0.579	8.815	8.633	0.500	0.684
Recurvina	*paenoblita*	P		3.00		26.412	2.725	0.585	9.064	8.548	0.667	0.684
Recurvina	*penita*					20.372	2.868	0.588	7.276	7.427	0.667	0.526
Recurvina	*recurva*	L/P	465	2.25	0.20	19.089	2.838	0.664	6.074	7.834	0.750	0.895
Recurvina?	*sp. A*					11.811	2.323	0.385	4.421	6.383	0.667	0.526
Spiralta	*elegans*	P	548	3.00	0.20	21.982	2.859	0.890	6.611	6.131	0.583	0.421
Spiralta	*maculosa*	P	505	3.00	0.17	17.082	3.547	0.787	6.455	4.955	0.750	0.421
Strombina	*cf. colinensis*					12.733	2.699	0.588	4.423	7.362	0.667	0.526
Strombina	*cf. zorritosensis*					16.364	1.747	0.466	5.260	5.876	0.667	0.632
Strombina	*colinensis*	P		3.00		16.196	2.623	0.614	5.385	7.250	0.667	0.632
Strombina	*cyphonotus*					15.807	2.254	0.462	5.752	6.094	0.667	0.368
Strombina	*lanceolata*	L/P	505	2.75	0.20	22.009	3.100	0.598	6.275	12.857	0.667	0.789
Strombina	*lesepsiana*	L/P	485	2.50	0.20	18.738	2.438	0.530	7.036	5.500	0.750	0.474
Strombina	*zorritosensis*					17.807	2.483	0.569	6.877	6.255	0.667	0.684

Note: n = 97. DEV refers to inferred developmental mode: D = direct; L/P = lecithotrophic or planktotrophic; P = planktotrophic. Numbers of protoconch characters correspond to definitions in table 9.2: 1 = DIAM, 2 = WHL, 3 = RATIO. Numbers of teloconch characters correspond to definitions in table 9.4: 1 SIZE, 2 = SHAPE, 3 = SPIRE, 4 = APARM, 5 = SHPAP, 6 = SZAP, 7 = SCLPT.

REFERENCES

Allmon, W. D. 1988. Ecology of Recent turritelline gastropods (Prosobranchia, Turritellidae): Current knowledge and paleontological implications. *Palaios* 3:259–84.

———. 1992. Role of temperature and nutrients in extinction of turritelline gastropods: Cenozoic of the northwestern Atlantic and northeastern Pacific. *Palaeogeogr., Palaeoclimatol., Palaeoecol.* 92:41–54.

———. 1993. Age, environment, and mode of deposition of the densely fossiliferous Pinecrest Sand (Pliocene of Florida): Implications for the role of biological productivity in shell bed formation. *Palaios* 8:183–201.

Allmon, W. D., S. D. Emslie, D. S. Jones, and G. S. Morgan. 1996. Late Neogene oceanographic change along Florida's west coast: Evidence and mechanisms. *J. Geol.* 104. In press.

Allmon, W. D., G. Rosenberg, R. W. Portell, and K. S. Schindler. 1993. Diversity of Atlantic coastal plain mollusks since the Pliocene. *Science* 260:1626–29.

Bayer, F. M., G. L. Voss, and C. R. Robins. 1970. *Bioenvironmental and radiological safety feasibility studies, Atlantic-Pacific interoceanic canal: Report on the marine fauna and benthic shelf-slope communities of the isthmian region.* Coral Gables, Fla.: University of Miami.

Beck, J. W., R. L. Edwards, E. Ito, F. W. Taylor, J. Recy, F. Rougerie, P. Joannot, and C. Henin. 1992. Sea-surface temperature from coral skeletal strontium/calcium ratios. *Science* 257:644–47.

Bhaud, M. 1989. Rôle de la dissémination larvaire en paleobiogéographie reévalue a la lumière des données concernant l'époque actuelle. *Bull. Soc. Geol. Fr.* 5:551–59.

———. 1993. Relationship between larval type and geographic range in marine species: Complementary observations on gastropods. *Oceanol. Acta* 16: 191–98.

Birkeland, C. 1977. The importance of the rate of biomass accumulation in early successional stages of benthic communities to the survival of coral recruits. *Proc. Third Int. Coral Reef Symp.* 1:15–21.

———. 1987. Nutrient availability as a major determinant of differences among coastal hard-substratum communities in different regions of the tropics. In *Comparison between Atlantic and Pacific tropical marine coastal ecosystems: Community structure, ecological processes, and productivity,* ed. C. Birkeland, 45–97. UNESCO Reports in Marine Science 46. Paris.

Borne, P. F., T. M. Cronin, and H. J. Dowsett. 1994. Microfaunal evidence of Late Pliocene–Early Pleistocene coastal oceanic upwelling in the Moin Formation, Costa Rica. *Geol. Soc. Am. Ann. Meeting, Abstr. Progr.* 26:A-170.

Budd, A. F., and K. G. Johnson. 1991. Size-related evolutionary patterns among species and subgenera in the *Strombina*-group (Gastropoda: Columbellidae). *J. Paleontol.* 65:417–34.

Budd, A. F., T. A. Stemann, and K. G. Johnson. 1994. Stratigraphic distributions of genera and species of Neogene to Recent Caribbean reef corals. *J. Paleontol.* 68:951–77.

Bush, M. B., D. R. Piperno, P. A. Colinvaux, P. E. De Oliveira, L. A. Krissek, M. C. Miller, and W. E. Rowe. 1992. A 14,300-yr paleoecological profile of a lowland tropical lake in Panama. *Ecol. Monogr.* 62:251–75.

Cipriani, R., and P. E. Penchaszadeh. 1993. How does *Strombina* reproduce? Evidence from two Venezuelan species (Prosobranchia: Columbellidae). *Veliger* 36:178–84.

Clarke, A. 1993. Temperature and extinction in the sea: A physiologist's view. *Paleobiology* 19:499–518.

Coates, A. G., J. B. C. Jackson, L. S. Collins, T. M. Cronin, H. J. Dowsett, L. M. Bybell, P. Jung, and J. A. Obando. 1991. Closure of the Isthmus of Panama: The near-shore marine record of Costa Rica and western Panama. *Geol. Soc. Am. Bull.* 104:814–28.

Collins, L. S., and A. G. Coates. 1993. Marine paleobiogeography of Caribbean Panama: Last Pacific influences before closure of the Tropical American Seaway. *Geol. Soc. Am. Ann. Meeting, Abstr. Prog.* 25:A-428–29.

Collins, L. S., D. H. Geary, and K. C. Lohmann. 1995. A test of the prediction of decreased Caribbean coastal upwelling caused by the emergence of the Isthmus of Panama, using stable isotopes of neritic Foraminifera. *Geol. Soc. Am. Ann. Meeting, Abstr. Progr.* 27:A-156.

Cubit, J. D., H. M. Caffey, R. C. Thompson, and D. M. Windsor. 1989. Meteorology and hydrography of a shoaling reef flat on the Caribbean coast of Panama. *Coral Reefs* 8:59–66.

D'Croz, L., J. B. Del Rosario, and J. A. Gomez. 1991. Upwelling and phytoplankton in the Bay of Panama. *Rev. Biol. Trop.* 39:233–41.

De Maintenon, M. J. 1994. Evolution of *Columbella* (Neogastropoda: Columbellidae) in the Neogene American tropics. *Geol. Soc. Am. Ann. Meeting, Abstr. Progr.* 26:A-53.

Diaz, H. F., and V. Markgraf. 1992. *El Niño: Historical and paleoclimatic aspects of the Southern Oscillation.* Cambridge: Cambridge University Press.

Dowsett, H. J., and Poore, R. Z. 1991. Pliocene sea surface temperatures of the North Atlantic ocean at 3.0 Ma. *Quatern. Sci. Revs.* 10:189–204.

Duque-Caro, H. 1990. Neogene stratigraphy, paleoceanography, and paleobiogeography in northwest South America and the evolution of the Panama seaway. *Palaeogeogr., Palaeoclimatol., Palaeoecol.* 77:203–34.

Felsenstein, J. 1985. Phylogenies and the comparative method. *Am. Nat.* 125:1–15.

Fortunato, H. 1995. Direct observations on the larval development and occlusion time in strombinid gastropods. *Larv. Biol. Meetings, Abstr. Progr.* 2:13.

Fortunato, H., and P. Jung. 1995. The *Strombina*-group (Neogastropoda: Columbellidae): a case study of evolution in the neotropics. *Geol. Soc. Am. Ann. Meeting, Abstr. Progr.* 27:A-52.

Fortunato, H., P. Penchaszadeh, P. Miloslavich, H. Alvarez. 1995. Observations on the reproduction of *Bifurcium bicanaliferum* (Sowerby, 1832) (Gastropoda: Columbellidae: *Strombina*-group) *Amer. Mac. Un. Ann. Meeting, Abstr.* 61:26.

Glynn, P. W. 1988. El Niño–Southern Oscillation, 1982–1983: Nearshore population, community, and ecosystem responses. *Ann. Rev. Ecol. Syst.* 19:309–45.

Glynn, P. W., and M. W. Colgan. 1991. Sporadic disturbances in fluctuating coral reef environments: El Niño and coral reef development in the eastern Pacific. *Am. Zool.* 32:707–18.

Guilderson, T. P., R. G. Fairbanks, and J. L. Rubenstone. 1994. Tropical temperature variations since 20,000 years ago: Modulating interhemispheric climate change. *Science* 263:663–65.

Hansen, T. A. 1978. Larval dispersal and species longevity in Lower Tertiary gastropods. *Science* 199:885–87.

———. 1980. Influence of larval dispersal and geographic distribution on species longevity in neogastropods. *Paleobiology* 6:193–207.

Haq, B. U., J. Hardenbol, and P. R. Vail. 1987. Chronology of fluctuating sea levels since the Triassic. *Science* 235:1156–66.

Harvey, P. H., and M. D. Pagel. 1991. *The comparative method in evolutionary biology.* Oxford: Oxford University Press.

Hasson, P. F., and A. G. Fischer. 1986. Observations on the Neogene of northwestern Ecuador. *Micropaleontology* 32:32–42.

Houbrick, R. S. 1983. A new *Strombina* species (Gastropoda: Prosobranchia) from the tropical western Atlantic. *Proc. Biol. Soc. Wash.* 96:349–54.

Jablonski, D. 1986. Larval ecology and macroevolution in marine invertebrates. *Bull. Mar. Sci.* 39:565–87.

Jablonski, D., and R. A. Lutz. 1983. Larval ecology of marine benthic invertebrates: Paleobiological implications. *Biol. Rev.* 58:21–89.

Jablonski, D., K. W. Flessa, and J. W. Valentine. 1985. Biogeography and paleobiology. *Paleobiology* 11:75–90.

Jackson, J. B. C. 1974. Biogeographic consequences of eurytopy and stenotopy among marine bivalves and their evolutionary significance. *Am. Nat.* 108:541–60.

———. 1988. Does ecology matter? *Paleobiology* 14:307–12.

———. 1994. Constancy and change of life in the sea. *Phil. Trans. Roy. Soc. Lond.* B-343:55–60.

Jackson, Jeremy B. C., Ann F. Budd, and John M. Pandolfi. 1996. The shifting balance of natural communities? In *Evolutionary paleobiology,* ed. D. Jablonski, D. H. Erwin, and J. Lipps. Chicago: University of Chicago Press.

Jackson, J. B. C., P. Jung, A. G. Coates, and L. S. Collins. 1993. Diversity and extinction of tropical American mollusks and emergence of the Isthmus of Panama. *Science* 260:1624–26.

Jackson, J. B. C., and F. K. McKinney. 1990. Ecological processes and progressive macroevolution of marine clonal benthos. In *Causes of evolution,* ed. R. M. Ross and W. D. Allmon, 173–209. Chicago: University of Chicago Press.

Jansen, E., and J. Sjøholm. 1991. Reconstruction of glaciation over the past six Myr from ice-borne deposits in the Norwegian Sea. *Nature* 349:600–603.

Jones, D. S., and P. F. Hasson. 1985. History and development of the marine invertebrate faunas separated by the Central American Isthmus. In *The Great American Biotic Interchange,* ed. F. G. Stehli and S. D. Webb, 325–55. New York: Plenum Press.

Joyce, J. E., L. R. C. Tjalsma, and J. M. Prutzman. 1990. High-resolution planktic stable isotope record and spectral analysis for the last 5.35 M. Y.: Ocean Drilling Program site 625 northeast Gulf of Mexico. *Paleoceanography* 5: 507–29.

Jung, P. 1989. Revision of the *Strombina*-group (Gastropoda: Columbellidae), fossil and living. *Schweiz. Paläontol. Abh.* 111:1–298.

Keen, A. M. 1971. *Sea shells of tropical West America*, 2d ed. Stanford: Stanford University Press.

Keigwin, L. D., Jr. 1982. Isotopic paleoceanography of the Caribbean and east Pacific: Role of Panama uplift in Late Neogene time. *Science* 217:350–53.

Keller, G., C. E. Zenker, and S. M. Stone. 1989. Late Neogene history of the Pacific-Caribbean gateway. *J. South Am. Earth Sci.* 2:73–108.

Knowlton, N., L. A. Weigt, L. A. Solorzano, D. K. Mills, and E. Bermingham. 1993. Divergence in proteins, mitochondrial DNA, and reproductive compatibility across the Isthmus of Panama. *Science* 260:1629–32.

Lessios, H. A. 1981. Divergence in allopatry: Molecular and morphological differentiation between sea urchins separated by the Isthmus of Panama. *Evolution* 35:618–34.

———. 1990. Adaptation and phylogeny as determinants of egg size in echinoderms from the two sides of the Isthmus of Panama. *Am. Nat.* 135:1–13.

Matthews, R. K. 1988. Technical comment on sea level history. *Science* 241: 597–99.

Olson, R. R., and M. H. Olson. 1989. Food limitation of planktotrophic marine invertebrate larvae: Does it control recruitment success? *Ann. Rev. Ecol. Syst.* 20:225–47.

Radwin, G. E. 1977. The family Columbellidae in the western Atlantic. *Veliger* 19:403–17.

Raymo, M. E., W. F. Ruddiman, J. Backman, B. M. Clement, and D. G. Martinson. 1989. Late Pliocene variation in Northern Hemisphere ice sheets and North Atlantic deep water circulation. *Paleoceanography* 4:413–46.

Richmond, R. H., and C. L. Hunter. 1990. Reproduction and recruitment of corals: Comparisons among the Caribbean, the tropical Pacific, and the Red Sea. *Mar. Ecol. Prog. Ser.* 60:185–203.

Robinson, D. G. 1993. The zoogeographic implications of the prosobranch gastropods of the Moin Formation of Costa Rica. *Am. Malacol. Bull.* 10: 251–55.

Rubinoff, I. 1968. Central American sea-level canal: Possible biological effects. *Science* 161:857–61.

Scheltema, R. S. 1989. Planktonic and non-planktonic development among prosobranch gastropods and its relationship to the geographic range of species. In *Reproduction, genetics, and distribution of marine organisms*, eds. J. S. Ryland and P. A. Tyler, 183–88. Fredensborg, Den.: Olsen and Olsen.

Shackleton, N. J., J. Backman, H. Zimmerman, D. V. Kent, M. A. Hall, D. G. Roberts, D. Schnitker, J. G. Baldauf, A. Desprairies, R. Homrighausen, P. Huddlestun, J. B. Keene, A. J. Kaltenback, K. A. O. Krumsiek, A. C. Morton, J. W. Murray, and J. Westberg-Smith. 1984. Oxygen isotope calibration of the onset of ice-rafting and history of glaciation in the North Atlantic region. *Nature* 307:620–23.

Sikes, E. L., L. D. Keigwin, Jr., and W. B. Curry. 1991. Pliocene paleoceanography: Circulation and oceanographic changes associated with the 2.4 Ma glacial event. *Paleoceanography* 6:245–57.

Stanley, S. M. 1986. Anatomy of a regional mass extinction: Plio-Pleistocene decimatioh of the western Atlantic bivalve fauna. *Palaios* 1:17–36.

———. 1988. Adaptive morphology of the shell in bivalves and gastropods. In *The Mollusca, vol. 2, Form and function*, ed. K. M. Wilbur. New York: Academic Press.

Stanley, S. M., and L. D. Campbell. 1981. Neogene mass extinction of western Atlantic molluscs. *Nature* 293:457–59.

Strathmann, R. R. 1985. Feeding and nonfeeding larval development and life-history evolution in marine invertebrates. *Ann. Rev. Ecol. Syst.* 16:339–61.

———. 1986. What controls the type of larval development? Summary statement for the evolution session. *Bull. Mar. Sci.* 39:616–22.

Thorson, G. 1950. Reproductive and larval ecology of marine bottom invertebrates. *Biol. Rev.* 25:1–45.

Vermeij, G. J. 1978. *Biogeography and adaptation*. Cambridge: Harvard University Press.

———. 1987. *Evolution and Escalation*. Princeton: Princeton University Press.

———. 1989. Interoceanic differences in adaptation: Effects of history and productivity. *Mar. Ecol. Prog. Ser.* 57:293–305.

———. 1993. The biological history of a seaway. *Science* 260:1603–4.

Vermeij, G. J., and E. J. Petuch. 1986. Differential extinction in tropical American molluscs: Endemism, architecture, and the Panama land bridge. *Malacologia* 27:29–41.

Webb, T., III, and P. J. Bartlein. 1992. Global changes during the last three million years: Climatic controls and biotic responses. *Ann. Rev. Ecol. Syst.* 23:141–73.

Whittaker, J. E. 1988. *Benthic Cenozoic Foraminifera from Ecuador*. London: British Museum (Nat. Hist.).

Woodring, W. P. 1966. The Panama land bridge as a sea barrier. *Proc. Am. Phil. Soc.* 110:425–33.

10

Diversity of Pliocene–Recent Mollusks in the Western Atlantic: Extinction, Origination, and Environmental Change

Warren D. Allmon, Gary Rosenberg, Roger W. Portell, and Kevin Schindler

INTRODUCTION

The richly fossiliferous Neogene deposits of the western Atlantic, particularly those of the U.S. Atlantic Coastal Plain, are probably as well known paleontologically and paleoenvironmentally as any other time period of any other large region in the world, and should therefore be an excellent source of insights into biotic and abiotic changes during this interval. Despite this potential, however, there is surprising lack of consensus about temporal and geographic patterns of diversity in molluscan faunas contained in these units. This is partly because only recently has adequate attention been paid to the detailed bio-, litho- and chronostratigraphy of the region, especially the correlation of Coastal Plain deposits to global microfossil zonations. There is also a feeling among many paleontologists that Atlantic Coastal Plain mollusks have long ago "been done" by workers such as Tuomey and Holmes (1857), Dall (1890–1903), Martin (1904), Glenn (1904), Clark (1906), and Gardner (1943–48). Consequently, systematic knowledge of these mollusks is much less complete than is generally imagined. Recently published species-level monographs are few, and even lists of species occurrences are lacking for many stratigraphic units. So although at least some information on patterns of diversity in these faunas is available, it is frequently difficult to test either the data or the conclusions derived from them. Furthermore, little attention has been given to patterns of origination and extinction underlying changes in diversity, and still less to the causal processes underlying origination and extinction.

In an effort to document and investigate causes of change in these faunas, we have compiled lists of gastropod species for individual faunal units in the Plio-Pleistocene of the Atlantic Coastal Plain, and for putatively synoptic "time slices" across the region. These lists are based on both the literature and on new collections deposited at the Florida Museum of Natural History. We have also compiled from the literature the first comprehensive list of Recent gastropod species from the western Atlantic, using a standard taxonomy corrected where possible for synonymies. We have compared patterns of faunal change in these gastropod data to patterns reported for bivalves from the same region and time period (Stanley and Campbell 1981; Stanley 1986).

Summary findings of our analysis (Allmon et al. 1993) suggested that (1) gastropod diversity has not changed in Florida since the Pliocene, despite approximately 70% extinction; (2) Recent gastropod diversity in the low-latitude western Atlantic is not demonstrably different than in the low-latitude eastern Pacific; and (3) high extinction rates in the western Atlantic must have been balanced by high origination rates. In this chapter, we will (1) discuss some of the advantages and limitations of the new data sets; (2) compare diversity measures and compositional changes for individual sites across the region; and (3) evaluate hypotheses for the cause of these changes, specifically hypotheses that changes in extinction and origination rates were due to climatic cooling or to changes in productivity caused by changes in paleocirculation.

STRATIGRAPHIC SETTING

The molluscan faunas considered here are contained in Pliocene and younger sediments exposed on the Atlantic Coastal Plain from Virginia to Florida (figs. 10.1, 10.2). Although these faunas have been studied for many years, their stratigraphic relationships have only recently become closely tied to global geochronologic timescales. Of particular importance is the previous assignment of several western Atlantic faunas now believed to be of Pliocene age to the Miocene, largely because of the uncritical use of Lyellian percentages to date them (e.g., Woodring 1928, 1957–82). Stanley (1986 and references therein) has argued convincingly that a high level of Pliocene faunal turnover in the western Atlantic led to this error; many Pliocene faunas of this region have fewer Recent species than they "should"—not because they are old (i.e., Miocene), but because an unusually high proportion have become extinct since the Pliocene.

Based on recent detailed litho-, chrono-, and biostratigraphic work (e.g., Ward and Blackwelder 1980, 1987; Blackwelder 1981b, c; Gibson 1983; Hazel 1983; Cronin et al. 1984; Lyons 1991; Ward et al. 1991), something approaching a consensus on at least the large-scale features of the Plio-Pleistocene stratigraphy of the U.S. Atlantic Coastal Plain has been reached (fig. 10.2).

Yet a number of serious problems remain, and there are considerable

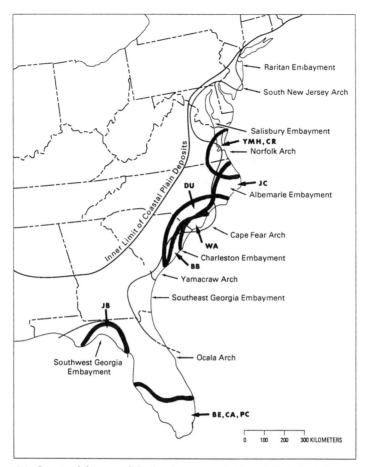

Fig. 10.1 Structural features of the Southeastern U.S. Coastal Plain, including outcrop area of the principal stratigraphic units (see fig. 10.2). Outlined embayments are major sites of sediment accumulation: YMH = Yorktown Formation, Moore House Member; CR = Chowan River Formation; JC = James City Formation; DU = Duplin Formation; WA = Waccamaw Formation; BB = Bear Bluff Formation; JB = Jackson Bluff Formation; BE = Bermont "formation"; CA = Caloosahatchee Formation; PC = Pinecrest Formation. (Base map modified from Ward 1992b; unit outcrop areas modified from Ward et al. 1991.

improvements needed in chronological control of Coastal Plain sediments before a really adequate understanding of the timing of all faunal changes can be obtained (or before these changes can be compared to others from other areas in the Caribbean Basin and beyond). Despite these uncertainties, however, we feel that stratigraphic control is sufficient to point out that the story of molluscan faunal change in the west-

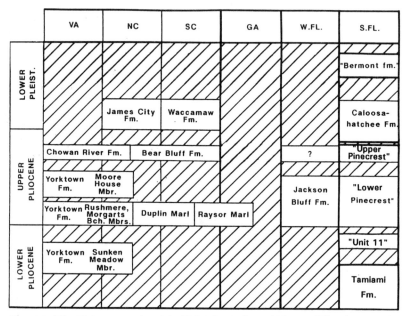

Fig. 10.2 Plio-Pleistocene units of the U.S. Atlantic Coastal Plain and Florida peninsula (modified from Hazel 1983; Ward and Blackwelder 1987; Ward et al. 1991; Ward 1992a; Allmon 1993). Dates for the "Bermont" are uncertain; it may be Lower or Middle Pleistocene in age.

ern Atlantic Neogene is more complex than previously believed. Three stratigraphic problems in particular are relevant to our discussion: the Pinecrest beds, the status of the Calosahatchee Formation, and the so-called Bermont formation.

Pinecrest. The molluscan fauna of the "Pinecrest beds" (fig. 10.2) of southern Florida has figured prominently in previous discussions of environmental and faunal change in this region (e.g., Stanley and Campbell 1981; Stanley 1986). As exposed in quarries near Sarasota, on the central west coast of Florida, the Pinecrest may consist of beds of at least two discrete ages; units 10–5 of Petuch (1982b) (the "lower Pinecrest") appear to date from 3.5 to 2.5 Ma and to be separated by a hiatus from the overlying units 4–2 (the "upper Pinecrest"), which appear to date from 2.5 to 2.0 Ma (Jones et al. 1991; Allmon 1993). This is important for the present analysis because the Pinecrest as a whole has usually been correlated biostratigraphically with the Yorktown Formation of Virginia (e.g., Olsson and Petit 1964; Stanley 1986). Units 4–2 at Sarasota contain typical Pinecrest mollusk and ostracode taxa, but they also contain some taxa with affinities to faunas of the Caloosahatchee and Chowan River Formations (e.g., Jones et al. 1991; Lyons 1991; Ward 1992a; Allmon 1993; Ketcher 1993), which are widely acknowledged to be younger than the Yorktown (Lyons 1991; Ward and Gilinsky 1993). Faunal changes at the top of the Yorktown thus may or may not be simultaneous with changes at the top of the Pinecrest; they may be synchronous with the top of the "lower Pinecrest," the top of the "upper Pinecrest," or some point in-between. This is discussed further below. The nomenclatural history and stratigraphy of the Pinecrest beds are further reviewed by Lyons (1991) and Allmon (1993).

Caloosahatchee. The status of the Caloosahatchee Formation also remains unclear, even though it is one of the few Plio-Pleistocene units in Florida to have been treated explicitly as a lithostratigraphic unit (e.g., Hunter 1978; Lyons 1991). On the basis of He-U coral dates, its age has been estimated as ranging from 2.5 to 1.8 Ma, placing it wholly within the Late Pliocene (Bender 1973; Lyons 1991). Faunally, however, it has traditionally been placed at least partly in the Early Pleistocene (e.g., DuBar 1958a, b; Hazel 1983). As in the Pinecrest, there is also the possibility of significant hiatuses within the unit. DuBar (1958a, b) and Brooks (1968), among others, recognized several "members" of different ages within the Caloosahatchee. Most recently, Hazel (1977, 1983) suggested the existence of a lower (i.e., Upper Pliocene) Caloosahatchee unit at St. Petersburg, containing much of the diverse mollusk fauna described by Olsson and Harbison (1953), and an upper (i.e., Lower Pleistocene) Caloosahatchee unit at the type area along the Caloosahatchee River, with a hiatus of approximately 0.4 my between the two. As pointed out by Lyons (1991), however, Hazel had used 2.0 Ma as the date of the Plio-Pleistocene boundary. Lyons (1991) believes that the Caloosahatchee is entirely of Late Pliocene age, and correlates it with the Chowan River Formation of North Carolina and Virginia, rather than with the James City and Waccamaw Formations.

Bermont. The "Bermont formation" was proposed by DuBar (1974) as an informal name for the fauna designated by Olsson (in Olsson and Petit 1964) as "Unit A" and by Vokes (1963), Hoerle (1970), and McGinty (1970) as the "Glades" formation or unit. While it has traditionally been considered as Middle Pleistocene in age, Lyons (1991) has argued that the Bermont is more probably Early Pleistocene (1.6–1.0 Ma) and correlative with the James City and Waccamaw Formations of the Carolinas and Virginia. Petuch (1990, 1991) has described a number of new stratigraphic units and paleoenvironments from the Plio-Pleistocene of southernmost Florida, some of which may be equivalent to the Bermont of previous authors. Most of these units, however, remain inadequately documented, and they are consequently not dealt with further here. If Petuch's work is verified by other workers, considerable additional litho- and biostratigraphic complexity may become apparent in this region. Finally, Ward and colleagues (1991) indicate that the Pleistocene James City and Waccamaw Formations in North and South Carolina may encompass two separate marine transgressions; it is possible that the second of these is at least in part equivalent to the Bermont.

Pending more definitive study of the Florida Plio-Pleistocene section, and as a temporary heuristic device, in the present chapter we use a simple, three-time-slice model of Coastal Plain stratigraphy (fig. 10.3), consisting of the Late Pliocene Yorktown–Duplin–Jackson Bluff–Pinecrest (YDJP), the Late Pliocene Chowan River–Bear Bluff (CB), and the latest Pliocene–Early Pleistocene James City–Waccamaw–Caloosahatchee (JWC). The YDJP and JWC are taken to represent roughly isochronous faunas living between Virginia and southern Florida, while the CB is assumed to be present only in North Carolina and Virginia. The Bermont fauna is restricted to Florida. It is treated as younger than the JWC fauna, and of Early or Middle Pleistocene age.

MOLLUSCAN DIVERSITY IN THE ATLANTIC COASTAL PLAIN

Recent

The Recent western Atlantic molluscan fauna has traditionally been viewed as less diverse than that of the eastern Pacific (e.g., Dall 1890–1903, 2:349–50; Olsson 1961; Woodring 1966; Keen 1971; Vermeij 1978, 1991; Stanley and Campbell 1981; Stanley 1986), having declined from approximate parity before the uplift of the Central American Isthmus to a depauperate condition at present. Stanley and Campbell (1981), for example, demonstrated that Pliocene mollusk faunas from the western Atlantic suffered as much as 70–80% extinction, versus 30% extinction of Pliocene faunas from California. Using earlier estimates (e.g., Olsson 1968; Campbell et al. 1975), they then noted that the Pinecrest beds of southern Florida may contain as many as 1000 species of mollusks, a number they claimed "approaches the species richness of the entire Caribbean region today" (1981, 457).

The chief problem with these previous discussions of western Atlantic mollusks is that no one has published a complete list of known Recent mollusk species for the entire tropical and subtropical western Atlantic. Most statements have been based on knowledge of only North

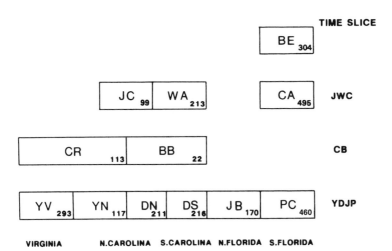

Fig. 10.3 Three putative synoptic "time slices" represented by macrofossiliferous sediments across the Atlantic Coastal Plain from Virginia to south Florida. In lower right corner of each stratigraphic unit is number of known gastropod species. Stratigraphic unit abbreviations: YV = Yorktown Formation (Virginia); YN = Yorktown Formation (North Carolina); DN = Duplin Formation (North Carolina); DS = Duplin Formation (South Carolina), including Raysor Formation; JB = Jackson Bluff Formation (north Florida); PC = "Pinecrest beds" (south Florida); CR = Chowan River Formation (North Carolina and Virginia); BB = Bear Bluff Formation (North Carolina); JC = James City Formation (North Carolina and Virginia); WA = Waccamaw Formation (North Carolina); CA = Caloosahatchee Formation (south Florida); BE = Bermont formation (south Florida). Time slice abbreviations: YDJP = Yorktown–Duplin–Jackson Bluff–Pinecrest; CB = Chowan River–Bear Bluff; JWC = James City–Waccamaw–Caloosahatchee.

American species, supplemented only by anecdotal data from elsewhere. Stanley and Campbell (1981), for example, cited Abbott 1974 for the Recent diversity of Caribbean mollusks, apparently having counted species listed from the Caribbean in that source. Much earlier, however, Van Hyning (1942) had reported the existence of "1213 species and subspecies of [Recent marine] mollusks well identified from Florida," and Nicol (1977) tallied the Florida molluscan fauna at 1137 species, including 1085 gastropods and bivalves.

   In an attempt to address this problem for gastropods, Gary Rosenberg has assembled a database, believed to be the first of its kind, of all described shelled marine gastropod species occurring between Cape Hatteras and Rio de Janeiro, including Bermuda (35°N–23°S). The database was compiled from Abbott 1974 and Turgeon et al. 1988 (North America), de Jong and Coomans 1988 (Netherlands Antilles), Vokes and Vokes 1983 (Yucatán), Ortiz-Corps 1985 and Warmke and Abbott

Table 10.1 Diversity of Recent Western Atlantic Shelled Gastropod Species

Region	Latitude	No. of species (total)	No. of species (< 30 m depth)	No. of species (< 30 m, > 5 mm size)
Cape Hatteras– Rio de Janeiro	35°N–23°S	2911		
Caribbean	30°N–8°N	2362		
Virginia–South Florida[1]	39°N–25°N	1542	1051	817
Florida	30°N–25°N	1171	827	645 (78%)
West Florida[2]		710	545	429 (78.7%)
East Florida[2]		778	617	490 (79.4%)
Western Atlantic[3]	31°N–6°S	2643		
Eastern Pacific	31°N–6°S	2475		

Source: Rosenberg data base.
[1]Excluding the Bahamas.
[2]Excluding the Florida Keys.
[3]Latitude chosen to match that treated by Keen (1971).

1961 (Puerto Rico), Rios 1985 (Brazil), Lyons 1989 (Florida), Altena 1975 (Surinam), and more than 1000 other published works.[1]

This western Atlantic database currently contains 2911 species of shell-bearing gastropods (table 10.1). More than 1100 of these species were not treated by Abbott (1974), who claimed to be comprehensive only for North America, not for the Caribbean or South America. Comparison of this database with the known fauna of the eastern Pacific (table 10.1; Keen 1971; Skoglund 1992) indicates that the gastropod faunas are equally diverse regardless of the latitudinal ranges compared. Within the range examined by Keen (1971) (31°N–6°S), for example, the western Atlantic contains 7% more species of shelled gastropods (table 10.1).

Within the western Atlantic, comprehensive data on species diversity at single localities ("alpha diversity," sensu Whittaker 1977), as opposed to the entire region, are surprisingly few (table 10.2). Based on these limited data, it appears that casual collecting in shallow water at a single locality in the Recent Caribbean Basin yields between 200 and 300 species of gastropods, and that the number of species encountered rises with every increase in exploratory effort (e.g., snorkeling, SCUBA, exposing cryptic habitats and greater depths), reaching around 1000 in at least one case.

1. See Rosenberg 1993a,b for details. This database can be accessed electronically via gopher on the Internet at erato.acnatsci.org.

Table 10.2 Diversity of Recent Western Atlantic Molluscan Communities

Location and Depth	No. of species
Florida Middle Grounds 23–36 m (Hopkins et al. 1977)	75 total 42 gastropods 24 bivalves
Southwest Florida (Phillips et al. 1990)	299 total
East Florida *Oculina* coral reefs (Reed and Mikkelsen 1987)	68 bivalves 155 gastropods
East Florida Hutchinson Island < 12 m (Lyons 1989)	181 gastropods
Yucatán (shallow) (Vokes and Vokes 1983)	533 gastropods
Puerto Rico (Ortiz-Corps 1985)	743 gastropods
Netherlands Antilles (de Jong and Coomans 1988)	723 gastropods
Brazil (Rios 1985)	740 gastropods
Bahamas (Worsfold collection)[1]	1200 gastropods

[1]Housed in the Academy of Natural Sciences of Philadelphia (ANSP).

Plio-Pleistocene

Data on Plio-Pleistocene gastropods of the southeastern U.S. Coastal Plain, including all faunas of this age from Virginia, the Carolinas, Georgia, and both coasts of Florida, were compiled from the database presented in Campbell et al. 1975, with the addition of data from Ward and Blackwelder 1987, Ward and Gilinsky 1993, and our own new collections from three Florida localities. We used the data from the published sources as given, assuming identifications and synonymies to be correct.

The three new Florida lists are based on identification of all gastropods larger than 5 mm in maximum dimension in stratigraphically collected bulk samples; specimens documenting the lists are housed in the

Florida Museum of Natural History (FLMNH). The localities are as follows:

1. "Pinecrest beds" (units 10–2 of Petuch 1982b) at APAC Quarry, Sarasota, Florida (see Stanley 1986; Jones et al. 1991; Allmon 1993)
2. Caloosahatchee Formation (unit 1 of Petuch 1982b) at APAC Quarry, Sarasota, Florida (see Lyons 1991, 1992)
3. Bermont "formation" at Leisey shell pit, Hillsborough County, Florida (see Webb et al. 1989; Hulbert and Morgan 1989; Portell et al. 1992; Portell et al. 1995)

These Plio-Pleistocene data are summarized in table 10.3.

Comparison of Fossil and Recent
Faunal differences between superposed stratigraphic units can be a result of evolutionary change, environmental change and associated tracking of habitats by faunas, or some combination of both. Although understanding of the environments of deposition of the densely macrofossiliferous units in Florida is still incomplete (Allmon 1993), they all appear to represent similar shallow-shelf settings. Indeed, they are often nearly identical lithologically (Scott and Allmon 1992; Allmon 1993), suggesting that differences in the faunas are results of evolutionary originations and extinctions rather than environmental preferences. Farther north on the Coastal Plain, although it is clear that profound lateral facies differences exist (e.g., Ward and Blackwelder 1980; Bailey and Tedesco 1986; Miller 1986), we have compared faunas from across areas much greater than single environments or even single embayments (fig. 10.1). Furthermore, although there are differences between Florida and Virginia Pliocene-Recent faunal patterns (see below), sufficient similarities exist to suggest common underlying causes rather than the vagaries of a series of independent local environmental shifts.

Although we believe these fossil data are useful, it is also necessary to recognize their limitations. Most published species records remain undocumented by detailed faunal monographs. We are thus unable to judge the reliability or comparability of the faunal records or species concepts in most of the literature we have used. Basic ecological data are also unavailable for most taxa, rendering it difficult to make generalizations about what types of species appear and disappear where and when they do.

With these qualifications in mind, we concentrate here on six aspects of the diversity patterns apparent in these data.

Table 10.3 Gastropod Diversity in Plio-Pleistocene of the Atlantic Coastal Plain

Unit	Locality	Time Slice	Reference	No. of species	No. of species restricted
Bermont	Belle Glade	BE	Hoerle 1970[1]	261	
Bermont	Leisey	BE	FLMNH[4]	97	
Bermont	total	BE		304	41
Caloosahatchee	Sarasota[2]	JWC	FLMNH	42	
Caloosahatchee	total	JWC	Campbell et al. 1975	493	209
James City	total	JWC	Ward and Gilinsky 1993	99	28
Waccamaw	total	JWC	Campbell et al. 1975	213	33
Bear Bluff	total	CB	Ward and Gilinsky 1993	22	0
Chowan River	total	CB	Ward and Gilinsky 1993	113	32
Duplin	North Carolina	YDJP	Campbell et al. 1975	211	49
Duplin	South Carolina	YDJP	Campbell et al. 1975	216	21
Jackson Bluff	total	YDJP	Campbell et al. 1975	170	44
Yorktown	North Carolina	YDJP	Campbell et al. 1975	117	15
Yorktown	Virginia	YDJP	Campbell et al. 1975	293	75
Pinecrest	Sarasota[3]	YDJP	FLMNH	333	
Pinecrest	total	YDJP	FLMNH + Campbell et al. 1975, etc.[5]	460	196

Note: See figure 10.3 for time-slice abbreviations.

[1]Hoerle's species count has been updated based on nomenclatural changes since 1970 and FLMNH collections.

[2]Unit 1 of Petuch 1982b at the APAC Quarry, northern Sarasota County; see Lyons 1991, 1992.

[3]Units 10–2 of Petuch 1982b at the APAC and Quality Aggregates quarries, northern Sarasota County; see Stanley 1986, Jones et al. 1991, and Allmon 1993.

[4]See Portell et al. 1992; Portell et al. 1995.

[5]From the FLMNH collections, 333 species identified; 39 species not in FLMNH identified by Campbell et al. (1975); 87 species not in FLMNH identified by Petuch (1982b, 1990, 1991); 1 species not in FLMNH identified by Tripp (1988). See Allmon et al. 1993.

Diversity of the Pinecrest fauna. The Pinecrest mollusk fauna of Florida has been cited as one of the most species-rich known (e.g., Olsson 1968; Petuch 1982b), yet there are no published data to support this. Our data total 460 species of gastropods larger than 5 mm in the Pinecrest beds, out of 333 total gastropods from quarries at Sarasota plus 127 from elsewhere (table 10.3). To obtain an estimate of diversity representing all size ranges, we added to these 460 species a proportion of species with maximum observed size no more than 5 mm that corresponds to the proportion present in the Recent Florida shallow-water fauna (22%; table 10.1). We thus obtain a total of about 590 species of gastropods. To obtain an estimate of the entire Pinecrest mollusk fauna, we first added the 211 bivalve species documented by Stanley (1986) from the Pinecrest. We then added a guess of 20% for species yet to be discovered, which would appear to be a safe estimate even given recent additions to the fauna (e.g., Petuch 1994; Stanley, unpublished data). We also added 5% for the two less diverse classes (chitons and scaphopods [Nicol 1977]). This yields a number of 1051 (590 + 211 + 250), closely matching the estimate of 1000 made by Campbell and others (1975), but below that of 1200 made by Olsson (1968).

Pinecrest bivalve diversity (Stanley 1986) can be used to check the reasonableness of our estimate of gastropod diversity. There are 320 species of bivalves known to occur in water depths of less than 30 m in Florida today (compiled from Florida references in Rosenberg 1993a), yielding a gastropod:bivalve ratio of 2.6:1 for the Recent Florida fauna (827:320). Applying this ratio to the 211 Pinecrest bivalves, we expect a total of 544 Pinecrest gastropods. Our estimate of 590 gastropods is within 8% of this number.

Diversity of Pinecrest and Recent faunas. Olsson (1968, 75) suggested that the Pinecrest fauna contains "approximately three times that of the living Florida fauna in the same depth range." Our estimates of Pinecrest and Recent gastropod diversity allow us to test this statement (table 10.1; fig. 10.4). The count of 460 fossil species larger than 5 mm is about 7% higher than the 429 species in this size range now living in western Florida; the estimate of 590 Pinecrest gastropod species of all sizes is only 8% greater than the 545 living species documented in western Florida. Similarly, numbers for the living fauna of eastern Florida are approximately 4% greater than those for the Pinecrest fauna.

Diversity of Pinecrest and Caloosahatchee faunas. Diversity in the Pinecrest as a whole does not differ substantially from that of the Caloosahatchee as a whole (table 10.3). The Caloosahatchee has a re-

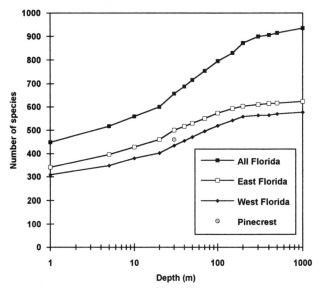

Fig. 10.4 Cumulative number of species by depth for Recent Floridian gastropods larger than 5 mm in maximum size. Each species is counted from its minimum known depth in the western Atlantic. The circle representing total Pinecrest diversity (table 10.3) is placed at 30 m, on the basis of independent paleoenvironmental data (Allmon 1993). (From Allmon et al. 1993).

ported fauna of 386 species of gastropods more than 5 mm in maximum size (table 10.4); this is about 16% fewer than the 460 Pinecrest species, a difference comparable to the 12% difference today between the eastern and western coasts of Florida (table 10.1). Reported diversity is considerably lower in the Bermont than in the underlying Caloosahatchee. The Bermont, however, was not recognized as containing a distinct fauna until the 1960s (Olsson in Olsson and Petit 1964; DuBar 1974), and has been undersampled relative to the longer-known Pinecrest and Caloosahatchee faunas. It is therefore not surprising that recent studies (e.g., Petuch 1988, 1990, 1991, 1994) indicate that the Bermont may contain a considerable number of undescribed or unreported species.

Comparison of diversities within restricted areas over time is more difficult because the Recent fauna has been sampled better geographically than have the fossil assemblages, whereas temporal scope of fossil faunas may be problematic. Stanley (1986, 1991, personal communication) has suggested that the entire Pinecrest section at Sarasota represents a single transgressive-regressive cycle. If this is true, the entire Sarasota Pinecrest gastropod fauna (333 species; table 10.3) lived essen-

Table 10.4 Molluscan Origination and Extinction in Plio-Pleistocene of the Atlantic Coastal Plain

	Number of Species			Apparent Extinction (%)	Apparent Origination (%)
Unit	Older Unit	Younger Unit	Shared		
Pinecrest-Caloosahatchee (gastropods)	460	386	173	62.4	55.2
Pinecrest-Caloosahatchee (bivalves)	211	150	110	47.9	26.7
YDJP-CB (gastropods)	600	100	53	91.1	47.0
CB-JWC (gastropods)	100	462	75	25.0	83.8

Source: Allmon et al. 1993.
Note: These calculations are based only on species larger than 5 mm in maximum size (table 10.3). Abbreviations for formations and time slices are the same as figure 10.3. Gastropod data are from table 10.3; bivalve data are from Stanley 1986.

tially contemporaneously and represents considerably higher local diversity than reported anywhere on the coast of Florida today (42–155 species; table 10.2). Based on taphonomic analysis and the existence of at least one significant hiatus within the Pinecrest section at Sarasota, however, it is apparent that this deposit formed over at least 10^5–10^6 years (Jones et al. 1991; Allmon 1993) and that its constituent beds contain taxa that did not necessarily co-occur. Documentation of species richness and composition of individual horizons at Sarasota is currently underway. Until this is completed we know only that the species richness of the entire Pinecrest at Sarasota is larger by an unknown degree than the species richness of any single time slice through the deposit. The 42 species of gastropods from the Caloosahatchee at Sarasota (see table 10.3), on the other hand, are taken from a single bed, and may be more representative of actual community diversity. Thus, at present we do not know whether local diversity has changed in Florida since the Late Pliocene.

Diversity changes farther north. If the total gastropod faunas of the YDJP and JWC time slices are compared, there appears to be little change in diversity with latitude (figure 10.3). If, however, the southern Florida Caloosahatchee and Pinecrest faunas are excluded, it becomes clear that the northern part of the Coastal Plain (Virginia and the Carolinas) experienced a decline in diversity in the Late Pliocene, while southern Florida did not. Without the Pinecrest, the YDJP contains 422 gastropod species, whereas without the Caloosahatchee the JWC contains only 241 species, a diversity decrease of 42%.

Patterns of extinction and origination. Estimates of extinction and origination based on the data considered here may be somewhat problematic because, in the absence of comprehensive published faunal monographs, taxonomic nomenclature remains formally unstandardized across the Coastal Plain. It should be noted, however, that the data we are using were compiled by a relatively small number of workers with broad experience in the Coastal Plain. Although there is currently no published source to verify it, we are confident that the patterns we report are at least qualitatively robust. Improvements could (and should!) certainly be made, but these are the best and most complete data at hand for an intensely studied region and time interval.

Of the 460 gastropod species we recorded from the Pinecrest, 173 (38%) are known to occur in the overlying Caloosahatchee Formation, yielding an extinction rate between the two faunas of 62% (table 10.4). Of the total gastropod species, only 150 survive today, a number reflecting an overall extinction rate of about 70% (cf. Stanley and Campbell 1981; Stanley 1986).

Whether these rates of extinction are higher than rates earlier in the Neogene is not clear, principally because the Miocene record in Florida is less well known than the Plio-Pleistocene. Stanley and Campbell (1981, 457) suggested that Late Miocene extinction rates in the better-known Chesapeake Bay region were "little, if any, above normal," suggesting a pattern of faunal stasis. In any case, a great deal of extinction did occur in the Late Pliocene, at rates higher than or equal to what they had been previously. At least in southern Florida, however, these extinction rates were approximately matched by high origination rates (table 10.4), and diversity did not decline significantly.

Farther north, extinction rates were approximately as high, but origination rates were lower. For example, Ward and Gilinsky (1993) found that of the 133 species of gastropods in the Moore House Member of the Yorktown Formation, 50 (38%) are known from the Chowan River Formation, giving an apparent extinction level of 62%. Of the 113 species present in the Chowan River, 65 (58%) are known from the James City Formation, giving an apparent extinction level of 43%. Of the 113 species in the Chowan River, 63 do not occur in the Moore House, giving an apparent origination level of 56%. Of the 99 species in the James City, 34 of them are not known from the Chowan River, giving an apparent origination level of 34%.

Timing of extinction. If the unconformity at the top of the "upper Pinecrest" is approximately the same age as that at the top of the Yorktown, it is possible that the high levels of extinction observed in the

Late Pliocene of Florida and Virginia occurred at approximately the same time (cf. Stanley 1986). The top of the Yorktown is traditionally dated at approximately 3.5–3.0 Ma (Blackwelder 1981b; Hazel 1983; Cronin et al. 1984). If, however, the top of the "upper Pinecrest" at Sarasota is considerably younger, approximately 2.5–2.0 Ma (Jones et al. 1991; Allmon 1993), then the extinction may have occurred earlier in the north than in the south.

An exactly opposite pattern may be indicated by Lyons's (1991) critique of Ward and Blackwelder's (1987) biostratigraphic interpretations of faunas from North Carolina. Ward and Blackwelder recognized four units (B–E in upward sequence) at the Lee Creek Mine, assigning unit B to the Chowan River and units C–E to the James City. Lyons points out that several gastropod species present in units D and E "do not occur in Florida deposits above the uppermost Caloosahatchee." This suggests "either that Units C–E are actually of late Pliocene age or that the taxa became extinct at the end of the Pliocene in Florida but survived into the early Pleistocene in North Carolina" (Lyons 1991, 148).

Diachroneity may extend farther than the Coastal Plain itself. McNeill and others (1988) have reported results of a detailed magneto-stratigraphic analysis of a core taken on San Salvador, Bahamas. The core contains molds and casts referable to molluscan taxa reported by Woodring (1928) from the Bowden fauna of Jamaica (Williams 1985). Based on the upper limit of this Bowden-type assemblage, McNeill and others conclude that the disappearance of the fauna from the Bahamas occurred between 2.7 and 2.6 Ma. Stanley (1991) has used this date to argue that the Pinecrest fauna must have become extinct at or before this date, since the "warm-adapted elements of the [Pinecrest] at Sarasota should have disappeared by this time because they were positioned at the margins of the tropics."

Recent reanalysis by McNeill of a larger number of cores across the Bahama Platform, however, indicates that the Bowden-type fauna disappeared at different times in different areas of the platform, and that the last appearance of the fauna on the platform was later than 2.4 Ma (McNeill 1989, personal communication). On Grand Bahama the Bowden fauna disappears in a reversed interval, interpreted as being before the Gauss-Gilbert chron boundary (i.e., earlier than 3.4 Ma). On Williams Island west of Andros, the fauna disappears in a normal interval, interpreted as being in the middle of the Gauss normal chron (i.e., earlier than 2.4 Ma). In both the San Salvador core and in a core from northeastern Andros, the fauna disappears in a reversed interval, interpreted as being the lower part of the Matuyama reversed chron and dated at 2.4–1.8 Ma (McNeill 1989, personal communication). Our analysis therefore shows that the Pinecrest is very species-rich, but not as species-rich as the highest previous estimates and probably no more species-rich than either the fauna of the Caloosahatchee Formation or the fauna living across a comparable range of space and environments in Florida today. Diversity thus appears not to have declined significantly in Florida since around 3.0 Ma. In the Carolinas and Virginia during this time, however, diversity does appear to have declined. Substantial extinction, therefore, did occur in the Late Pliocene–Early Pleistocene across the Atlantic Coastal Plain. In the south this extinction was replaced by new originations, whereas in the north it was not.

COMPARISON WITH OTHER FAUNAL PATTERNS

If the patterns of faunal turnover described here were limited to mollusks of the southeastern U.S. Coastal Plain, it might be argued that they could be explained by strictly regional, or even a series of coordinated local environmental changes. Comparison with other nonmolluscan faunal data, however, suggests that such explanations are insufficient.

Budd and others (1993, this volume, chap. 7) have demonstrated that between 4.0 and 2.0 Ma, rates of both origination and extinction of Caribbean coral species were greatly accelerated, resulting in an episode of major faunal turnover. During this interval, overall species richness remained roughly the same. A subsequent interval of only accelerated extinction between 2.0 and 1.0 Ma created the modern scleractinian fauna of the Caribbean, which is depauperate relative to the Indo-Pacific. Budd and colleagues further note that shallow-grass flat corals, characterized by small, short-lived, free-living colonies, were more susceptible to extinction in both episodes. Coastal areas and offshore carbonate platforms, however, were equally affected.

Similarly, diversity of Late Miocene to Recent mollusks from the Caribbean coast of Costa Rica and Panama, as measured by number of genera and subgenera, has increased in this region since the Miocene, despite high levels of extinction in the Late Pliocene (Jackson et al. 1993). Increased origination in these taxa coincided with increased extinction in the Late Pliocene. In contrast, high rates of extinction were not balanced by origination in some "paciphile" taxa (Jackson et al. 1993; Jackson et al., this volume, chap. 9).

A

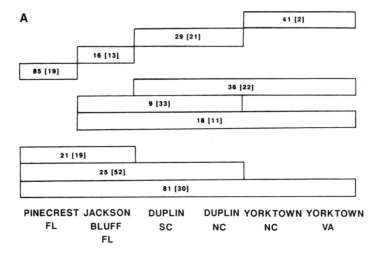

41 [2]

29 [21]

16 [13]

85 [19]

36 [22]

9 [33]

18 [11]

21 [19]

25 [52]

81 [30]

PINECREST FL	JACKSON BLUFF FL	DUPLIN SC	DUPLIN NC	YORKTOWN NC	YORKTOWN VA

B

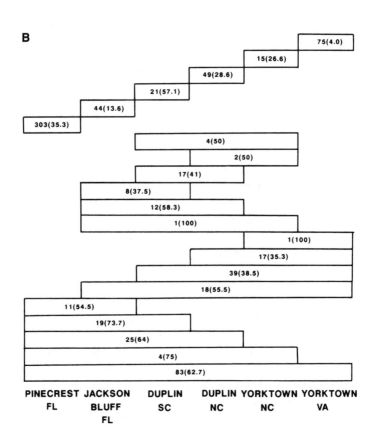

75(4.0)

15(26.6)

49(28.6)

21(57.1)

44(13.6)

303(35.3)

4(50)

2(50)

17(41)

8(37.5)

12(58.3)

1(100)

1(100)

17(35.3)

39(38.5)

18(55.5)

11(54.5)

19(73.7)

25(64)

4(75)

83(62.7)

PINECREST FL	JACKSON BLUFF FL	DUPLIN SC	DUPLIN NC	YORKTOWN NC	YORKTOWN VA

Thus mollusks in the southwestern Caribbean and corals throughout the Caribbean appear to show a pattern of turnover similar to that shown by gastropods (and probably bivalves) in the southern U.S. Atlantic Coastal Plain. Late Pliocene "regional mass extinction" did not greatly reduce diversity of these groups in the western Atlantic because extinction was approximately balanced by origination.

CORRELATION WITH PALEOENVIRONMENTAL CHANGE

Stanley (Stanley 1986; Stanley and Campbell 1981; see also Blackwelder 1981a; Ward et al. 1991) argued that high extinction rates among western Atlantic bivalves around 3.0 Ma were due almost completely to cooling temperatures. His principal evidence was biogeographic—the higher survivorship (as represented by Lyellian percentages) into the Pleistocene of Pliocene species with larger geographic ranges (Stanley 1986; fig. 10.5A). Similar analysis of our gastropod data (fig. 10.5B) gives mixed results. Species with the greatest geographic range (those that occur in the Pinecrest, Jackson Bluff, Duplin, and Yorktown) show relatively high survivorship (63%) into JWC time, but not the highest, and the species with the narrowest geographic range do not have the lowest survivorship. Even if total range size in degrees plotted against survivorship does show a positive relationship (fig. 10.6), however, this does not necessarily require a primary role for temperature tolerance in survival. Large geographic range itself can confer resistance to extinction (Jablonski 1987, 1989, and references therein), independent of temperature effects. A plot of range midpoint in degrees north against survivorship should be more revealing of temperature effects, for it indicates the actual latitudinal position (and therefore a first approximation temperature preference) of different cohorts of species, rather than just range size. Such a plot for our gastropod data (fig. 10.7) yields only a very weak relationship.

Stanley also cited as evidence for a primary role for temperature decline in late Neogene mollusk extinctions the initiation of continental

Fig. 10.5 Geographic range and survivorship in late Pliocene mollusk species from the YDJP time slice: (A) bivalves (data from Stanley 1986); (B) gastropods (our data). Width of each bar represents geographic range across formations listed underneath the bar; first number is total number of species occurring in the formations. In A, second number in brackets is Lyellian percentage (percentage of species surviving to the Recent); in B, second number in parentheses is percentage of these species surviving into the later JWC time slice.

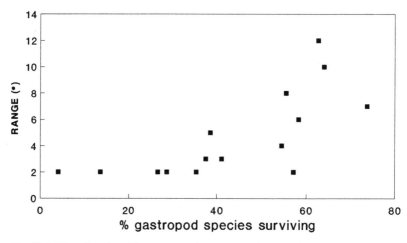

Fig. 10.6 Plot of survivorship vs. geographic range (in degrees of latitude) for gastropod species from the Late Pliocene YDJP time slice (data from fig. 10.5). Simple linear regression of these data yields an r^2 value of .50 ($p = .25$).

glaciation in the Northern Hemisphere at 3.2 Ma, and suggested that conditions at this time were similar to those that prevailed during the last glacial maximum 18 thousand years ago. It is no longer clear that this conclusion is valid.

Based on data on ice-rafted debris, there is evidence of significant Northern Hemisphere ice at 3.4–3.0 Ma (e.g., Jansen and Sjøholm 1991), but this evidence suggests very small ice volume compared to the period of ice buildup that began 2.8–2.6 Ma (Jansen et al. 1990). Other data are consistent with the view that Northern Hemisphere glaciation between 3.5 and 3.0 Ma was of a lesser magnitude than the glaciation that commenced at 2.8–2.6 Ma (e.g., Keigwin 1986; Curry and Miller 1989; Cronin 1990; Joyce et al. 1990; Sikes et al. 1991). A growing body of data suggests that sea surface temperatures in the North Atlantic between 3.5 and 3.0 Ma were higher than at present (e.g., Cronin 1991; Dowsett and Poore 1991; Crowley 1991; Cronin and Dowsett this volume, chap. 4). These warm temperatures may have occurred on both sides of the North Atlantic at this time (Jenkins and Houghton 1987; Jenkins et al. 1988; Wilkinson 1980). It is therefore between 3.0 and 2.0 Ma, and not between 3.5 and 3.0 Ma, that we should look for faunal changes to be correlated with glaciation or its major effects.

As discussed above, significant molluscan extinction, associated with a reduction in species diversity, did occur at higher latitudes (i.e., north of Cape Hatteras) at approximately the end of Yorktown deposition

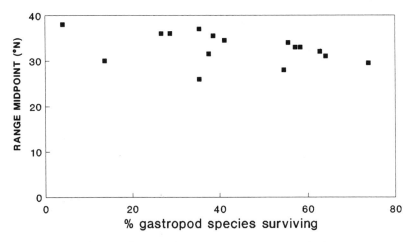

Fig. 10.7 Plot of survivorship vs. midpoint of geographic range (in degrees north latitude) for gastropod species from the Late Pliocene YDJP time slice (data from fig. 10.5). Simple linear regression of these data yields an r^2 value of .17 (p = .30).

(3.5–3.0 Ma; Cronin et al. 1984). Faunal changes at lower latitudes (i.e., south of Cape Hatteras, especially Florida) may have occurred somewhat later, as late as 2.5–2.0 Ma (between the Pinecrest and Caloosahatchee faunas). The change at higher latitudes would thus not appear to coincide with the onset of glaciation, but the lower latitude change might. It is not clear, however, whether this event actually coincided with cooler temperatures at these latitudes.

Although 3–2 Ma was a time of major ice buildup in the Northern Hemisphere, it is not clear that low-latitude sea surface temperatures declined. In the eastern Atlantic at low latitudes, for example, the 2.4 Ma $\delta^{18}O$ event was only 50–60% as great as "full glacial" (i.e., 18 Ka) conditions (Curry and Miller 1989). Elsewhere in the Atlantic, evidence of cooling at low latitudes at a time that could be roughly coincident with the Pinecrest-Caloosahatchee transition (i.e., 2.5–2.0 Ma) is ambiguous (Cronin 1990, 1991; Krantz 1990; Dowsett and Poore 1991; Cronin and Dowsett, this volume, chap. 4; Jones and Allmon 1995a,b): Deep Sea Drilling Project/Ocean Drilling Program sites 502 (11.5°N), 667 (4.5°N), and 672 (15.5°N) all show no decline in temperature based on foraminiferal faunas at or around 2.4 Ma (Wiggs and Poore 1991; Dowsett and Poore 1991; Foley and Dowsett 1992; Wiggs and Dowsett 1992; Dowsett, personal communication), but site 665A (2°N) does show a decline (Sikes et al. 1991). At high latitudes, DSDP sites 552 (56°N) and 548 (49°N) show foraminiferal evidence of cooling and/or ice-rafted detritus at around 2.4 Ma (Dowsett and Poore 1990; Dowsett and Loubere 1992).

Recent studies (Anderson and Webb 1994; Guilderson et al. 1994) have thrown into doubt earlier conclusions that low-latitude temperatures did not decline during Late Pleistocene glacial intervals. Even if these revisionist results are confirmed, however, they have no necessary bearing on Late Pliocene climates, which may or may not have been significantly different from those of the Late Pleistocene (R. G. Fairbanks, personal communication).

Evidence for temperature decline at times of Pliocene faunal change is thus problematic. At least enough doubt exists about the timing and magnitude of cooling in low latitudes to support exploration of additional environmental factors that might be coincident with faunal change. In fact, even if significant low-latitude cooling in the Late Pliocene is eventually demonstrated, there is considerable evidence for changes in another environmental variable—productivity—at around the time of formation of the Central American Isthmus (Allmon et al. 1996).

As discussed by Stanley (1986, 19), several authors (e.g., Woodring 1966; Vermeij 1980, 1987, 1989, 1990) have suggested changes in the productivity or nutrient regimes in the western Atlantic as an explanation for the extinction of Neogene mollusks in the region. Good circumstantial evidence exists that (at least) local upwelling and associated high productivity existed prior to around 3.0 Ma in the low-latitude western Atlantic and then declined (Allmon et al. 1996); in contrast, upwelling and productivity appear to have changed little in the low-latitude eastern Pacific. Evidence for upwelling in the western Atlantic prior to 3.0 Ma includes (1) local areas of cooler temperatures in an otherwise warm Late Pliocene (Cronin and Dowsett 1990; Cronin 1991); (2) vertebrate and invertebrate faunal indicators of cool waters and/or high productivity amid otherwise subtropical temperatures in the Late Pliocene Pinecrest beds of Florida (Allmon et al. 1996); (3) carbon and oxygen isotopic evidence (Jones and Allmon 1996); (4) widespread phosphogenesis in Florida and the Carolinas throughout the Miocene and into the Early Pliocene (Riggs 1984; Riggs and Sheldon 1991); (5) high Plio-Pleistocene extinction in the western Atlantic in at least one group of mollusks (turritellid gastropods) that today occur in highest abundance in high-productivity environments (Allmon 1992), with survival of relict Caribbean pockets in areas of upwelling (Petuch 1976, 1981, 1982a; Vermeij and Petuch 1986; Allmon 1988); (6) prior to 11–10 Ma, accumulation of biogenic silica was much more common in the Atlantic than at present, indicating higher productivity (Keller and Barron 1983); (7) diversity and productivity of planktonic foraminifera were higher in the western Atlantic before about 3.0 Ma (Ma'alouleh and Moullade 1987; Keller et al. 1989).

After formation of the isthmus, these patterns of productivity appear to have broken down in the Caribbean more than in the eastern Pacific. Gartner and colleagues (1987, 266), for example, note increased "instability" in plankton in the Caribbean and Gulf of Mexico in the Late Miocene and Early Pliocene, but no such changes in the Pacific. Both empirical data and ocean circulation modeling indicate that upwelling in the eastern Pacific did not change significantly after formation of the isthmus (Hayes et al. 1989; Maier-Reimer et al. 1990). Collectively, these data suggest that productivity was higher in the western Atlantic prior to closure of the Central American Isthmus than it was afterward (Allmon et al. 1996).

Stanley (1986, 27) has criticized the hypothesis that a reduction in productivity could have been the primary cause of increased extinction in the western Atlantic on the basis of "the fact that the extinction occurred throughout the Gulf of Mexico and Caribbean, even in areas never characterized by upwelling." This argument may not be valid, however, because it remains to be documented exactly where in the western Atlantic upwelling did and did not occur during the Pliocene.

CONCLUSIONS

The Late Pliocene was thus a time of high extinction *and* high origination for gastropods of the southeastern U.S. Atlantic coast, particularly in Florida. This conclusion broadens our view of faunal change in the Late Neogene beyond a "regional mass extinction" and suggests that causes should be sought not only for extinction but for speciation as well.

It is not clear that temperature was the primary cause of this faunal change. Temperature decrease may have been to blame for some of the Late Pliocene extinctions north of Cape Hatteras, but south of this area additional causal factors for both extinction and origination should be sought. There is growing evidence that a change in productivity, associated in some way with the formation of the Central American Isthmus, may have been at least partially responsible for some of these faunal patterns. Diachroneity of faunal turnover appears to be a real phenomenon across at least this part of the western Atlantic at this time, and may provide a model for resolution of this question on a case by case basis.

ACKNOWLEDGMENTS

We are grateful to Alexandra Collazos, Kathleen Ketcher, and Stephen Schellenberg for assistance in data analysis, to Ann Budd, Thomas Cronin, Thomas Crowley, Harry Dowsett, Jeremy Jackson, Douglas Jones, and Edward Joyce for discussion, and to Ann Budd, Jeremy Jackson, Kenneth Johnson, William Lyons, Geerat Vermeij, and an anonymous reviewer for comments on earlier drafts. This chapter is contribution to paleobiology number 416 from the University of Florida.

REFERENCES

Abbott, R. T. 1974. *American seashells.* 2d ed. New York: Van Nostrand Reinhold.

Allmon, W. D. 1988. Ecology of Recent turritelline gastropods (Prosobranchia: Turritellidae): current knowledge and paleontological implications. *Palaios* 3:259–84.

———. 1992. Role of temperature and nutrients in extinction of turritelline gastropods: Cenozoic of the northwestern Atlantic and northeastern Pacific. *Palaeogeogr., Palaeoclimatol., Palaeoecol.* 92:41–54.

———. 1993. Age, environment, and mode of deposition of the densely fossiliferous Pinecrest Sand (Pliocene of Florida): Implications for the role of biological productivity in shell bed formation. *Palaios* 8:183–201.

Allmon, W. D., S. D. Emslie, D. S. Jones, and G. S. Morgan. 1996. Late Neogene oceanographic change along Florida's west coast: Evidence and mechanisms. *J. Geol.* 104:143–62.

Allmon, W. D., G. Rosenberg, R. W. Portell, and K. S. Schindler. 1993. Diversity of Atlantic Coastal Plain mollusks since the Pliocene. *Science* 260:1626–29.

Altena, C. O. van Regteren. 1975. The marine Mollusca of Suriname (Dutch Guiana), Holocene and Recent. Part 3, Gastropoda and Cephalopoda. *Zool. Verhandl.* no. 139:1–104.

Anderson, D. S., and R. S. Webb, 1994, Ice-age tropics revisited. *Nature* 367:23–24.

Bailey, R. H., and S. A. Tedesco. 1986. Paleoecology of a Pliocene coral thicket from North Carolina: An example of temporal change in community structure and function. *J. Paleontol.* 60:1159–76.

Bender, M. L. 1973. Helium-uranium dating of corals. *Geochim. Cosmochim. Acta* 37:1229–47.

Blackwelder, B. W. 1981a. Late Cenozoic marine deposition in the United States Atlantic Coastal Plain related to tectonism and global climate. *Palaeogeogr., Palaeoclimatol., Palaeoecol.* 34:87–114.

———. 1981b. Late Cenozoic stages and molluscan zones of the U.S. Middle Atlantic Coastal Plain. *J. Paleontol.* 55, no. 5 (suppl.):1–35.

———. 1981c. Stratigraphy of Upper Pliocene and Lower Pleistocene marine and estuarine deposits of northeastern North Carolina and southeastern Virginia. *U.S. Geol. Surv. Bull.* 1502-B:1–16.

Brooks, H. K. 1968. The Plio-Pleistocene of Florida, with special reference to the strata outcropping on the Caloosahatchee River. In *Late Cenozoic stratigraphy of southern Florida—a reappraisal*, comp. R. D. Perkins, 3–42. Miami: Miami Geological Society.

Budd, A. F., K. G. Johnson, and T. A. Stemann. 1993. Plio-Pleistocene extinctions and the origin of the modern Caribbean reef coral fauna. Paper presented at the Colloquium on Global Aspects of Coral Reefs: Health, Hazards, and History, Miami, Fla.

Campbell, L. D., S. Campbell, D. Colquhoun, and J. Ernissee. 1975. Plio-Pleistocene faunas of the central Carolina Coastal Plain. *S.C. Geol. Notes* 19:51–124.

Clark, W. B. 1906. Mollusca. In *Pliocene and Pleistocene*, 176–209. Baltimore: Maryland Geological Survey, Johns Hopkins University Press.

Cronin, T. M. 1990. Evolution of Neogene and Quaternary Marine Ostracoda, United States Atlantic Coastal Plain. Part 4, Evolution and speciation in Ostracoda. *U.S. Geol. Sur. Prof. Pap.* 1367-C:1–43.

———. 1991. Pliocene shallow water paleoceanography of the North Atlantic Ocean based on marine ostracodes. *Quatern. Sci. Revs.* 10:175–88.

Cronin, T. M., L. M. Bybell, R. Z. Poore, B. W. Blackwelder, J. C. Liddicoat, and J. E. Hazel. 1984. Age and correlation of emerged Pliocene and Pleistocene deposits, U.S. Atlantic Coastal Plain. *Palaeogeogr., Palaeoclimatol., Palaeoecol.* 47:21–51.

Cronin, T. M., and H. J. Dowsett. 1990. A quantitative micropaleontologic method for shallow marine paleoclimatology: Application to Pliocene deposits of the western North Atlantic Ocean. *Mar. Micropaleontol.* 16:117–47.

Crowley, T. J. 1991. Modeling Pliocene warmth. *Quatern. Sci. Revs.* 10:275–82.

Curry, W. B., and K. G. Miller. 1989. Oxygen and carbon isotopic variation in Pliocene benthic foraminifers of the equatorial Atlantic. *Proc. ODP, Sci. Results* 108:157–66.

Dall, W. H. 1890–1903. Contributions to the Tertiary fauna of Florida, with especial reference to the Miocene silex-beds of Tampa and the Pliocene of the Caloosahatchie river. *Trans. Wagner Free Inst. Sci., Phil.* 3, part 1 (1890); part 2 (1892); part 3 (1895); part 4 (1898); part 5 (1900); part 6 (1903).

de Jong, K. M., and H. E. Coomans. 1988. *Marine gastropods from Curaçao, Aruba, and Bonaire*. Studies on the Fauna of Curaçao and Other Caribbean Islands, vol. 69, no. 214. Amsterdam: Found. Sci. Res. Neth. Ant.

Dowsett, H. J. and P. Loubere. 1992. High resolution Late Pliocene sea-surface temperature record from the northeast Atlantic Ocean. *Mar. Micropaleontol.* 20:91–105.

Dowsett, H. J. and R. Z. Poore. 1990. A new planktic foraminifer transfer function for estimating Pliocene-Holocene paleoceanographic conditions in the North Atlantic. *Mar. Micropaleontol.* 16:1–23.

———. 1991. Pliocene sea surface temperatures of the North Atlantic Ocean at 3.0 Ma. *Quatern. Sci. Revs.* 10:189–204.

DuBar, J. R. 1958a. Neogene stratigraphy of southwestern Florida. *Trans. Gulf Coast Assoc. Geol. Soc.* 8:129–55.

————. 1958b. Stratigraphy and paleontology of the Late Neogene strata of the Caloosahatchee River area of southern Florida. *Fla. Geol. Surv. Bull.* 40: 1–267.

————. 1974. Summary of the Neogene stratigraphy of southern Florida. In *Post-Miocene stratigraphy of central and southern Atlantic Coastal Plain*, ed. R. Q. Oaks and J. R. Dubar, 206–31. Logan: Utah State University Press.

Foley, K. M., and Dowsett, H. J. 1992. Pliocene planktic foraminifer census data from Ocean Drilling Program holes 667 and 659A. Open File Report 92–434, U.S. Geological Survey, Reston, Va.

Gardner, J. A. 1943–48. Mollusca from the Miocene and Lower Pliocene of Virginia and North Carolina. Part 1, Pelecypoda. Part 2, Scaphopoda and Gastropoda. *U.S. Geol. Surv. Prof. Pap.* 199-A (1943):1–178; 199-B (1948): 179–310.

Gartner, S., J. Chow, and R. J. Stanton, Jr. 1987. Late Neogene paleoceanography of the eastern Caribbean, the Gulf of Mexico, and the eastern equatorial Pacific. *Mar. Micropaleontol.* 12:255–304.

Gibson, T. G. 1983. Stratigraphy of Miocene through Lower Pleistocene strata of the United States central Atlantic Coastal Plain: In *Geology and paleontology of the Lee Creek Mine, North Carolina*, vol. 1. ed. C. E. Ray, 35–80. Smithsonian Contributions to Paleobiology, no. 53. Washington, D.C.

Glenn, L. C. 1904. Pelecypoda. In *Miocene*, ed. W. B. Clark, 274–401. Baltimore: Maryland Geological Survey.

Guilderson, T. P., R. G. Fairbanks, and J. L. Rubenstone. 1994. Tropical temperature variations since 20,000 years ago: Modulating interhemispheric climate change. *Science* 263:663–65.

Hayes, P. E., N. G. Pisias, and A. K. Roelofs. 1989. Paleoceanography of the eastern equatorial Pacific during the Pliocene: A high-resolution radiolarian study. *Paleoceanography* 4:57–74.

Hazel, J. E. 1977. Distribution of some biostratigraphically diagnostic ostracodes in the Pliocene and Lower Pleistocene of Virginia and northern North Carolina. *U.S. Geol. Surv. J. Res.* 5(3): 373–88.

————. 1983. Age and correlation of the Yorktown (Pliocene) and Croatan (Pliocene and Pleistocene) Formations at the Lee Creek Mine. In *Geology and paleontology of the Lee Creek Mine, North Carolina*, ed. C. E. Ray, 81–200. Smithsonian Contributions to Paleobiology, no. 53. Washington, D.C.

Hoerle, S. E. 1970. Mollusca of the "Glades" unit of southern Florida. Part 2, List of molluscan species from the Belle Glade Rock Pit, Palm Beach County, Florida. *Tulane Stud. Geol. Paleontol.* 8:56–68.

Hopkins, T. S., D. R. Blizzard, and D. K. Gilbert. 1977. The molluscan fauna of the Florida Middle Grounds, with comments on its zoogeographical affinities. *NE Gulf Sci.* 1(1): 39–47.

Hulbert, R. C., Jr., and G. S. Morgan. 1989. Stratigraphy, paleoecology, and vertebrate fauna of the Leisey Shell Pit local fauna, Early Pleistocene (Irvingtonian) of southwestern Florida. *Pap. Fla. Paleontol.*, no. 2:1–19.

Hunter, M. E. 1978. What is the Caloosahatchee Marl? In *Hydrogeology of south-central Florida*, ed. M. P. Brown, 61–88. Twenty-second Field Conference. Southeastern Geological Society Publication 20.

Jablonski, D. 1987. Heritability at the species level: Analysis of geographic ranges of Cretaceous mollusks. *Science* 238:360–63.

———. 1989. The biology of mass extinction: A paleontological view. *Phil. Trans. Roy. Soc. Lond.* B-325:357–68.

Jackson, J. B. C., P. Jung, A. G. Coates, and L. S. Collins. 1993. Diversity and extinction of tropical American mollusks and the emergence of the Isthmus of Panama. *Science* 260:1624–26.

Jansen, E., and J. Sjøholm. 1991. Reconstruction of glaciation over the past 6 myr from ice-borne deposits in the Norwegian Sea. *Nature* 349:600–603.

Jansen, E., J. Sjøholm, U. Bleil, and J. A. Erichsen. 1990. Neogene and Pleistocene glaciations in the Northern Hemisphere and Late Miocene–Pliocene global ice volume fluctuations: Evidence from the Norwegian Sea. In *Geological history of the polar oceans: Arctic versus Antarctic*, ed. U. Bleil and J. Thiede, 677–705. NATO ASI Series. Dordrecht: Kluwer.

Jenkins, D. G., D. Curry, B. M. Funnell, and J. E. Whittaker. 1988. Planktonic Foraminifera from the Pliocene Coralline Crag of Suffolk, eastern England. *J. Micropaleontol.* 7:1–10.

Jenkins, D. G., and S. D. Houghton. 1987. Age, correlation, and paleoecology of the St. Erth beds and the Coralline Crag of England. *Mededl. Werkgr. Tert. Kwart. Geol.* 24(1/2): 147–56.

Jones, D. S., and W. D. Allmon. 1995. Records of upwelling, seasonality, and growth in stable isotope profiles of Pliocene mollusk shells from Florida. *Lethaia* 28:61–74.

———. 1996. Pliocene marine temperatures on the west coast of Florida. In *The Pliocene*, ed. J. Wrenn. Dallas: American Association of Stratigraphic Palynologists. In press.

Jones, D. S., B. J. MacFadden, S. D. Webb, P. M. Mueller, D. A. Hodell, and T. M. Cronin. 1991. Integrated geochronology of a classic Pliocene fossil site in Florida: Linking marine and terrestrial biochronologies. *J. Geol.* 99: 637–48.

Joyce, J. E., L. R. C. Tjalsma, and J. M. Prutzman. 1990. High-resolution planktic stable isotope record and spectral analysis for the last 5.35 M.Y.: Ocean Drilling Program site 625, Northeast Gulf of Mexico. *Paleoceanography* 5: 507–29.

Keen, A. M. 1971. *Sea shells of tropical west America*. Stanford: Stanford University Press.

Keigwin, L. D., Jr. 1986. Pliocene stable-isotope record of Deep Sea Drilling Project site 606: Sequential events of ^{18}O enrichment beginning at 3.12 Ma. *Init. Repts. DSDP* 94:911–20.

Keller, G., and J. A. Barron. 1983. Paleoceanographic implications of Miocene deep-sea hiatuses. *Geol. Soc. Am. Bull.* 94:590–613.

Keller, G., C. E. Zenker, and S. M. Stone. 1989. Late Neogene history of the Pacific-Caribbean gateway. *J. South Am. Earth Sci.* 2(1): 73–108.

Ketcher, K. M. 1993. Preliminary molluscan biozonation of Plio-Pleistocene fossiliferous deposits in southwestern Florida. M.S. thesis, University of South Florida, Tampa, Fla.

Krantz, D. E. 1990. Mollusk-isotope records of Plio-Pleistocene marine paleoclimate, U.S. Middle Atlantic Coastal Plain. *Palaios* 5:317–35.

Lyons, W. G. 1989. *Nearshore marine ecology at Hutchinson Island, Florida, 1971–1974.* Part II, *Mollusks.* Florida Marine Research Institute, Publication no. 47.

———. 1991. Post-Miocene species of *Latirus* Montfort, 1810 (Mollusca: Fasciolariidae) of southern Florida, with a review of regional marine biostratigraphy. *Bull. Fla. Mus. Nat. Hist., Biol. Sci.* 35(3): 131–208.

———. 1992. Caloosahatchee-age and younger molluscan assemblages at APAC Mine, Sarasota County, Florida. In *The Plio-Pleistocene stratigraphy and paleontology of southern Florida,* ed. T. M. Scott and W. D. Allmon, 133–60. Florida Geological Survey Special Publication 36. Tallahassee, Fla.

Ma'alouleh, K., and M. Moullade. 1987. Biostratigraphic and paleoenvironmental study of Neogene and Quaternary planktonic foraminifers from the lower continental rise of the New Jersey margin (western North Atlantic), Deep Sea Drilling Project leg 93. *Init. Repts. DSDP* 93 (603): 481–91.

Maier-Reimer, E., U. Mikolajewicz, and T. J. Crowley. 1990. Ocean General Circulation Model sensitivity experiment with an open Central American Isthmus. *Paleoceanography* 5:349–66.

Martin, G. C. 1904. Gastropoda. In *Miocene,* ed. W. B. Clark, 131–269. Baltimore: Maryland Geological Survey.

McGinty, T. L. 1970. Mollusca of the "Glades" unit of southern Florida. Part 1, Introduction and observations. *Tulane Stud. Geol. Paleontol.* 8(2): 53–56.

McNeill, D. F. 1989. Magnetostratigraphic dating and magnetization of Cenozoic platform carbonates from the Bahamas: Ph.D. diss., University of Miami, Coral Gables, Fla.

McNeill, D. F., R. N. Ginsburg, S.-B. R. Chang, and J. L. Kirschvink. 1988. Magnetostratigraphic dating of shallow-water carbonates from San Salvador, Bahamas. *Geology* 16:8–12.

Miller, W., III. 1986. Community replacement in estuarine Pleistocene deposits of eastern North Carolina. *Tulane Stud. Geol. Paleontol.* 19(3/4): 97–122.

Nicol, D. 1977. Mollusk fauna of Florida. *Nautilus* 91:4–10.

Olsson, A. A. 1961. *Mollusks of the tropical Eastern Pacific, particularly from the southern half of the Panamic-Pacific faunal province: Panamic-Pacific Pelecypoda.* Ithaca, N.Y.: Paleontological Research Institution.

———. 1968. A review of Late Cenozoic stratigraphy of southern Florida. In *Late Cenozoic stratigraphy of southern Florida—a reappraisal,* comp. R. D. Perkins, 66–82. Miami: Miami Geological Society.

Olsson, A. A., and A. Harbison. 1953. *Pliocene Mollusca of southern Florida.* Academy of Natural Sciences of Philadelphia, Monograph 8. Philadelphia.

Olsson, A. A., and R. E. Petit. 1964. Some Neogene Mollusca from Florida and the Carolinas. *Bulls. Am. Paleontol.* 47:509–75.

Ortiz-Corps, E. 1985 (1983). *An annotated checklist of the Recent marine Gastro-*

poda (Mollusca) from Puerto Rico. Memorias del Sexto Simposio de la Fauna de Puerto Rico y el Caribe. Humacao: University of Puerto Rico.

Petuch, E. J. 1976. An unusual molluscan assemblage from Venezuela. *Veliger* 18:322–25.

———. A relict-Neogene caenogastropod fauna from northern South America. *Malacologia* 20:307–48.

———. 1982a. Geographical heterochrony: Contemporaneous coexistence of Neogene and Recent molluscan faunas in the Americas. *Palaeogeogr., Palaeoclimatol., Palaeoecol.* 37:277–312.

———. 1982b. Notes on the molluscan paleoecology of the Pinecrest beds at Sarasota, Florida with the description of *Pyruella*, a stratigraphically important new genus (Gastropoda: Melongenidae). *Proc. Acad. Nat. Sci. Phil.* 134:12–30.

———. 1988. *Neogene history of tropical American mollusks: Biogeography and evolutionary patterns of tropical western Atlantic Mollusca.* Charlottesville, Va.: Coastal Education and Research Foundation.

———. 1990. New gastropods from the Bermont Formation (Middle Pleistocene) of the Everglades Basin. *Nautilus* 104:96–104.

———. 1991. *New gastropods from the Plio-Pleistocene of southwestern Florida and the Everglades Basin.* W. H. Dall Paleontological Research Center Special Publication 1. Boca Raton: Florida Atlantic University.

———. 1994. *Atlas of Florida fossil shells: Pliocene and Pleistocene marine gastropods.* Evanston, Ill.: Chicago Spectrum Press.

Phillips, N. W., D. A. Gettleson, and K. D. Spring. 1990. Benthic biological studies of the southwest Florida coast. *Am. Zool.* 30:65–75.

Portell, R. W., K. S. Schindler, and G. S. Morgan. 1992. The Pleistocene molluscan fauna from Leisey Shell Pit 1, Hillsborough County, Florida. In *The Plio-Pleistocene stratigraphy and paleontology of southern Florida*, ed. T. M. Scott and W. D. Allmon, 181–94. Florida Geological Survey Special Publication 36. Tallahassee, Fla.

Portell, R. W., K. S. Schindler, and D. Nicol. 1995. Biostratigraphy and paleoecology of the Pleistocene invertebrates from the Leisey Shell Pits, Hillsborough County, Florida. *Bull. Fla. Mus. Nat. Hist.* 37:127–64.

Reed, J. K., and P. M. Mikkelsen. 1987. The molluscan community associated with the scleractinian coral *Oculina varicosa. Bull. Mar. Sci.* 40:99–131.

Riggs, S. R. 1984. Paleoceanographic model of Neogene phosphorite deposition, U.S. Atlantic continental margin. *Science* 223:123–31.

Riggs, S. R., and R. P. Sheldon. 1991. Paleoceanographic and paleoclimatic controls on the temporal and geographic distribution of Upper Cenozoic continental margin phosphorites. In *Neogene*, vol. 3 of *Phosphate deposits of the world*, ed. W. C. Burnett and S. R. Riggs, 207–22. Cambridge: Cambridge University Press.

Rios, E. C. 1985. *Seashells of Brazil.* Río Grande: Empresas Ipiranga.

Rosenberg, G. 1993a. A database approach to studies of molluscan taxonomy, biogeography, and diversity, with examples from western Atlantic marine gastropods. *Am. Malacol. Bull.* 10(2): 257–66.

————. 1993b. Malacolog 1.0: An electronic database of western Atlantic Mollusca. Distributed by the author.

Scott, T. M., and W. D. Allmon, eds. 1992. *The Plio-Pleistocene stratigraphy and paleontology of southern Florida*. Florida Geological Survey Special Publication 36. Tallahassee, Fla.

Sikes, E. L., L. D. Keigwin, and W. B. Curry. 1991. Pliocene paleoceanography: Circulation and oceanographic changes associated with the 2.4 Ma glacial event. *Paleoceanography* 6:245–57.

Skoglund, C. 1992. Pacific mollusk update. *Festivus* 24 (suppl.): 1–60.

Stanley, S. M. 1986. Anatomy of a regional mass extinction: Plio-Pleistocene decimation of the western Atlantic bivalve fauna. *Palaios* 1:17–36.

————. 1991. Evidence from marine deposits in Florida that the Great American Interchange of mammals began prior to 3 Ma. *Geol. Soc. Am. Abstr. Progr.* 23:A-405.

Stanley, S. M., and L. D. Campbell. 1981. Neogene mass extinction of western Atlantic molluscs. *Nature* 293:457–59.

Tripp, J. J. 1988. A new species of fossil *Busycon (Busycotypus)* from the lower Pliocene Buckingham Formation of Florida (Gastropoda: Melongenidae). *Ann. Carn. Mus.* 57:259–66.

Tuomey, M., and F. S. Holmes. 1857. *Pleiocene fossils of South-Carolina*. Charleston, S.C.: James and Williams; reprint, Ithaca, N.Y.: Paleontological Research Institution, 1974.

Turgeon, D. D., A. E. Bogan, E. V. Coan, W. K. Emerson, W. H. Lyons, W. L. Pratt, C. F. E. Roper, A. Scheltema, F. G. Thompson, and J. D. Williams. 1988. *Common and scientific names of aquatic invertebrates from the United States and Canada: Mollusks*. American Fisheries Society Special Publication 16.

Van Hyning, T. 1942. The Van Hyning collection of Florida shells. *Nautilus* 55:106–7.

Vermeij, G. J. 1978. *Biogeography and adaptation*. Cambridge: Harvard University Press.

————. 1980. Molluscan extinction at the Panama Isthmus: Predation or food? *Geol. Soc. Am. Abstr. Progr.* 12:541.

————. 1987. Interoceanic differences in architecture and ecology: The effects of history and productivity. In *Comparison between Atlantic and Pacific tropical marine coastal ecosystems: Community structure, ecological processes, and productivity*, ed., C. Birkeland, 105–25. UNESCO Reports in Marine Science 46. Paris.

————. 1989. Interoceanic differences in adaptation: Effects of history and productivity. *Mar. Ecol. Prog. Ser.* 57:293–305.

————. 1990. Tropical Pacific pelecypods and productivity: A hypothesis. *Bull. Mar. Sci.* 47:62–67.

————. 1991. When biotas meet: Understanding biotic interchange. *Science* 253:1099–1104.

Vermeij, G. J., and E. J. Petuch. 1986. Differential extinction in tropical American molluscs: Endemism, architecture, and the Panama land bridge. *Malacologia* 27:29–41.

Vokes, E. H. 1963. Cenozoic Muricidae of the Western Atlantic region. Part 1, *Murex sensu stricto. Tulane Stud. Geol. Paleontol.* 1(3): 93–123.

Vokes, H. E., and E. H. Vokes. 1983. *Distribution of shallow water marine Mollusca, Yucatán, Mexico.* Middle American Research Institute Publication 54. New Orleans: Tulane University.

Ward, L. W. 1992a. Diagnostic mollusks from the APAC Pit, Sarasota, Florida. In *The Plio-Pleistocene stratigraphy and paleontology of southern Florida*, ed. T. M. Scott and W. D. Allmon, 161–66. Florida Geological Survey Special Publication 36. Tallahassee, Fla.

———. 1992b. *Molluscan biostratigraphy of the Miocene, Middle Atlantic Coastal Plain of North America.* Virginia Museum of Natural History Memoir 2. Martinsville, Va.

Ward, L. W., and B. W. Blackwelder. 1980. Stratigraphic revision of Upper Miocene and Lower Pliocene beds of the Chesapeake Group, Middle Atlantic Coastal Plain. *U.S. Geol. Surv. Bull.* 1482-D:1–61.

———. 1987. Late Pliocene and Early Pleistocene Mollusca from the James City and Chowan River Formations at the Lee Creek Mine. In *Geology and paleontology of the Lee Creek Mine, North Carolina*, vol. 2, ed. C. E. Ray, 113–283. Smithsonian Contributions to Paleobiology, no. 61. Washington, D.C.

Ward, L. W., and N. L. Gilinsky. 1993. *Molluscan assemblages of the Chowan River Formation.* Part A, *Biostratigraphic analysis of the Chowan River Formation (Upper Pliocene) and adjoining units, the Moore House Member of the Yorktown Formation (Upper Pliocene) and the James City Formation (Lower Pleistocene).* Virginia Museum of Natural History Memoir 3. Martinsville, Va.

Ward, L. W., R. H. Bailey, and J. G. Carter. 1991. Pliocene and Early Pleistocene stratigraphy, depositional history, and molluscan paleobiogeography of the Coastal Plain. In *The geology of the Carolinas*, ed. J. W. Horton and V. A. Zullo, 274–89. Knoxville: University of Tennessee Press.

Warmke, G., and R. T. Abbott. 1961. *Caribbean seashells.* New York: Dover Publications.

Webb, S. D., G. S. Morgan, R. C. Hulbert, Jr., D. S. Jones, B. J. MacFadden, and P. A. Mueller, 1989. Geochronology of a rich Early Pleistocene vertebrate fauna, Leisey Shell Pit, Tampa Bay, Florida. *Quatern. Res.* 32:96–110.

Whittaker, R. H. 1977. Evolution of species diversity in land communities. *Evol. Biol.* 10:1–67.

Wiggs, L. B., and H. J. Dowsett. 1992. Pliocene planktic foraminifer census data from Deep Sea Drilling Project hole 396 and Ocean Drilling Program hole 672. Open File Report 92-414, U.S. Geological Survey, Reston, Va.

Wiggs, L. B., and R. Z. Poore. 1991. Pliocene planktic foraminifer census data from Deep Sea Drilling Project holes 502A,B,C. Open File Report 91-325, U.S. Geological Survey, Reston, Va.

Wilkinson, I. P. 1980. Coralline Crag Ostracoda and their environmental and stratigraphical significance. *Proc. Geol. Assoc.* 91:291–306.

Williams, S. C. 1985. Stratigraphy, facies evolution, and diagenesis of Late Cenozoic limestones and dolomites, Little Bahama Bank. Bahamas. Ph.D. diss., University of Miami, Coral Gables, Fla.

Woodring, W. P. 1928. *Miocene mollusks from Bowden, Jamaica*. Part 2, *Gastropods and discussion of results*. Carnegie Institution of Washington Publication 385. Washington, D.C.

————. 1957–1982. Geology and paleontology of Canal Zone and adjoining parts of Panama. *U.S. Geol. Surv. Prof. Pap.* 306-A–F. 743 pp.

————. 1966. The Panama land bridge as a sea barrier. *Proc. Am. Phil. Soc.* 110:425–33.

11

Molecular Comparisons of Transisthmian Species Pairs: Rates and Patterns of Evolution

Timothy Collins

Hence, marine fishes are about as good as geologists
in telling us when the Isthmus of Panama was formed.
R. K. SELANDER, "PHYLOGENY"

INTRODUCTION

The continuum of memory is punctuated by key events which frame the chronology of our past, a birth or death, a marriage or a move. In the absence of such markers, our recollections are compressed or jumbled, and the pattern of our lives difficult to discern. In John Steinbeck's words, "eventlessness collapses time" (1962). Steinbeck's insight has an analog in studies of the evolutionary process. The absence of discrete tie points in the past collapses time to the present, limiting our understanding of this historical process. Historical events are the stepping stones which allow us to view evolution from a temporal perspective. This volume focuses on one such stepping stone, the Pliocene emergence of the Central American Isthmus, and attempts to analyze the ecological and evolutionary consequences of this event.

The purpose of this chapter is to review and synthesize the results of studies dealing with the consequences of the isthmian emergence at the molecular level on marine animals isolated by this event, particularly in regard to the importance of these studies to the notion of a molecular clock of evolution for informational macromolecules. Other significant aspects, such as adaptation at the molecular level (Graves et al. 1983), will not be discussed, except as they relate to the question of regularity of rates of molecular evolution. It is intended for those who, while not directly involved in molecular studies, have an interest in the intersection of their disciplines with molecular biology regarding phylogenetic, evolutionary, and biogeographic questions, and as as aid in comparing and critically evaluating the congruence and conflict of results from these fields. For example, geologists may want to know if in fact molecules from marine fishes are as good as geologists in dating the emergence of the Isthmus of Panama, and what methods are used to derive dates from molecules. I will begin by discussing transisthmian protein (allozyme) comparisons, which make up the bulk of the work published to date, followed by some discussion of recent mitochondrial DNA (mtDNA) restriction endonuclease and sequence analyses. I have prefaced this with discussions of the significance of molecular rate calibrations for evolutionary studies, the history of transisthmian comparisons, the important issue of recognition of geminate (twin) species across the isthmus, and the unique evidential contribution of paleontology and geology to studies of rates of molecular evolution.

RATES OF MOLECULAR EVOLUTION

Molecular data are employed with increasing frequency in evolutionary studies. Aside from their obvious role in elucidating the mechanistic bases of processes such as heredity and development, molecular data have made contributions to the understanding of such perennial questions as the extent of continuity among populations within species, the nature of hybrid zones, and the factors involved in the process of speciation. Molecular data have also proven useful as a complement to traditional characters used in mapping out relationships among organisms. This has been true at very high taxonomic levels, among kingdoms and phyla, for example, where few morphological characteristics are shared among the groups being compared, and at very low taxonomic levels, for example, cryptic or sibling species, which may often appear morphologically indistinguishable.

Many molecular biologists have noted a correlation between the time elapsed since species diverged from a common ancestor and the amount of molecular difference between them (e.g., Nei 1987; Gillespie 1991). Early on (Zuckerkandl and Pauling 1962), this observation led to the proposal that amino-acid replacements and nucleotide substitutions might occur at a stochastically constant rate; a biological analog to the geological stochastic clocks of decaying radionuclides. This concept, termed the molecular clock, has been one of the most exciting and contentious developments in molecular biology. There are many interrelated questions concerning this concept. Is the variance in rates of molecular evolution sufficiently constrained to justify the regularity and linearity that the term "molecular clock" implies? Is there a single clock for all molecules and organisms or are there gene-specific and taxon-specific rates? Finally, if there are clocks, what are their calibrations?

Evolutionary biologists using molecular techniques are increasingly aware of the importance of calibrating rates of molecular evolution and of defining the limits of variation in these rates. Suppositions about the tempo and degree of uniformity of rates have particular significance for phylogenetic analysis and the dating of cladogenetic events. Methods of phylogenetic reconstruction may make rate-dependent clustering or rooting assumptions (Hillis 1987) or assumptions about the distribution of rates among the branches of lineages (Felsenstein 1978). Resulting phylogenies and dates of cladogenetic events will be unreliable to the extent that these assumptions are violated. Confidence intervals of phylogenies may be dramatically affected by the levels of rate variation allowed (Felsenstein 1988).

Information about absolute rates of evolution for different portions of the genome (nuclear, mitochondrial, protein-encoding, ribosomal, pseudogene, etc.) and among the different types of sites within genes is useful in determining characters to be analyzed at a given phylogenetic level (Hillis 1987; Simon 1991). Rate calibrations derived for one taxonomic group are often applied to other groups for which no calibrations are currently available. These transferred calibrations are then used to search for temporal congruence or incongruence of dates derived from molecular sequence divergence values and independently (usually geologically/paleontologically) derived dates for events in the history of the clade. Finally, our views on some of the most basic questions concerning evolutionary processes, such as the relative importance of stochastic versus selective processes, are affected by considerations of rate (Nei 1987; Gillespie 1991). For these reasons it is necessary to understand how rates of molecular evolution vary, both within and among groups.

Two methods are commonly used to determine levels of variation in rates of molecular evolution, relative rates and absolute rates (calibration from independently dated events in the histories of clades). The relative rates test uses variation in observed differences between a reference taxon and each of several ingroup members as an estimate of uniformity in rates (Sarich and Wilson 1973; Wilson et al. 1977) (fig. 11.1A). If the distances between the reference and each ingroup member are similar, rates of evolution are considered to have been regular. If there is a great disparity in these distances, rate variation is inferred. The great advantage of this test is its convenience. It is not necessary to know the times of divergence of lineages, only that the relationship of the reference taxon as an outgroup to the ingroup taxa be unequivocally known. Unfortunately, there are also several disadvantages. The relative rates test is predisposed to finding equality of rates (Fitch 1976; Ayala

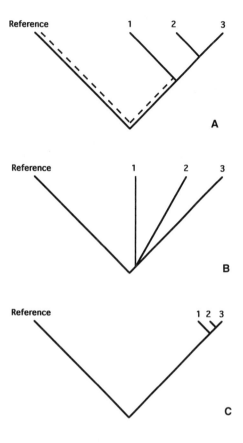

Fig. 11.1 The relative rates test compares the amount of divergence from each member of the ingroup (numbered individuals) to a reference taxon. Similarity of pairwise differences between the reference taxon and each ingroup member is taken to indicate regularity of evolutionary rate. However, as the length of the dashed portion of the tree increases (in A representing the period of shared history of the ingroup members and the distance to the reference taxon from the common ancestor of the ingroup), the likelihood of finding similar distances increases. This can be seen more easily in B and C. B represents the ideal relative rates situation, where the reference taxon is very closely related to the ingroup, and the ingroup radiated simultaneously following the split with the reference taxon. The period of shared history of the ingroup members relative to the outgroup has been minimized. A less appropriate, but probably more common, situation for a relative rates test is seen in C.

1985; Scherer 1989, 1990), a problem which becomes exacerbated as the time since divergence from the outgroup and length of shared ancestry of the ingroup members increase. A related problem, particularly in the analysis of sequence data, is that the pairwise divergence values between the reference taxon and each ingroup member are not independent, making these values unsuitable for statistical evaluation (Felsenstein 1988). These difficulties may be most easily understood by considering an analogy with geographic distances. Imagine that you find yourself in Dodge City, Kansas, with four cars and drivers. You want to determine the average speed and variation in speed of these cars and their drivers. You decide on an experimental design in which one car, the reference car, will drive east, and the other three cars, the sample cars, will drive west. At a specific time, you measure the distance between the reference car heading east and each of the sample cars heading west. The distances between each sample car and the reference taxon are taken to be measures of the relative rates of speed of those cars. Very similar distances among cars would be considered evidence of uniform rates; very different distances would indicate rate variation. This is the basic model of the relative rates test (fig. 11.1B). Actual comparisons, however, are typically more complicated than this simplified case. To continue our analogy, imagine that you discover that the three sample cars heading west were loaded onto one truck. At some point (unknown to you) one car was unloaded and began to drive west under its own power. Later, the other two cars were unloaded, and also began their drive. Your confidence in the independence of your measures will obviously be dramatically affected by how long the cars were traveling together on the truck. What appears to be three independent measures of average speed may actually be one measure, if the cars spent most of the trip west together on the truck. This represents the problem of shared history of the ingroup members (fig. 11.1C). Now consider one final complication. As the reference and sample cars get closer to the East and West Coasts, they move more and more slowly. They are limited by the ocean. The coasts therefore represent the levels of saturation of the samples with regard to the reference. There is a limit to how different they can become. Therefore, as the cars approach the coasts, their distances will converge on a similar value regardless of the differences in speed that may have obtained earlier in the trip. As the distance of the ingroup members from the reference taxon increases, equality of rates becomes more and more likely. This analogy illustrates some of the difficulties inherent in the relative rates test, although actual situations typically will be even more complicated than the case outlined above. Finally, it should be noted that absolute time calibrations are not possible with a relative rates test, precluding meaningful comparisons across higher taxa, particularly on short time scales.

The second commonly used approach involves measuring rates of molecular evolution with reference to independently dated events, usually fossil occurrences or vicariance episodes such as the separation of Africa and South America by the opening of the South Atlantic. This allows calibration of rates on an absolute time scale, and if based on several such events representing successively deeper branches in a clade, permits the determination of linearity or nonlinearity of rates on different time scales (Gingerich 1986). There is, of course, an error associated with the independent dating of each event. In some cases these errors may be quite large.

Fossils give the minimum time of divergence based on the earliest appearance of specimens with characters which indicate that a particular cladogenetic event has occurred (Marshall 1990b; Norell 1992; Smith 1992). It is not possible to determine whether the earliest occurrence is actually the time of divergence, although confidence intervals may be calculated if certain assumptions are accepted (Marshall 1990a,b). Given the one-sided nature of information from the fossil record, rates derived will be overestimates to the extent that cladogenetic events predate first occurrences.

Calibrations based on vicariant events in the histories of regional biotas, such as the isthmian emergence, afford the opportunity for absolute calibrations and comparisons across higher taxa on the same time scale. These comparisons, across taxa for the same vicariant event, are not subject to artifacts of temporal scaling (Gingerich 1983, 1986). However, one important concern in calibrating rates of molecular evolution with vicariant events or fossil occurrences is the age of the event relative to the rate of molecular evolution. If the age is great compared to the rate of molecular evolution, severe underestimates of rate may result. This can be most easily understood by considering pairwise nucleotide sequence divergence with increasing time. Imagine the bifurcation of a population, in which all individuals have the identical nucleotide sequence for a particular gene, into two subpopulations. Early in the divergence of these isolated subpopulations virtually every new substitution fixed in one of the subpopulations adds to the pairwise sequence difference between populations for that gene. But as more and more substitutional events occur, the chance of a site which has already undergone one substitution undergoing another substitution increases. This second substitution may result in the reversion of the site to the ancestral state at that site, it may parallel a substitution that has already occurred in the other subpopulation, or it may change to a new state. In any case, two substitutions will have occurred, whereas a pairwise comparison of the sequences will suggest that only one, if any, substitution has occurred. There are many different methods to augment pairwise difference values to account for this problem of multiple substitutions at a site, each based on a different model of the substitutional process (e.g. Nei 1987). Most of these methods give very similar values for recent divergences, but can give increasingly divergent and unreliable values for more ancient splits. In the case of transisthmian comparisons, studies to date suggest that the time since isolation of the pairs of taxa is of a recency such that augmentations of pairwise distances for superimposed evolutionary events are often unnecessary or are in the range where most of the different methods give similar values for many of the comparisons made. In addition, the values derived for different taxon pairs are independent, making these data amenable to statistical analysis of rate variation (Felsenstein 1985, 1988). Error in this type of calibration arises from error in the dating of the biogeographic event and some measure of the abruptness and efficacy with which the barrier isolated taxa. It also requires the assumption that the sister taxa on either side of the barrier were isolated by the vicariant event and not some other mechanism. This leads us to a discussion of the particulars of the isthmian vicariant event.

THE ISTHMIAN VICARIANT EVENT

The Isthmus of Panama has been recognized as a unique opportunity for the study of evolutionary processes since the observation by Günther in the mid nineteenth century (Günther 1869) that similarities of Central American fish faunas of the western Atlantic and eastern Pacific implied a relatively recent marine connection. Since that time, the isthmian region has been used as a testing ground for a variety of theories. It has served as a paradigm of geographic speciation (Jordan 1908; Mayr 1954; Jones 1972). Its importance in understanding biogeographic patterns of marine organisms has been discussed by Ekman (1967) and Vermeij (1978). The distribution of representatives of monophyletic groups on either side of the isthmus has been cited as a "spectacular" example of a marine vicariant generalized track (Humphries and Parenti 1986). The marine species on either side of the isthmus have been cited as evidence in support of a punctuated model of evolution (Stanley 1979; Vrba 1980; Cronin 1985) and have been studied to understand changing patterns in shell architecture and extinction within the Mollusca (Vermeij 1978; Vermeij and Petuch 1986). The contemplated construction of a sea-level canal across the isthmus has stimulated research on the potential impact of large-scale mixing of biotas (Voss 1972; Vermeij 1991). This sampling of the research utilizing the Isthmus of Panama is far from complete but gives one an impression of the variety of questions which may be addressed within this system.

The reason that the isthmian region has proven such fertile ground for research is that we have some information on several important parameters which are typically missing or poorly delimited in evolutionary studies. The final emergence of the isthmus is well constrained temporally (see discussion below) compared to many biogeographic events. Thus, we have a good, independent estimate of the minimum length of time that the populations have been isolated. We also know that the isthmus represents an absolute barrier to migration among most shallow-water tropical marine species. These two features alone set the isthmus apart from the majority of evolutionary studies, where the precise time of isolation is rarely known, and the isolating barrier is often permeable.

The biogeographic ranges of the extant geminate pairs suggest that the emergence of the isthmus centrally divided the broad geographic range of the ancestral species into two large subpopulations, rather than the creation of a small peripheral isolate (Mayr 1967; Lessios 1979a), and that accelerated evolution due to founder effects is therefore unlikely. Paleobiogeographic studies of mollusks support this conclusion. During the latter portion of the Tertiary period, the biogeographic provinces which we recognize as the Panamic in the eastern Pacific and the Caribbean in the western Atlantic were united into one large province centered on the isthmus. This province has been referred to as the Middle Miocene Caribbean Province (Woodring 1965, 1974), the Tertiary Caribbean Province (Woodring 1966), and the Gatunian Province (Petuch 1982). The limits of this province suggest that the emergence of the isthmus did in fact divide the province into two approximate halves, beginning the differentiation that resulted in the distinct faunas that we see today. The lack of differentiation of the western Atlantic and eastern Pacific faunas prior to the shoaling of the isthmus suggests that, in general, the divergences that we are studying are unlikely to have preceded the emergence. Woodring (1966) writes: "Practically identical middle Miocene [molluscan] faunas are found along the Caribbean coast of central Panamá, in northern Columbia, in the Atrato trough itself, and also near the Pacific coast of southwestern Columbia, Darién, and Chiriquí."[1] The generalized amphi-American vicariant track figured in Rosen 1976, derived from nonmolluscan taxa (terrestrial isopods, brachyuran crabs, atyid shrimps, stomatopods, and marine fishes), corresponds closely to the limits of the molluscan paleoprovince and further supports the notion of the bisection of a previously undifferentiated faunal range. However, recent, more refined analyses of temporal patterns of diversity and differentiation in mollusks, particularly strombinid gastropods (Coates et al. 1992; Jackson et al. 1993), suggest that at least some groups may have begun to differentiate into Caribbean and eastern Pacific elements prior to the final closure of the isthmus. The implications

---

1. What Woodring identifies as Middle Miocene is now generally considered to be Late Miocene or Early Pliocene (Stanley and Campbell 1981, references this volume, personal communication, Panama Paleontology Project).

of this will be discussed below, but they ultimately depend on the degree to which the differentiation of faunas was a drawn-out affair. Keeping this caution in mind, a fair summary of the current evidence is that we know the approximate time at which an unequivocal barrier to further genetic exchange bisected the ranges of a large number of species. In the next two sections, I review what I consider to be important caveats to this summary statement with regard to molecular studies of transisthmian species.

RECOGNITION OF GEMINATE SPECIES

David Starr Jordan coined the term "geminate species" for "twin species—each one representing the other on opposite sides of some form of barrier" (1908, 75). Jordan considered geminates to be the result of reproductive isolation following the genesis of a barrier, and the species on either side of the Central American Isthmus as the exemplar of his "law" of geminate species. In modern terms and in the isthmian context, these geminate species are sister taxa that diverged as a result of the final emergence of the isthmus. But as in many other areas, definition does not necessarily assist in recognition. For any particular pair of species, it is difficult to imagine how it could be that the divergence is the result of the emergence of the isthmus. The conviction that the morphologically similar sister taxa on either side of the isthmus date from the emergence of the isthmus is derived from first principles of evolutionary theory and biogeography, as well as examination of patterns, both taxonomic and biogeographic, in the marine species of the eastern Pacific and western Atlantic. For groups with an adequate fossil record, it would be possible to demonstrate that a divergence occurred before the emergence of the isthmus. If broad geographic sampling and phylogenetic analysis demonstrated characters which uniquely defined the living species, and fossils predating the emergence of the isthmus were found to display these characters, it would suggest that the divergence of the species predated the emergence of the isthmus. Groups without an adequate fossil record lack the potential for this falsification.

The referral of geminate status to species pairs on either side of the isthmus depends on general considerations as well as on specific criteria. The most important is the recognition that geographic isolation is the predominant cause of speciation among organisms. This proposition is accepted both by those working within the framework of the modern evolutionary synthesis (e.g., Mayr 1963) or challenging it (e.g., Eldredge and Gould 1972). It is also a basic assumption of the school of vicariance biogeography (Kluge 1988). Vicariance biogeographers further assume that if related taxa are isolated by a barrier, the taxa and the barrier between them are the same age (Humphries and Parenti 1986).

If one accepts even the more general of these propositions, the distribution of morphologically similar sister taxa on either side of a recently developed geographic barrier suggests a causal relationship. This pattern becomes more compelling as it is recognized as being more general, that is, numerically common within a group and persistent across a range of taxa. It is further strengthened if ranges of the various taxa are concordant (Lessios 1979b) and if the ranges of the taxa are restricted to areas adjacent to the barrier (particularly if the taxa are not cosmopolitan). If these features are present, the most parsimonious explanation is that the taxa have diverged as a result of the erection of a geographic barrier. Alternative explanations are less parsimonious, as they require many extra events and coincidences to produce the same result.

In the case of the Central American Isthmus, such morphologically similar sister taxa are common in most of the groups that have been studied, including the sea stars, brittle stars, sea urchins, brachyuran crabs, and pycnogonids (Ekman 1967; Chesher 1972; Bayer et al. 1970), gastropods, bivalves, chitons, cephalopods, and polychaetes (Vermeij 1978; Voight 1988), trematodes and sponges (Briggs 1974), stomatopods (Manning 1969) and fishes (Jordan 1908; Rosenblatt 1963, 1967; Rubinoff 1963). The concordance of the ranges of many of these species pairs argues for a common cause (Lessios 1979b). The pattern in marine geminates is also supported by transisthmian elements of the terrestrial fauna (Rosen 1976). The coincidence of ranges of species pairs on either side of the Central American Isthmus belonging to genera endemic to the Americas are difficult to explain in any way other than as the result of their isolation following erection of the isthmian barrier.

In practice, for any specific pair of species, geminate status has been determined by status as sister taxa, biogeographic distribution, and morphological similarity. There is no quantitative measure of what degree of morphological differentiation is sufficient to determine that species on either side of the isthmus are not geminate, although this determination has been made for some echinoderms (Lessios 1979a), mollusks (Vermeij 1978), and based on morphometric distances, isopods (Weinberg and Starczak 1989). Insofar as morphological evolution is directly related to molecular evolution, this criterion will tend to limit geminate distances to lower values, truncating the range of variation and therefore potentially biasing results in favor of reduced variance in rates of molecular evolution. A conservative approach is to examine first those species pairs that are morphologically so similar that there is broad consensus on their geminate status (the extreme of this is the use of conspecific populations on either side of the isthmus), realizing that this involves the bias discussed above. The results from these species will determine the necessity of testing species pairs which are morphologically more divergent.

In some cases, there may be more than one species in an ocean which is a potential twin to the species in the other ocean (Rubinoff and Rubinoff 1971; Lessios 1979a,b, 1981; Vawter et al. 1980; West 1980; Weinberg and Starczak 1989; Lessios and Cunningham 1990; Knowlton et al. 1993). Most of the confusion regarding the choice of geminates in these cases is due to the lack of a hypothesis of relationships for the taxa in question. The combination of a phylogenetic hypothesis and the biogeographic distributions of the species in question should clarify the geminate relationships. It should be remembered that phylogenetic hypotheses based on molecular distances may potentially bias measures of geminate distances to lower values. Geminates may in some cases be represented by more than two species, if one member of a geminate pair has speciated subsequent to the isthmian isolation.

Finally, it should be remembered that, given all of the congeneric species on either side of the Isthmus of Panama, some have likely diverged owing to events other than the Pliocene emergence of the isthmus and thus share their transisthmian distribution with true geminates by coincidence. An important point that can be made in this regard is that divergences between taxa which are much older than the Pliocene are likely to have resulted in enough morphological divergence to make their geminate status questionable using the conservative criteria outlined above, and if one does accept a predominant role of geographic isolation in speciation, the isolation of faunas across the isthmus prior to the Pliocene was at the latest between the Cretaceous and the Paleogene periods (Jones and Hasson 1985; Stehli and Webb 1985).

It is also worthwhile to keep in mind the distinctions between (*a*) applying independently derived molecular clocks to species pairs to infer times of divergence, (*b*) calibrating molecular clocks with geminate species, and (*c*) testing for molecular clocks with geminate species. These different goals and methods are often blended in ways which result in circularity or bias (see discussion below). One may (*a*) simply assume a molecular clock and an independent calibration, apply it to species pair divergence values, and given the range of estimates for the time of the final emergence, draw conclusions as to which of the species pairs are geminates and which predate the final isthmian emergence. It is also possible to (*b*) measure divergence values between species pairs and, based on results and some a posteriori criteria, choose the most likely candidates to have been isolated by the final isthmian emergence and calibrate rates of molecular evolution for these geminates. Finally, one may (*c*) decide on geminate status prior to generating divergence values and use the results to estimate the variance in rates of molecular evolution between or among groups. The latter method is the most robust test of the molecular clock. The first method assumes a molecular clock and therefore cannot test it. Divergence values which are below the average divergence values for congeneric species or which give time divergence values which are reasonable by any independent calibration of molecular rates of evolution cannot be rejected a posteriori, by reference to a particular calibration of the molecular clock in tests of the molecular clock. The second method may be a reasonable approach for calibrating rates, but it is less useful for testing for variation in rates of molecular evolution because it is biased at two levels in favor of finding limited rate variation: in the choice of morphologically similar sister species as discussed above, and then at a second level in the a posteriori rejection of species pairs as geminates based on molecular distances. It may also be difficult in practice to develop criteria for these a posteriori judgments, which do not involve assumptions that favor the finding of a molecular clock. It is more typical to bias experimental design, as far as possible, against the favored hypothesis, and for this reason method (*c*) may be preferable. If one is willing to accept Panamanian geminates as a test of the molecular clock, one should be willing to accept the results of these tests. The results of these tests should be sound if based on many geminate pairs across a broad range of taxa, in order for the true pattern of geminate divergences to be discerned from the possibly confounding influence of the occasional false geminate pair. However, to the extent that accumulating evidence confirms suggestions (Coates et al. 1992; Jackson et al. 1993; Knowlton et al. 1993) that the isolation of species pairs on either side of the isthmus was a protracted affair spanning many millions of years before final closure, the use of geminate species for (*c*), testing the reality of and variance in molecular clocks, will be vitiated.

TIMING AND TERM OF THE ISTHMIAN EMERGENCE

For the reasons mentioned above, an understanding of the timing and term of the isthmian emergence is critical to the use of transisthmian geminates for testing and calibration of molecular rates. A detailed review of the evidence that bears on this question is beyond the scope of this chapter. There are several recent reviews (Stehli and Webb 1985; Collins 1989) and considerable current research on this problem (Coates et al. 1992; L. Collins 1993, this volume, chap. 6). In this brief discussion, I consider the kinds of evidence that are most pertinent to the determination of the timing and duration of the final closure of the Central American Seaway. Conclusions concerning the exact timing and duration of shoaling and emergence of the Central American Isthmus are based primarily on biogeographic patterns and diversity trends as well as isotopic and physical (usually sedimentological) data. The arguments developed generally depend on the assumption of a correlation between an occurrence or trend and the emergence of the isthmus. The validity of the assumptions made varies widely, because there are often events or processes other than the isthmian emergence which could as easily be responsible for the patterns observed, particularly if a trend is measured on only one side of the isthmus. In addition, some assumptions are faulty in a logical sense. The difficulty may be traced to the fact that, at least for the marine environment, the evidence collected to date is often based on the absence of fossils. Geologists and paleontologists have looked for indications that the western Atlantic and eastern Pacific Oceans were *not* sharing a connection. The evidence typically takes the form of inferring the time of the emergence of the isthmus as the point at which a faunal element (usually microfaunal) is found to be present in one basin and absent in the other. Given the limited amount of core material and land-based section examined for microfossils to date (particularly in the eastern Pacific), these types of arguments must be regarded as tentative, although the results of the Panama Paleontology Project and studies in this volume are allowing more confident conclusions. In any case, inferences based on the absence of fossils will always be weaker than those based on the presence of fossils. In contrast, the terrestrial biogeographic evidence is usually based on the presence of fossils. The point at which the North and South American faunas share land-bridge dispersed faunal elements previously restricted to one continent sets an upper limit for the time of the emergence. It is likely, however, that isolation of the marine faunas preceded these terrestrial dispersal events by some interval. In addition, the fossil record of terrestrial environments is generally poor compared to marine environments. To further refine the timing of the isthmian emergence, positive biogeographic evidence from the marine environment must be collected. These data would indicate the latest point at which the western Atlantic and eastern Pacific were connected and therefore set a lower limit for the timing of the emergence of the isthmus. The latest first appearance datums (FADs) for New World tropical endemics found in both the western Atlantic and eastern Pacific would mark the most conservative and reliable estimate of the latest time of the isthmian seaway's existence. The timing of the final emergence would then be bracketed by the sharing of tropical New World endemic marine animals from the Caribbean and eastern Pacific at the lower end and the dispersal of terrestrial animals between North and South America at the upper end. A secondary line of evidence would consist of the occurrence of new lineages of highly vagile tropical species on one side of the isthmus, without dispersal to the other side. This is particularly true of Caribbean species that do not disperse into the Pacific, as it is likely that the Atlantic North Equatorial Current flowed through the isthmian portals into the Pacific (Berggren and Hollister 1974, 1977; Holcombe and Moore 1977; Haq 1984). This type of biogeographic evidence requires the additional assumptions that dispersal must occur whenever there is a connection between the ocean basins, and that absence from a sample implies absence from the basin and not simply a preservational or sampling bias. These assumptions may not be valid, particularly if environmental conditions are locally variable. Trends (usually isotopic values or species abundance and diversity) may be measured on either side of the isthmus. Concordance or covariation in trends between basins, followed by deviation and discordance, may be indicative of isolation of the basins. The basins are usually considered to have become isolated at the point at which trends are no longer concordant across the isthmus. Finally, concordance in all of these different lines of evidence may be taken as supporting the inferred time of emergence. The best current evidence seems to suggest that the Central American Isthmus was complete by approximately 3.5 Ma. The degree to which this process was drawn out remains to be seen, but as mentioned above, it will ultimately determine the utility of geminate comparisons in tests of variation in rates of molecular evolution.

THE ISTHMIAN EMERGENCE AS A TEST OF THE PROTEIN MOLECULAR CLOCK

> The protein clock theory has clearly and unduly molded
> the interpretation of data, rather than vice versa.
> J. C. AVISE AND C. F. AQUADRO, "A COMPARATIVE
> SUMMARY OF GENETIC DISTANCE IN THE VERTEBRATES"

Use of the emergence of the Isthmus of Panama as a test of the molecular clock began with a series of papers using protein electrophoretic data (Gorman et al. 1976; Gorman and Kim 1977; Lessios 1979a,b, 1981;

West 1980; Vawter et al. 1980; Laguna 1987) and continues to the present (Bermingham and Lessios 1993; Knowlton et al. 1993; see also Rosenblatt and Waples 1986; Wilson and Waples 1984). Table 11.1 is a compilation of transisthmian protein electrophoretic comparisons from these studies. These data are summarized by taxonomic group in figure 11.2. One of the species, *Coryphaena hippurus*, is a circumtropical pelagic fish that ranges to latitude 35°S (Masuda et al. 1984; Smith and Heemstra 1986). Another fish species, *Coryphaenoides armatus*, is a cosmopolitan abyssal species (Wilson and Waples 1984). These cosmopolitan species, by virtue of the potential for continued gene flow between populations on either side of the Central American Isthmus, are poor candidates for a test based on the assumption that populations of the species were irrevocably isolated by the emergence of this barrier. For this reason, cosmopolitan species cannot be considered an appropriate test of the molecular clock in this context. This reasoning also applies to two published interocean comparisons of mitochondrial DNA, in the pelagic skipjack tuna *Katsuwonas pelamis* (Graves et al. 1984) and the cosmopolitan migratory Green turtle *Chelonia mydas* (Bowen et al. 1989).

The D values (Nei's genetic distance) for the remaining species display a tenfold difference in amounts of protein divergence among echinoids, a sixfold difference among crabs, and a threefold difference in divergence among fishes. The total range of D values found encompasses a thirtyfold difference from the smallest to the largest value, and span almost a twentyfold range, even if the outlying *Euraphia* geminate value is dropped from the comparison (table 11.1, fig. 11.2). Some of this variation may be due to (1) the use of different sets of proteins in comparisons, resulting in variation due to proportion of more rapidly evolving versus slowly evolving proteins, (2) variation in laboratory protocols, and (3) sampling errors (Sarich 1977; Lessios 1981; Nei 1987). Proportion of fast versus slow loci and laboratory technique variation have been shown to be inadequate in explaining variation of a much smaller magnitude than found in the present survey (Avise and Aquadro 1982), because most surveys have similar proportions of fast versus slow loci and because large differences in genetic distance remained after standardization for different proportions of these loci. Much of the variation has been found within urchins, using largely the same proteins and laboratory protocols (Lessios 1979a,b; Bermingham and Lessios 1993). The potential error due to sampling is unlikely to be responsible for the differences found in transisthmian comparisons. Lessios (1981) demonstrated that, given the expected difference based on the molecular clock, the probability of missing differences was actually quite small

Table 11.1 Transisthmian Protein Electrophoretic Comparisons of Marine Species

Genus	Altantic species	Pacific species	D	No. of loci	No. of individuals	T_1	T_2	R_1	R_2	Study
Diodon	*holocanthus*	*holocanthus*	.192	41	4, 4	3.6	1.0	1.2	0.3	5
Diodon	*hystrix*	*hystrix*	.361	30	4, 7	6.8	1.8	2.3	0.6	5
Gerres	*cinereus*	*cinereus*	.282	34	5, 8	5.3	1.4	1.8	0.5	5
Haemulon	*steindachneri*	*steindachneri*	.131	35	4, 1	2.5	0.7	0.8	0.2	5
Scorpaena	*plumieri*	*plumieri*	.206	28	9, 6	3.9	1.0	1.3	0.3	5
Abudefduf	*taurus*	*concolor*	.194	27	1, 3	3.7	1.0	1.2	0.3	5
Abudefduf	*saxatilis*	*troschelii*	.320	28	26, 23	6.0	1.6	2.0	0.5	2
Rypticus	*bicolor*	*saponaceus*	.142	39	5, 5	2.7	0.7	0.9	0.2	5
Mulloidicthys	*martinicus*	*dentatus*	.168	23	5, 12	3.2	0.8	1.1	0.3	5
Bathygobius	*soporator*	*andrei*	.146	26	24, 24	2.8	0.7	0.9	0.2	1
Bathygobius	*soporator*	*ramosus*	.420	26	20, 24	7.9	2.1	2.6	0.7	1
Coryphaena	*hippurus*	*hippurus*	<.01	36	15, 6	0.2	.05	—	—	7
Coryphaenoides	*armatus*	*armatus*	.03	27	40, 29	0.6	.15	—	—	8
Priacanthus	*cruentatus*	*cruentatus*	.140	36	9, 1	2.6	0.7	0.9	0.2	7
Diadema	*antillarum*	*mexicanum*	.038	34	+34, +34	0.7	0.2	0.2	0.06	3, 10
Eucidaris	*tribuloides*	*thouarsi*	.312	25	+34, +34	5.9	1.6	2.0	0.5	3, 10
Echinometra	*lucunter*	*vanbrunti*	.341	31	+34, +34	6.4	1.7	2.1	0.6	3, 10

Genus			D			T₁	T₂	R₁	R₂	
Echinometra	_viridis_	_vanbrunti_	.521	31	+34, +34	9.8	2.6	3.3	0.9	3, 10
Grapsus	_grapsus_	_grapsus_	.079	24	59, 58	1.5	0.4	0.5	0.1	4
Aratus	_pisonii_	_pisonii_	.064	19	78, 76	1.2	0.3	0.4	0.1	4
Goniopsis	_cruentata_	_pulchra_	.064	21	93, 78	1.2	0.3	0.4	0.1	4
Ocypode	_quadrata_	_occidentalis_	.380	22	82, 65	7.1	1.9	2.4	0.6	4
Euraphia	_rhizophorae_	_eastropacensis_	.950	17	22, 22	18.0	4.8	6.0	1.6	6
Alpheus	_paracrinitus_ b	_rostratus_	.028	15	19, 19	0.5	0.1	0.2	0.05	9
Alpheus	_paracrinitus_ a	_paracrinitus_	.114	15	19, 19	2.2	0.6	0.7	0.2	9
Alpheus	_formosus_ a	_panamensis_	.109	16	19, 19	2.1	0.5	0.7	0.2	9
Alpheus	_cylindricus_	_cylindricus_	.121	16	19, 19	2.3	0.6	0.8	0.2	9
Alpheus	_simus_	_saxidomus_	.177	15	19, 19	3.3	0.9	1.1	0.3	9
Alpheus	_nuttingi_	_canalis_ b	.188	16	19, 19	3.6	0.9	1.2	0.3	9
Alpheus	_cristulifrons_	_cristulifrons_	.272	16	19, 19	5.1	1.4	1.7	0.5	9

Note: D is Nei's genetic distance. The number of individuals from the western Atlantic is followed by the number of individuals from the eastern Pacific. Numbers of individuals from study 4 are averages for loci. Numbers of individuals from studies 3 and 10 ranged from 34 to 104 per species. Numbers of individuals from study 9 are the mean sample size per locus averaged across all species. The genetic distance from the _Alpheus cristulifrons_ pair is between .232 and .272 (Knowlton et al. 1993). T_1 is divergence time estimated using calibration of _D_ of 1 = 18.9 my (Sarich 1977; Carlson et al. 1978). T_2 is divergence time estimated using calibration _D_ of 1 equal to 5 my (Chakraborty and Nei 1974; Nei 1987). R_1 and R_2 are ratios created by dividing T_1 and T_2 by 3 Ma, a conservative estimate for the time of the final isthmian emergence. Data are from (1) Gorman et al. 1976; (2) Gorman and Kim 1977; (3) Lessios 1979b; (4) West 1980; (5) Vawter et al. 1980; (6) Laguna 1987; (7) Rosenblatt and Waples 1986; (8) Wilson and Waples 1984; (9) Knowlton et al 1993; (10) Bermingham and Lessios 1993.

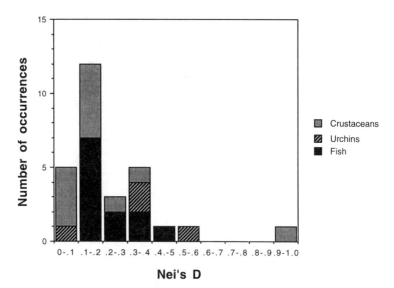

Fig. 11.2 Nei's genetic distances for geminate pairs by taxonomic group. Data are from table 11.1, excluding *Coryphaena* and *Coryphaenoides* geminates. The outlier is the *D* value for the barnacle geminate pair in the genus *Euraphia* (Laguna 1987).

when as few as eighteen loci were assayed. In addition, Bermingham and Lessios (1993), essentially doubled the number of isozymes assayed for the urchin geminates to thirty-four without an appreciable change in the genetic distance between these geminates. In short, the variation in *D* values cannot be largely due to the possible sources of error listed above.

These *D* values may be converted to estimates of time since divergence with two commonly used calibrations of the protein electrophoretic clock: T_1 (*D* of 1 = 18.9 my [Sarich 1977; Carlson et al. 1978]) and T_2 (*D* of 1 = 5 my [Chakraborty and Nei 1974; Nei 1987]). Echinoid divergence estimates range between 0.7 my and 9.8 my for T_1 and 0.2 my to 2.6 my for T_2. Crustacean divergence estimates span the range from 0.5 my to 7.1 my for T_1 to 0.1 my to 1.9 my for T_2 (In these comparisons, the *Euraphia* geminate pair has not been included. In excluding it, I am not suggesting that the *Euraphia* species pair is not geminate, but rather that the conclusions drawn remain the same, even if this outlier is excluded). Fish divergence estimates range between 2.5 my and 7.9 my for T_1 and 0.7 my to 2.1 my for T_2. The relationship between protein divergence estimates for geminate pairs and a conservative geological estimate for the timing of the final isthmian emergence is shown in figure 11.3. This range, summarized from the discus-

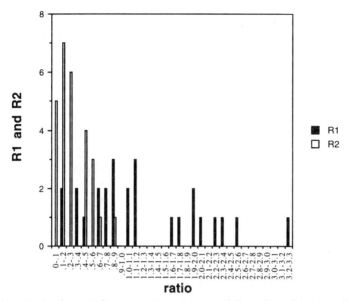

Fig. 11.3   Ratio of protein divergence time estimates and the geologically estimated time of emergence of the Central American Isthmus. Ratios close to 1 indicate close correspondence between protein divergence time estimate (R_1) and a conservative geological estimate for the time of the final isthmian emergence (R_2). Ratios less than 1 indicate a divergence younger than 3 my while those greater than 1 indicate an emergence older than 3 my (for R_1 and R_2, see note to table 11.1). *Euraphia* ratios not plotted.

sion above, suggests that the molecular clock hypothesis would be rejected for these data. However, only a minority of the scientists comparing transisthmian protein divergences have come to this conclusion. The arguments developed in these papers, as well as discussion of these results by other authors, are informative for two reasons. They trace the progression of arguments concerning what constitutes a valid test of the molecular clock and demonstrate that there has been a significant degree of bias toward finding regular rates of evolution, as alluded to in the quote by Avise and Aquadro at the beginning of this section.

The criticism of many of these arguments is that they are, to a greater or lesser extent, circular. Specifically, there is a tacit assumption of clocklike behavior of proteins underlying many of the assumptions made. For example, Gorman and colleagues (1976), studied transisthmian representatives of the fish genus *Bathygobius: B. soporator* (western Atlantic) and *B. ramosus* and *B. andrei* (eastern Pacific). According to Rubinoff and Rubinoff (1971), these three species " are [morphologically] very similar and appear to have diverged almost equally from each

other. Certainly the magnitude of the differences is too small to pre-
sume a closer phyletic relationship between any two species." The D
values found for these species were .146 for *B. soporator/B.andrei* and
about .419 for *B. soporator/B. ramosus* and *B. ramosus/B. andrei*. Once the
D values are reported, it is presumed that the geminate pair is *B. sopora-
tor/B. andrei*, the pair with the lowest D value. This is the presumption
that one would make if one had accepted a molecular clock, but it is
not an acceptable assumption when one is testing a molecular clock.
Alternative scenarios are not discussed. The possibility that rates of evolution might be variable and therefore distance data might not be phy-
logenetically informative, or that the species might represent a tri-
chotomy, with accelerated evolution in the *B. ramosus* lineage, is not
considered.

The *B. ramosus/B. soporator* comparison is dropped without comment
from the Vawter compilation (Vawter et al. 1980) of transisthmian comparisons. The point here is not that *B. soporator* and *B. andrei* are not the
geminate pair; independent studies of other characters may well demonstrate their geminate status. The point is that it may not be assumed
on the basis of protein electrophoretic distance when one is testing the
protein electrophoretic clock. The choice of *B. soporator/B. andrei* as the
cognate pair in turn determined the calibration of the molecular clock
for transisthmian comparisons.

Having accepted *B. soporator/B. andrei* as the geminate pair, Vawter
and colleagues calculated the estimated divergence times derived from
this D value using two of the commonly employed calibrations of the
molecular clock (D of $1 = 18.9$ my and D of $1 = 4$–4.5 my). The former
extreme calibration was accepted because it gave a "reasonable" estimate; that is, it gave a value similar to the time estimate based on geological evidence. Given this, statements such as "the predicted divergence time is about 2.5 million years, which is in accord with the
geological evidence" (Gorman et al. 1976), or Selander's comment that
begins this chapter, are hardly surprising. As Avise and Aquadro point
out, calibrations from D of $1 = 0.8$ my to D of $1 = 18.9$ my have been
used by various authors, and in each case they "concluded that molecular based estimates of absolute divergence time were compatible with
geological or fossil evidence" and that "given the huge range of time
estimates available from different clocks that might be chosen, it is hard
to imagine a genetic distance estimate that would not be compatible
with almost any fossil or geologic data" (Avise and Aquadro 1982).

This circularity is important because it sets the stage for the discussions concerning acceptable falsifiers of the molecular clock in trans-

isthmian comparisons. In the original paper Gorman and colleagues (1976) state that "a value of less than 2 million years would cause us to question the model." However, given the range of possible calibrations, a D value of from .106 to 1.6 could represent 2 my of divergence. Table 11.1 and fig. 11.3 show that even if we use the extreme calibration of the molecular clock, D values from five of the geminate pairs yield estimated times of divergence considerably below 2 my. If we pick an intermediate calibration for the molecular clock (D of 1 = 12 my), almost half of the geminates listed give estimated divergence times below 2 my. Gorman and Kim (1977), in discussing conditions for rejection of the clock, state: "The hypothesis can be falsified if a pair of populations shows little or no genetic differentiation. The hypothesis cannot be falsified if there is too much differentiation. This merely implies that they are not true geminates." Vawter and colleagues (1980) concur: "High D values do not test the clock hypothesis as differentiation may have predated the rise of the isthmus."

But how much is "too much"? For these authors, the answer is, too much based on the calibration of the molecular clock chosen. This is the problem with these arguments: they are based on an acceptance not only of the molecular clock, but on a specific calibration of that clock. They are thus unacceptable in a test of the molecular clock. The key to the acceptance or rejection of the molecular clock in transisthmian comparisons is the range of values found for the taxa examined. If the overall pattern is of little variation, the clock is supported; if the range is great among a variety of taxa, the clock is rejected. Vawter and colleagues (1980) reject Lessios's finding of a D of .549 to .638 (the values reported in Lessios 1979a,b) among *Echinometra* geminates because they give values (10.4–12.0 my) which are too large based on the calibration they have chosen. However, using the commonly employed calibration of Chakraborty and Nei (1974), these D values give divergence estimates of 2.8 to 3.2 my, very close to current estimates of the timing of the emergence of the isthmus. This is the calibration that Nei (1987) uses in order to reach the conclusion that the average time estimate from D values for echinoderm geminates "is consistent with the one from geological data." The *B. soporator/B. ramosus* comparison also becomes "consistent" if this calibration is used.

The combination of choosing an extreme calibration of the protein clock and a rejection of any values representing "too much" divergence has restricted the possible range of variation in D values, making rejection of the clock improbable. A conservative estimate for the time of the final emergence of the isthmus based on geological/paleontological data is 3 my, corresponding to a D of .16 using D of $1 = 18.9$ my. This D value is equal to what is considered to be the upper limit of D values for conspecific populations (Nei 1987). The upper limit, as noted previously, is based on the acceptance of a molecular clock with a calibration of D of $1 = 18.9$ my. Vawter and colleagues (1980) reject Lessios's D values between *Echinometra* geminates on this basis. *Echinometra* D values (about 0.6) are below the average D values between congeneric species for fish, amphibia, reptiles, and mammals (Avise and Aquadro 1982). Thus, by squeezing down the predicted D value into the range of conspecific populations and rejecting values as too large which are at or below the mean for many congeneric species pairs, rejection of the clock is made unlikely.

In summary, review of the literature concerning the Isthmus of Panama has shown that the circularity of assuming a protein electrophoretic clock while supposedly testing it has affected virtually every level of the analysis. These circularities include (1) circularity in the choice of geminates where there is more than one potential twin on the basis of genetic distance, (2) circularity in the claimed congruence between estimated divergence times based on genetic distance and estimated divergence times based on geological estimates when a calibration is selected from a wide range of possible calibrations to coincide with the geological time estimate, (3) circularity in the rejection of certain genetic distances as "too much" divergence based on the acceptance of the molecular clock and an extreme calibration of that clock. One cannot assume a molecular clock when one is testing, or seeking support, for the same. The success or failure of tests of the molecular clock in the context of the Central American Isthmus depends on the range of values found for the geminate pairs examined.

MITOCHONDRIAL DNA COMPARISONS IN GEMINATE SPECIES

The focus of transisthmian molecular comparisons has recently shifted to analysis of mtDNA divergence (Collins 1989; Martin et al. 1992; Bermingham and Lessios 1993; Knowlton et al. 1993; Collins, Bermingham, and Lessios, unpublished data). The potential advantage of this approach lies in the ability to detect many evolutionary changes which would be missed by standard protein electrophoretic techniques, especially over the evolutionarily recent time scale of the isthmian emergence. Protein electrophoresis detects those substitutions that affect the electrophoretic mobility of a protein. Direct DNA analyses by restriction endonuclease assay or di-deoxy sequencing allow one to examine nucleotide substitutions that either do not alter the amino acid sequence or that result in amino acid replacements that do not alter protein mobility during electrophoresis. Nucleotide sequence analyses also offer the potential to study the evolutionary dynamics of different types of nucleotide substitution. For example, transitional substitutions (purine to purine, or pyrimidine to pyrimidine) occur at a much higher rate in the mitochondrial genomes of animals than transversions (purine to pyrimidine or vice versa) (Moritz et al. 1987). Analysis at the nucleotide level permits the study of rates and rate variation for particular types of substitutions. The discrete nature of sequence data also makes it more easily analyzed using parsimony techniques than electrophoretic data, where the hierarchical relationships of the many different alleles are difficult to determine from electrophoretic mobility. In addition, the conservation of the mitochondrial genome across the metazoa gives us confidence that comparisons of mtDNA of distantly related organisms are truly homologous, making rate comparisons meaningful (Moritz et al. 1987).

Results from these studies suggest that the variation in transisthmian mtDNA divergences is appreciable, though not as broad as that found in allozyme comparisons (Lessios and Bermingham 1993; Knowlton et al. 1993). Variation in rates among geminates was reduced for mtDNA compared to allozymes. This was true for analyses of variation throughout the urchin mtDNA molecule with restriction endonucleases (Bermingham and Lessios 1993) and for sequence analysis within the Cytochrome Oxidase I gene within alpheid shrimp (Knowlton et al. 1993).

Knowlton and others (1993) found strong correlations between divergence estimates based on allozymes and mtDNA sequences, and a measure of reproductive incompatibility, for most species pairs in a study of seven pairs within the snapping shrimp genus *Alpheus*. The authors interpreted the correlations among these factors as indicating staggered isolation of pairs rather than correlated rates of evolution within species. Other molecular support for the notion that some isthmian geminate species may have been isolated prior to 3.5 my comes from a study of rates of mitochondrial cytochrome *b* evolution in the temperate North Pacific and North Atlantic caenogastropod genus *Nucella* (Collins et al. n.d.). If a molecular clock for cytochrome *b* transversions, based on the phylogeny and fossil record of *Nucella*, is assumed and applied to divergence values for snail geminates, a Miocene isolation is suggested. We should keep in mind, however, the possibility that the correlations between different molecular measures of divergence

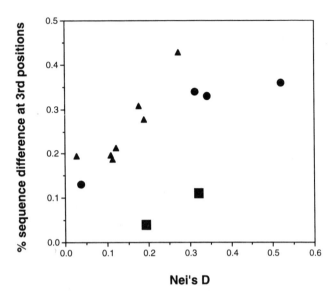

Fig. 11.4 Nei's genetic distance (*D*) plotted against sequence difference at third positions of mitochondrial genes for alpheid shrimp geminates (triangle), ray-finned fish geminates (square), and urchin geminates (circle). *D* values are from the sources cited in table 11.1. Third codon position differences for mitochondrial genes are from an approximately 650 base pair portion of the Cytochrome Oxidase I gene in Alpheid shrimp (Knowlton et al. 1993), and from a 302 base pair portion of the cytochrome *b* gene in fishes and urchins (Collins, Bermingham, and Lessios, unpublished data).

could indicate correlated rate variation between these measures rather than staggered isolation. These molecules have all been through similar population histories, and factors which accelerate rates in one molecule may have similar effects on others. In figure 11.4, I have plotted the Nei's genetic distance and pairwise sequence difference at third positions for the alpheid shrimp data from Knowlton et al. 1993. Similar measures for two geminate pairs of ray-finned fishes and three pairs of urchins (table 11.1; Collins, Bermingham, and Lessios unpublished data) broaden the taxonomic range of comparison. The addition of these taxa emphasizes two points. First, overall, there is no obvious pattern to suggest that some species are geminates while others are not. Second, there appears to be a taxonomic patterning to the distribution of points, with shrimp having higher sequence divergence values, and the fish having low sequence divergences and Nei's distances. This suggests that the fish geminate pairs may have been isolated later than the urchins or shrimp, or that they may have been isolated simultaneously, with correlated lower rates of evolution between two measures of divergence in the fish. Unfortunately, compatibility measures comparable to those gathered by Knowlton and colleagues are not available for the fish and urchin geminates sampled here.

Recent analyses of mitochondrial nucleotide sequences (Martin et al. 1992; Knowlton et al. 1993; Collins, Bermingham, and Lessios; unpublished data) suggest a strong possibility that rates may vary much more broadly across major taxonomic groups than within groups. Cytochrome *b* sequences in one pair of bonnethead sharks, five pairs of bony fishes, and three pairs of urchins (Martin et al. 1992; Collins, Bermingham, and Lessios n.d.) reveal that cytochrome *b* in cartilaginous and bony fishes is evolving by almost an order of magnitude more slowly than the values suggested for the same sequence in mammals (Irwin et al. 1991). The average value for the urchins is much closer to the rates commonly cited for mammals. There is a tenfold variation in amounts of sequence divergence for these geminates. Preliminary results from sequence analysis of snail geminates for the same region of cytochrome *b* indicate that values for these species are similar to those found in urchins (Collins, unpublished data). The overall pattern is one of great variation in divergence values among all geminates, with a much reduced variance when looking within a taxonomic group. The most striking result is the extremely low rate of sequence divergence in teleost fishes. Restriction endonuclease analysis of mtDNA in Shad (Bentzen 1989) and salmonid fishes (Thomas and Beckenbach 1989; Smith 1992) and sequence analysis in cichlids (Kocher et al. 1989), neopterygian fishes (Normark et al. 1991), and sharks (Martin et al. 1992) support the finding of reduced rates of nucleotide sequence evolution in fishes.

CONCLUSIONS

The future of transisthmian rate studies will be based on analysis at the sequence level and should focus on the search for taxon-specific rates and possible explanations for the characteristic rates that have been found in some groups. Factors such as compositional effects (unequal frequencies of the four nucleotides in sequences) and the proportions of sites free to change (Fitch 1976; Palumbi 1989), which have been found to affect rate estimates, need to be considered. A "brute force" approach, with analysis of a large number of geminates from each taxonomic group for the same genes will likely yield the most satisfying results and, with the advent of techniques such as the polymerase chain reaction (PCR), are no longer prohibitively time consuming. Comparisons of nuclear versus mitochondrial genes in geminates may give us some insight into the intrinsic factors controlling rates of evolution, and the analysis of closely related geminates from different habitats may help us to understand which extrinsic factors are the most important to accelerating rates of divergence. To sum up, molecular distances between geminate pairs could potentially be used by molecular biologists to (a) determine the time of isolation of geminate species (and hence the time of the final isthmian emergence, (b) calibrate molecular clocks with geminate species, and (c) test for the regularity of molecular rates. I would argue that molecular biologists have made very little contribution to determining the timing of the isthmian emergence, primarily because of the lack of reliable independent calibrations. Apparent confirmations of geological estimates are largely traceable to the choice of calibration to coincide with geological or paleontological estimates. Calibration of molecular clocks with geminate pairs has been more successful, but there is still considerable ambiguity about the criteria for recognizing "false geminates" that may have been isolated prior to the final isthmian emergence. The best approach outlined to date for recognizing these false geminates is in the comparisons of various measures of divergence. Pairs which have large amounts of divergence in all of these measures are considered to be false geminates. The success of this strategy will depend on our ability to specify what we mean by large, and to determine that these various measures should not be correlated for reasons other than similar time of isolation. As data accumulate, it is possible that clusters of divergence values for different taxonomic groups will become apparent, suggesting taxon-specific clusters of fairly uniform rate, with the occasional outlier indicating a "false geminate" or atypical rate. Alternatively, the pattern of variation might prove to be continuous, suggesting a long drawn-out isolation of the faunas on either side of the isthmus, limiting the utility of transisthmian comparisons for tests of variance in molecular clocks, but still permitting a possible calibration of rates, based on the geminates most likely to have been isolated by the final isthmian emergence. Finally, the evidence from transisthmian comparisons to date, argue strongly against a universal molecular clock, either for proteins or nucleotide sequences, although taxon-specific rates appear to be less variable. To return to the introductory quote, there is a chance that fishes may some day be able to tell time as well as geologists. If this proves to be the case, it will be because geologists, paleontologists, and systematists helped them calibrate their timepieces.

ACKNOWLEDGMENTS

I thank Eldredge Bermingham, Ann Budd, Laurel Collins, Rick Harrison, Jeremy Jackson, Nancy Knowlton, Haris Lessios, and Geerat Vermeij for valuable comments on various drafts of this chapter, as well as discussions of work in progress. Part of this work was carried out during a postdoctoral fellowship at the Smithsonian Tropical Research Institute. I thank María Carles for assistance in the laboratory and the field.

REFERENCES

Avise, J. C., and C. F. Aquadro. 1982. A comparative summary of genetic distance in the vertebrates: Patterns and correlations. *Evol. Biol.* 15: 151–85.

Ayala, F. J. 1985. Review of *Molecular evolutionary genetics. Mol. Biol. Evol.* 4:96–198.

Bayer, F. M., G. L. Voss, and C. R. Robins. 1970. *Bioenvironmental and radiological safety feasibility studies Atlantic-Pacific interoceanic canal: Report on the marine fauna and benthic shelf-slope communities of the isthmian region*, Coral Gables, Fla.: University of Miami.

Bentzen, P. 1989. Mitochondrial DNA polymorphism in American shad (*Alosa sapidissima*) and its implications for population structure. Ph.D. diss., McGill University, Montreal.

Berggren, W. A., and C. D. Hollister. 1974. Paleogeography, paleobiogeography, and the history of circulation in the Atlantic Ocean. *SEPM Spec. Pub.* 20:126–85.

———. 1977. Plate tectonics and paleocirculation—commotion in the ocean. *Tectonophysics* 38:11–48.

Bermingham, E., and H. A. Lessios. 1993. Rate variation of protein and mitochondrial DNA evolution as revealed by sea urchins separated by the Isthmus of Panama. *Proc. Nat. Acad. Sci.* 90:2734–38.

Bowen, B. W., A. B. Meylan, and J. C. Avise. 1989. An odyssey of the green sea turtle: Ascension Island revisited. *Proc. Nat. Acad. Sci.* 86:573–76.

Briggs, J. C. 1974. *Marine zoogeography*. New York: McGraw Hill.

Carlson, S. S., A. C. Wilson, and R. D. Maxson. 1978. Do albumin clocks run on time? Reply to Radinsky. *Science* 200:1182–85.

Chakraborty, R., and M. Nei. 1974. Dynamics of gene differentiation between incompletely isolated populations of unequal sizes. *Theor. Popul. Biol.* 5: 460–69.

Chesher, R. H. 1972. The status of knowledge of Panamanian echinoids, 1971, with comments on other echinoderms. *Bull. Biol. Soc. Wash.* 2:139–58.

Coates, A. G., J. B. C. Jackson, L. S. Collins, T. M. Cronin, H. J. Dowsett, L. M. Bybell, P. Jung, J. A. Obando. 1992. Closure of the Isthmus of Panama: The near-shore marine record of Costa Rica and western Panama. *Geol. Soc. Am. Bull.* 104:814–28.

Collins, T. M. 1989. Rates of mitochondrial DNA divergence in transisthmian geminate species. Ph.D. diss., New Haven: Yale University.

Collins, T. M., K. Fraser, A. R. Palmer, G. J. Vermeij, and W. M. Brown. N.d. Evolutionary history of Northern Hemisphere Nucella (Gastropoda, Muricidae): Molecules, morphology, ecology, and fossils. *Evolution.* In press.

Cronin, T. M. 1985. Speciation and stasis in marine Ostracoda: Climatic modulation of evolution. *Science* 227:60–63.

Ekman, S. 1967. *Zoogeography of the sea.* London: Sidgwick and Jackson.

Eldredge, N., and S. J. Gould. 1972. Punctuated equilibria, an alternative to phyletic gradualism. In *Models in paleobiology,* ed. T. J. M. Schopf, 82–115. San Francisco: W. H. Freeman.

Felsenstein, J. 1978. Cases in which parsimony or compatibility methods will be positively misleading. *Syst. Zool.* 27:401–10.

———. 1985. Phylogenies and the comparative method. *Am. Nat.* 125:1–15.

———. 1988. Phylogenies from molecular sequences: Inference and reliability. *Ann. Rev. Gen.* 22:521–65.

Fitch, W. M. 1976. Molecular Evolutionary Clocks. In *Molecular evolution,* ed. F. J. Ayala, 160–78. Sunderland, Mass.: Sinauer.

Gillespie, J. H. 1991. *The causes of molecular evolution.* New York: Oxford University Press.

Gingerich, P. D. 1983. Rates of evolution: Effects of time and temporal scaling. *Science* 222:159–61.

———. 1986. Temporal scaling of molecular evolution in primates and other mammals. *Mol. Biol. Evol.* 3:205–21.

Gorman, G. C., and Y. J. Kim. 1977. Genotypic evolution in the face of phenotypic conservativeness: *Abudefduf* (Pomacentridae) from the Atlantic and Pacific sides of Panama. *Copeia,* 694–97.

Gorman, G. C., Y. J. Kim, and R. Rubinoff. 1976. Genetic relationships of three species of *Bathygobius* from the Atlantic and Pacific sides of Panama. *Copeia,* 361–64.

Graves, J. E., R. H. Rosenblatt, and G. N. Somero. 1983. Kinetic and electrophoretic differentiation of lactate dehydrogenase of teleost species-pairs from the Atlantic and Pacific coasts of Panama. *Evolution* 37:30–37.

Graves, J. E., S. D. Ferris, and A. E. Dizon. 1984. Close genetic similarity of Atlantic and Pacific skipjack tuna (*Katsuwonas pelamis*) demonstrated with restriction endonuclease analysis of mitochondrial DNA. *Mar. Biol.* 79:315–19.

Günther, A. 1869. An account of the fishes of the states of Central America, based on the collections made by Capt. J. M. Dow, F. M. Godman, Esq., and O. Salvin, Esq. *Trans. Zool. Soc. Lond.* 6:377–402.

Haq, B. U. 1984. Paleoceanography: A synoptic overview of 200 million years of ocean history. In *Marine geology and oceanography of Arabian Sea and Coastal Pakistan,* ed. B. U. Haq and J. D. Milliman, 200–31. New York: Van Nostrand Reinhold.

Hillis, D. M. 1987. Molecular versus morphological approaches to systematics. *Ann. Rev. Ecol. Syst.* 18:23–42.

Holcombe, T. L., and W. S. Moore. 1977. Paleocurrents in the eastern Caribbean: Geologic evidence and implications. *Mar. Geol.* 23:35–56.

Humphries, C. J., and L. R. Parenti. 1986. *Cladistic biogeography*. Oxford: Clarendon Press.

Irwin, D. M., T. D. Kocher, and A. C. Wilson. 1991. Evolution of the cytochrome *b* gene of mammals. *J. Mol. Evol.* 32:128–44.

Jackson, J. B. C., P. Jung, A. G. Coates, and L. S. Collins. 1993. Diversity and extinction of tropical American mollusks and emergence of the Isthmus of Panama. *Science* 260:1624–26.

Jones, D. S., and P. F. Hasson. 1985. History and development of the marine invertebrate faunas separated by the Central American Isthmus. In *The Great American Biotic Interchange*, eds. F. G. Stehli and S. D. Webb, 325–55. New York: Plenum Press.

Jones, M. L. 1972. Introduction to a symposium on the biological effects of a sea level canal across Panama. *Bull. Biol. Soc. Wash.*, vi–viii.

Jordan, D. S. 1908. The law of geminate species. *Am. Nat.* 42:73–80.

Kluge, A. G. 1988. Parsimony in vicariance biogeography: A quantitative method and a greater Antillean example. *Syst. Zool.* 37:315–28.

Knowlton, N., L. A. Weight, L. A. Solórzano, D. K. Mills, and E. Bermingham. 1993. Divergence in proteins, mitochondrial DNA, and reproductive compatibility across the Isthmus of Panama. *Science* 260:1629–31.

Kocher, T. D., W. K. Thomas, A. Meyer, S. V. Edwards, S. Pääbo, F. X. Villablanca, and A. C. Wilson. 1989. Dynamics of mitochondrial DNA evolution in animals: Amplification and sequencing with conserved primers. *Proc. Nat. Acad. Sci.* 86:6196–6200.

Laguna, J. E. 1987. *Euraphia eastropacensis* (Cirripedia, Chthamaloidea), a new species of barnacle from the tropical eastern Pacific: Morphological and electrophoretic comparisons with *Euraphia rhizophorae* (deOliveira) from the tropical western Atlantic and molecular evolutionary implications. *Pac. Sci.* 41:132–40.

Lessios, H. A. 1979a. Molecular, morphological, and ecological divergence of shallow water sea urchins separated by the Isthmus of Panama. Ph.D. diss., Yale University, New Haven, Conn.

———. 1979b. Use of Panamanian sea urchins to test the molecular clock. *Nature* 280:599–601.

———. 1981. Divergence in allopatry: Molecular and morphological differentiation between sea urchins separated by the Isthmus of Panama. *Evolution* 35:618–34.

Lessios, H. A., and C. W. Cunningham. 1990. Gametic incompatibility between species of the sea urchin Echinometra on the two sides of the Isthmus of Panama. *Evolution* 44:933–41.

Manning, R. B. 1969. Stomatopod Crustacea of the western Atlantic. *Stud. Trop. Oceanogr.* 8:1–380.

Marshall, C. R. 1990a. Confidence interval on stratigraphic ranges. *Paleobiology* 16:1–10.

———. 1990b. The fossil record and estimating divergence times between lineages: Maximum divergence times and the importance of reliable phylogenies. *J. Mol. Evol.* 30:400–408.

Martin, A. P., G. J. P. Naylor, and S. R. Palumbi. 1992. Rates of mitochondrial

DNA evolution in sharks are slow compared with mammals. *Nature* 357:153–55.

Masuda, H., et al. 1984. *The fishes of the Japanese archipelago.* Tokyo: Tokyo University Press.

Mayr, E. 1954. Geographic speciation in tropical echinoids. *Evolution* 8:1–18.

———. 1963. *Animal species and evolution.* Cambridge: Harvard University Press.

———. 1967. Population size and evolutionary parameters. In *Mathematical challenges to the Neo-Darwinian interpretation of evolution,* ed. P. Moorhead and M. M. Kaplan, 47–58. Wistar Institute Symposium, mon. 5. Philadelphia: Wistar Institute Press.

Moritz, C., T. E. Dowling, and W. M. Brown. 1987. Evolution of animal mitochondrial DNA: Relevance for population biology and systematics. *Ann. Rev. Ecol. Syst.* 18:269–92.

Nei, M. 1987. *Molecular evolutionary genetics.* New York: Columbia University Press.

Norell, M. A. 1992. Taxic origin and temporal diversity: The effect of phylogeny. In *Extinction and phylogeny,* ed. M. J. Novacek and Q. D. Wheeler, 89–118. New York: Columbia University Press.

Normark, B. B., A. R. McCune, and R. G. Harrison. 1991. Phylogenetic relationships of Neopterygian fishes, inferred from mitochondrial DNA sequences. *Mol. Biol. Evol.* 8:819–34.

Palumbi, S. R. 1989. Rates of molecular evolution and the fraction of nucleotide positions free to vary. *J. Mol. Evol.* 29:180–87.

Petuch, E. J. 1982. Geographical heterochrony: Contemporaneous coexistence of Neogene and Recent molluscan faunas in the Americas. *Paleogeogr., Paleoclimatol., Paleoecol.* 37:277–312.

Rosen, D. E. 1976. A vicariance model of Caribbean biogeography. *Syst. Zool.* 24:431–64.

Rosenblatt, R. H. 1963. Some aspects of speciation in marine shore fishes. In *Speciation in the sea,* ed. J. P. Harding and N. Tebble, 171–80. London: Systematics Assoc.

———. 1967. The zoogeographic relationships of the marine shore fishes of tropical America, *Stud. Trop. Oceanogr.* 5:579–92.

Rosenblatt, R. H., and R. S. Waples. 1986. A comparison of allopatric populations of shore fish species from the eastern and central Pacific Ocean: Dispersal or vicariance? *Copeia,* 275–84.

Rubinoff, I. 1963. Morphological comparisons of shore fishes separated by the Isthmus of Panama. Ph.D. diss., Harvard University, Cambridge, Mass.

Rubinoff, R. W., and I. Rubinoff. 1971. Geographic and reproductive isolation in Atlantic and Pacific populations of Panamanian *Bathygobius. Evolution* 23:88–97.

Sarich, V. M. 1977. Rates, sample sizes, and the neutrality hypothesis for electrophoresis in evolutionary studies. *Nature* 265:24–28.

Sarich, V. M., and A. C. Wilson. 1973. Generation time and genomic evolution in primates. *Science* 179:1144–47.

Scherer, S. 1989. The relative-rate test of the molecular clock hypothesis: A note of caution. *Mol. Biol. Evol.* 6:436–41.

———. 1990. The protein molecular clock: Time for a reevaluation. *Evol. Biol.* 24:83–106.

Selander, R. K. 1982. Phylogeny. In *Perspectives on evolution*, ed. R. Milkman, 32–59. Sunderland, Mass.: Sinaeur Assoc.

Simon, C. 1991. Molecular systematics at the species boundary: Exploiting conserved and variable regions of the mitochondrial genome of animals via direct sequencing from amplified DNA. In *Molecular techniques in taxonomy*, ed. G. M. Hewitt, A. W. B. Johnston, and J. P. W. Young, 33–71. NATO ASI series, H:57. Berlin: Springer Verlag.

Smith, G. R. 1992. Introgression in fishes: Significance for paleontology, cladistics, and evolutionary rates. *Syst. Biol.* 41:41–57.

Smith, M. M., and P. C. Heemstra. 1986. *Smith's sea fishes.* Berlin: Springer Verlag.

Stanley, S. M. 1979. *Macroevolution: Pattern and process.* San Francisco: W. H. Freeman.

Stanley, S. M., and L. D. Campbell. 1981. Neogene mass extinction of western Atlantic molluscs. *Nature* 293:457–59.

Stehli, F. G., and S. D. Webb. 1985. *The Great American Biotic Interchange.* New York: Plenum Press.

Steinbeck, J. 1962. *Travels with Charley in search of America.* New York: Viking Press.

Thomas, W. K., and A. T. Bechenbach. 1989. Variation in salmonid mitochondrial DNA: Evolutionary constraints and mechanisms of substitution. *J. Mol. Evol.* 29:233–245.

Thorpe, J. P. 1982. The molecular clock hypothesis: Biochemical evolution, genetic differentiation, and systematics. *Ann. Rev. Ecol. Syst.* 13:139–68.

Vawter, A. T., R. Rosenblatt, and G. C. Gorman. 1980. Genetic divergence among fishes of the eastern Pacific and the Caribbean: Support for the molecular clock. *Evolution* 34:705–11.

Vermeij, G. J. 1978. *Biogeography and adaptation.* Cambridge: Harvard University Press.

———. 1991. When biotas meet: Understanding biotic interchange. *Science* 253:1099–1104.

Vermeij, G. J., and E. J. Petuch. 1986. Differential extinction in tropical American molluscs: Endemism, architecture, and the Panama land bridge. *Malacologia* 27:29–41.

Voight, J. R. 1988. Trans-panamanian geminate octopods (Mollusca: Octopoda), *Malacologia* 29:289–93.

Voss, G. L. 1972. Biological results of the University of Miami deep-sea expedition 93. Comments concerning the University of Miami's marine biological survey related to the Panamanian sea-level canal. *Bull. Biol. Soc. Wash.* 2: 49–58.

Vrba, E. S. 1980. Evolution, species, and fossils: How does life evolve? *S.A. J. Sci.* 76:61–84.

Weinberg, J. R., and V. R. Starczak. 1989. Morphological divergence of eastern Pacific and Caribbean isopods: Effects of a land barrier and the Panama canal. *Mar. Biol.* 103:143–52.

West, D. A. 1980. Genetic variation in transisthmian geminate species of brachyuran crabs from the coasts of Panamá. Ph.D. diss., Yale University, New Haven, Conn.

Wilson, A. C., S. S. Carlson, and T. J. White. 1977. Biochemical Evolution. *Ann. Rev. Biochem.* 46:573–69.

Wilson, R. R., Jr., and R. S. Waples. 1984. Electrophoretic and biometric variability in the abyssal grenadier *Coryphaenoides armatus* of the western North Atlantic, eastern South Pacific, and eastern North Pacific Oceans. *Mar. Biol.* 80:227–37.

Woodring, W. P. 1965. Endemism in Miocene Caribbean molluscan faunas. *Science* 148:961–73.

———. 1966. The Panama land bridge as a sea barrier. *Proc. Am. Phil. Soc.* 110:425–33.

———. 1974. The Miocene faunal province and its subprovinces. *Verhandl. Naturforsch. Ges. Basel.* 84:209–13.

Zuckerkandl, E., and L. Pauling. 1962. Molecular disease, evolution, and genic heterogeneity. In *Horizons in biochemistry*, ed. M. Kasha and B. Pullman. New York: Academic Press.

12

Late Cenozoic Evolution of the Neotropical Mammal Fauna

S. David Webb and Alceu Rancy

INTRODUCTION

By most measures tropical America holds the world's richest terrestrial biota (De Onis 1992; Wilson 1992). The living mammal fauna of the Neotropical Region is no exception, as the rainforests alone hold about 500 mammalian species, including some 300 endemics (Emmons 1990). By examining the rich fossil record of Late Cenozoic mammals in both American continents, it is possible to outline the evolution of the neotropical mammal fauna.

There are several reasons for focusing on land mammals. First, they have left a more complete fossil record than any other major group of terrestrial organisms in the American tropics. Furthermore, they have evolved rapidly, moved widely, and diversified substantially within the two American continents. They provide the key terrestrial record associated with the emplacement of the isthmian land bridge and the subsequent biotic events known as the Great American Interchange.

The history of land mammals reciprocally crossing through the isthmian connection embodies what has been called the "Central American paradox" (Webb 1978). During the Tertiary, Central America was an adjunct of North America, as demonstrated by exclusively North America faunas in Honduras, El Salvador, and Panama (Stehli and Webb 1985). Yet the modern mammal fauna pertains decidedly to the Neotropical Region, with its peak diversity in South America. As Alfred Russell Wallace observed, "The portion of North America that lies within the tropics (Mexican Subregion) closely resembles the [Brazilian Subregion] in general zoological features" (Wallace 1876, 2:5). How did Central America shift its zoogeographic affiliation so completely in only two million years?

Fossil evidence from the American tropics suggests that this "Central American paradox" is not a true paradox, but rather a history that resolves itself into two phases characterized by substantially different ecological patterns. We illustrate two ecologically distinct phases by separately considering the Early Quaternary and Late Quaternary records of the Great American Interchange. A third phase was imposed by mass extinctions of mammalian megafauna in the latest Pleistocene. These three phases provide an outline of how the neotropical mammal fauna evolved.

Previous Studies of Interamerican Land Mammals

Alfred Russell Wallace was the first to recognize the impact and approximate timing of the interamerican mammal dispersal. (Darwin's work came too early to give him a perspective on the isolation and later merger of South American with North American fossil mammals). By comparing the "wonderful extinct fauna ... discovered in North America, with what was previously known from South America," Wallace found "unmistakable evidence of an extensive immigration from South into North America, not very long before the Glacial epoch" (Wallace 1876, 1:131). Wallace also appreciated the importance of the previous isolation of the South American mammal and bird faunas. "Richness combined with isolation is the predominant feature of Neotropical zoology, and no other region can approach it in the number of its peculiar family and generic types" (Wallace 1876, 2:5).

Tertiary records of American mammals, as one can see from Wallace's remarks, still left the direction of many dispersals ambiguous. By the turn of the century, however, mammalian paleontologists had sorted out the directions in which diverse interchange groups had migrated. For example, tapirs and llamas, contrary to Wallace's view, were determined to have northern ancestry. Matthew (1915) unduly stressed the high-latitude origin of most land mammal taxa. Because the predominant records came from the pampas in Argentina and the high plains in the United States, vertebrate paleontologists tended to ignore the role of equatorial latitudes. Indeed, Scott attributed to the American tropics a negative role: "Between the two warm-temperate zones extended the Tropics, which acted as a vast sieve holding back most mammals and allowing but a relatively small number of climatically adaptable species to pass through." (1937, 263).

The classic work of Simpson (1940) was the first to fully recognize the broad scope of the Late Pliocene reciprocal immigration episode, with at least a dozen families moving in each direction. Unfortunately there was still little direct fossil evidence of tropical mammals, and the plate tectonics mechanisms producing the isthmian land bridge were not appreciated.

Work in the tropics is sometimes discouraged as being unproductive.

For example, Raup stated that "the fossil record in present-day rain forest areas is notoriously poor because of the paucity of good rock exposures from which collections can be made" and that land animals there "have very low fossilization potentials" (1988, 56). In recent decades, however, increasing numbers of mammalian paleontologists have noted the need and sought to produce land mammals from the American tropics. In the western Amazon Basin, for example, extensive fossiliferous exposures are accessible during the dry season and yield especially rich samples of Middle Miocene and Late Pleistocene vertebrate fossils. Well-timed expeditions there have proven immensely successful (e.g., Simpson and Paula Couto 1981; Frailey 1986; Rancy 1991).

Even so, paleontological work in the American tropics has only begun. The nearly complete absence of small terrestrial vertebrates, especially rodents, birds, and squamates, is particularly frustrating because these taxa often provide a more sensitive register of environmental conditions than the large mammals which are more easily collected. Intensified efforts to screenwash productive sediments will surely produce further significant advances in this important research frontier.

CHRONOLOGY OF THE INTERCHANGE

Emplacement of the isthmian land bridge is chronicled both in the marine record by separation between Caribbean and Pacific biotas and in the continental record by biotic interchange of terrestrial animals between the Americas. This latter evidence dates final emplacement of the isthmian link in Central America at about 2.5 Ma (Webb 1985).

At present, the most reliable data registering the earliest interamerican immigrants in continental sequences occur at temperate latitudes in the western United States. In the Vallecito Creek badlands of southern California and in the San Pedro Valley and 111 Ranch sections in southern Arizona, the first major cohort of immigrant edentates and rodents from South America is well documented. The early records occur in the latest part of the Gauss magnetochron, earlier than 2.47 Ma (Galusha et al. 1984; Webb 1985).

In South America, the age of the first immigrants is subject to a wider range of interpretations. Reig (1980) claimed that cricetid rodents reached South America earlier in the Pliocene than other immigrant land mammals and attributed six such genera to the Chapadmalalan stage (stratigraphically near the base of the famous Barranca de Los Lobos section south of Mar del Plata, Argentina) and thus probably from the Gilbert magnetochron at least as early as 3.6 Ma.

The main cohort of northern immigrants (Canidae, Mustelidae,

Equidae, and Camelidae) appears in the Uquian stage (specifically, its earliest substage the Barrancolobian), and this mammal age most probably correlates with the early Matuyama magnetochron (Tonni et al. 1992). In this view, the first diverse cohort of immigrants in South America correlates closely with the arrival of the reciprocal cohort in western North America. Orgeira (1990), on the other hand, suggested that the Uquian immigrant-rich fauna at Barranca de Los Lobos was from the late Gilbert magnetochron and therefore that immigrant mammals from North America had traversed the isthmian land bridge earlier than 3.6 Ma. This would place the Uquian in correlation with the Chapadmalalan, which seems improbable (Tonni et al. 1992). The same objection arises from magnetic work in Bolivia, where Inchasi, a rich Chapadmalalan mammal fauna, is placed most probably within the late Gilbert and early Gauss magnetochrons, between about 4.0 and 3.3 Ma by MacFadden and colleagues (1993). Inchasi lacks northern immigrants and thus helps demonstrate that the Great American Interchange probably took place during the Uquian, after the Gauss magnetochron, and thus later than 3.0 Ma.

In Argentina, a still larger cohort of immigrant land mammal genera appears in the Ensenadan (Early Pleistocene) than in the preceding Uquian (Tonni et al. 1992). In North America, a number of additional immigrant groups appear in the Early Irvingtonian (Early Pleistocene), but the larger numbers had already arrived in the Late Blancan. Thus, the north temperate and south temperate records are not exactly symmetrical. The most probable general conclusion is that the interchange had begun in both directions by about 2.5 Ma and continued most actively for about a million years. In north and south temperate latitudes, there were few new appearances of interamerican immigrants after the Early to Middle Quaternary. Within South America, however, immigrant groups from the north continued to diversify strongly right through the Quaternary (Marshall et al. 1982; Webb 1985).

The ages of most Quaternary mammal faunas in the Americas are determined by biostratigraphic dating. The utility of such dates depends on the fact that mammals generally evolved rather rapidly in the course of the Pleistocene. In most taxa, Early Pleistocene species are readily distinguished from later Pleistocene forms. *Smilodon gracilis*, for example, an Early Pleistocene species of New World sabercats, has a much less rounded cranial shape and differs in tooth formula and many other details of its dentition from such Late Pleistocene species as *S. floridanus* in North America or *S. populator* in South America. Also it attained only about half the size of any of the later forms on either

American continent. On such bases, most faunas can be assigned readily either to the earlier or the later half of the Pleistocene.

In the latest Pleistocene, one ordinarily relies on carbon dating. Unfortunately, however, in the American tropics the sediments are often seasonally submerged and deeply weathered, rendering many samples unreliable. One of several problems that can arise in the use of carbon dates is illustrated by a recent report of a new species of ground sloth, genus *Nothropus*, from the Holocene of the western Amazon Basin (Frailey 1986). The associated carbon date led the investigator to assign it to the Holocene and to compare the new sloth with the typical Quaternary genus *Nothropus*. Subsequent analyses of the sloth show that this species probably represents a new genus, near *Pronothrotherium*, a Middle Miocene genus best known in Argentina. The accompanying fauna supports a Middle Miocene age (probably Huayquerian) (Rancy, unpublished data). Evidently, the Holocene carbon date resulted from recent water permeating older peaty sediments along the riverbanks. Another example of misleadingly young carbon dates is associated with the Mera flora in the Amazon Basin of Ecuador, where older sediments were probably contaminated by water seepage (Heine 1994). That is why most tropical mammal faunas must still rely on biostratigraphic dating.

METHODS

We divide our analysis of neotropical mammal evolution chronologically into three phases. First, we briefly treat the early interchange patterns of the Late Pliocene and Early Pleistocene. Second, we consider the later Pleistocene patterns in greater geographic detail, as warranted by the improved tropical record. And third, we summarize the devastating effects of the mass extinction of the mammal megafauna at the end of the Pleistocene.

Table 12.1 presents the basic data for land mammal families that participated in the Great American Interchange by Early Pleistocene time. This table summarizes data mainly from temperate latitudes, and has expanded only modestly since the results reported by Simpson (1940).

Table 12.2 provides Late Quaternary records of land mammal genera, including many new data from the American tropics. We use these data to distinguish patterns of mammalian evolution within the American tropics. We did not tabulate small mammal genera (here defined as those with body mass less than one kilogram), because their occurrences remain rare and thus have limited value in biogeographic summaries.

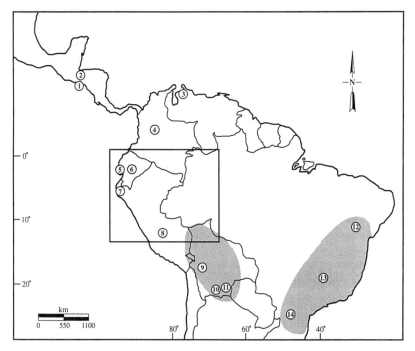

Fig. 12.1 Principal sites for Late Pleistocene mammals in tropical America. *Transandean lowlands:* (1) El Hormiguero, El Salvador; (2) Orillas del Humuya, Honduras; (3) Muaco, Venezuela; (4) Bogotá, Colombia; (5) La Carolina, Ecuador; (6) Talara, Peru. *Andes and altiplano:* (7) Punin and Guamote, Ecuador; (8) Pikimachay, Peru; (9) Ulloma, Bolivia; (10) Tarija, Bolivia; (11) Nuapua, Bolivia. *Eastern Brazilian lowlands:* (12) Janauba and Toco dos Ossos, Bahia, Brazil; (13) Lagoa Santa, Minas Gerais, Brazil; (14) Touro Passo, Rio Grande do Sul, Brazil. Insert: *Western Amazon* sites detailed in figure 12.2.

We relied primarily on recent reviews of Quaternary mammal data by Marshall (Marshall et al. 1984) for South America, by Tonni (Tonni et al. 1992) for Buenos Aires province, by Hoffstetter (1986) for the Andes, by Soruco (1991) for Bolivia, by Rancy (1991) for the western Amazon, by Cartelle (1994) for the Atlantic coastal region of Brazil, and by Webb and Perrigo (1984) for Central America.

We divided our tabulation of Late Quaternary mammal distribution into four broad geographic regions within tropical America. We compared the Amazon Basin, which still has the least complete samples, with the Atlantic coastal region of Brazil, the equatorial Andes and altiplano, and the transandean lowlands of northern South America and Central America. These regions and the principal Late Pleistocene mammal sites in each are indicated in Figures 12.1 and 12.2.

Our Late Pleistocene faunal regions follow along the lines of various

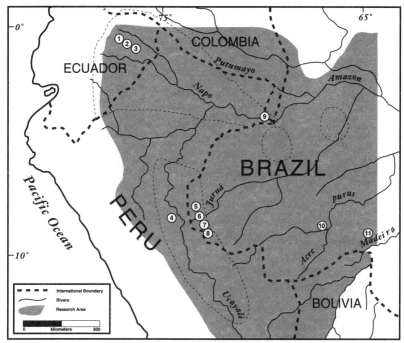

Fig. 12.2 Principal sites for Late Pleistocene mammals in western Amazon Basin (after Rancy 1991). (1–3) Upper Napo River, Ecuador; (4) Ucayali River, Peru; (5–8) Upper Juruá River, Brazil; (9) Lower Napo River, Ecuador; (10) Acre River, Brazil; (11) Madeira River, Brazil.

modern biogeographic regions. In delimiting groups of Late Pleistocene localities, the most difficult line to draw was that separating the Andean from the Amazonian province. We followed the elevational boundary of 700 meters selected by modern mammalogists; as shown by Eisenberg (1989), this represents the sharpest break in the upslope decline of tropical bats, marsupials, and rodents. As noted in our discussion, however, the Quaternary separation between Amazonian and Andean large mammal faunas may have been nearer 2000 meters.

In order to draw ecological significance from the faunal lists in table 12.2, we assigned to each land mammal genus an estimated range of environmental preferences using the following four kinds of evidence: (1) present ecology of living species or sister taxon, (2) significant morphological adaptations, (3) analogous taxa from other continents, and (4) authoritative opinions of other paleobiologists.

As a coarse summary of these ecological data, we weighed the relative importance of forest (closed environments) versus savanna (more open

Table 12.1 Mammal Families of the Early Interchange

Legions of the North	Legions of the South
Order Rodentia	Order Marsupialia
Cricetidae	Didelphidae
Order Carnivora	Order Edentata
Canidae	Dasypodidae
Mustelidae	Chlamytheriidae
Ursidae	Glyptodontidae
Order Proboscidae	Myrmecophagidae
Gomphotheriidae	Megalonychidae
Order	Megatheriidae
Perissodactyla	Order Rodentia
Tapiridae	Erethizontidae
Equidae	Hydrochoeridae
Order Artiodactyla	Order Sirenia
Tayassuidae	Trichechidae
Camelidae	Order Notoungulata
Cervidae	Toxodontidae

environments), and assigned each genus to one or both. Summary interpretations (F for forest, S for savanna, or F/S for both) were tabulated
with the taxonomic lists in tables 12.2 and 12.3.

EARLY QUATERNARY (SAVANNA) FAUNAS

Table 12.1 lists the land mammal families that by the Late Quaternary
had crossed through the tropics into temperate latitudes of the other
American continent. As pointed out by Webb (1976, 1978), the vast
majority of both the northern and the southern cohorts represent savanna (or open country) habitat preferences.

A few Late Pliocene and Early Pleistocene sites have been discovered
in the American tropics. For example, El Golfo on the Gulf of California in northern Mexico produced a mammal fauna of Irvingtonian
(Early Pleistocene) age, including the first North American fossil record
of the giant anteater, *Myrmecophaga tridactyla*. Living members of this
species rely exclusively on termite mounds in savanna settings. This record extends the geographic range of *Myrmecophaga* some 3000 km
north of its present Central American range (Shaw and McDonald
1987).

From El Salvador, Webb and Perrigo (1985) described a small fauna
(Barranca del Sisimico) of Irvingtonian age, most notable for the occurrence of two distinct megalonychid sloth genera, *Megalonyx* and the
very large endemic form, *Meizonyx*. Except for the novel sympatric oc-

currence of two megalonychid sloth genera, the fauna resembles other Central American large mammal faunas and evidently samples a savanna setting.

In South America, the most important Early Quaternary tropical fauna is the classic Tarija fauna from southern Bolivia. Tarija comes from moderate elevation (just below 2000 meters) and is of Middle Pleistocene (Ensenadan) age. In Hoffstetter's words, "this beautiful fauna (26 families, 52 genera) is assuredly the most important one known to this day from the Andean Pleistocene" (1986, 225). The fauna includes about equal numbers of northern and southern genera and generally indicates a savanna habitat. Most of the northern immigrants, such as *Onohippidion*, llamas, and several kinds of deer, are later characteristic of the Andes, whereas most South American elements subsequently become characteristic of the Amazon. Among subsequently Amazonian families and genera that occur at Tarija, Hoffstetter (1986, 242) cites the following: *Pampatherium, Neothoracophorus, Hoplophorus, Panochthus, Lestodon,* Hydrochoeridae, *Myocastor, Coendou, Toxodon, Canis dirus, Chrysocyon,* and *Notiomastodon*. (We omit *Stegomastodon* from this and other South American faunal lists because we are convinced that such records pertain to *Notiomastodon*, and that true *Stegomastodon*, typified in Blancan deposits of the western United States, never reached South America.) The first nine of these genera are of South American origin, whereas only the last three are of northern origin. Thus Tarija provides a strategic equatorial site from which to view the Middle Pleistocene aftermath of the interchange.

LATE QUATERNARY (MESIC) FAUNAS

Late Pleistocene large mammal faunas from Central America and the transandean lowlands of equatorial South America are surprisingly homogeneous, and that is why we combined them into one faunal region in table 12.2. Key local faunas from the coastal plain of northern South America are tar seeps at La Carolina in Ecuador and Talara in Peru, and also thick mud bogs at Muaco in Venezuela. The Central American faunas in Honduras and El Salvador are distinctive mainly because they include the southern limits of several families and genera from North America, notably Elephantidae (genus *Mammuthus*), Mammutidae (genus *Mammut*), Bovidae (genus *Bison*), and Camelidae (genus *Camelops*). These same faunas record the apparent northern limit of *Mixotoxodon*, the only representative of the order Notoungulata to reach beyond South America. It is astonishing how ubiquitously and abundantly one

Table 12.2 Ecological Valences of Late Quaternary Mammal Genera

	Atlantic	Amazon	Andes	Central America	Ecological Valence
Marsupialia					
Didelphis	x	.	.	.	S/F
Xenarthra-Pilosa					
Megalonyx	x	x	.	x	S
Megatherium	x	x	x	.	S
Eremotherium	x	x	x	x	S
Ocnopus	x	x	.	.	S
Valgipes	x	.	.	.	S
Nothrotherium	x	.	.	.	S
Glossotherium	x	x	x	x	S
Mylodon	x	x	.	.	S
Lestodon	x	x	.	.	S
Scelidotherium	x	x	x	x	S
Scelidodon	x	.	.	.	S
Xenarthra-Cingulata					
Myrmecophaga	.	.	.	x	S
Propraopus	x	x	x	x	S/F
Dasypus	x	x	.	.	S/F
Euphractus	.	x	x	.	S
Cabassous	x	.	.	.	S/F
Pampatherium	x	x	.	x	S
Hoplophorus	.	x	x	.	S
Panochthus	.	x	x	.	S
Parapanochthus	x	x	.	.	S
Neuryurus	x	.	.	.	S
Glyptodon	x	x	x	.	S
Glyptotherium	.	.	.	x	S
Doedicurus	.	x	.	.	S
Rodentia					
Coendu	x	.	.	.	F
Cavia	.	x	.	.	F
Lagidium	.	.	x	.	S
Dinomys	.	x	.	.	S/F
Hydrochoerus	x	x	x	x	S
Neochoerus	.	.	.	x	S
Tetrastylus	.	x	.	.	S/F
Myocaster	x	.	.	.	F
Dicolpomys	.	x	.	.	F
Agouti	.	x	.	.	F
Proechimys	.	.	.	x	S/F
Lagomorpha					
Sylvilagus	.	.	.	x	S/F
Carnivora					
Procyon	x	.	.	.	S/F
Arctodus	x	.	.	x	S
Canis	x	.	.	x	S/F
Protocyon	x	.	.	.	S/F
Dusicyon	.	.	x	.	S/F

Table 12.2 *continued*

	Atlantic	Amazon	Andes	Central America	Ecological Valence
Felis (Puma)	x	.	x	x	S/F
Leo (Jaguarius)	x	.	x	x	F
Smilodon	x	.	x	x	S/F
Conepatus	.	.	x	x	S/F
Eira	x	x	.	.	F
Notoungulata					
Toxodon	x	x	.	.	S
Mixotoxodon	.	x	x	x	S
Trigonodopsis	x	.	.	.	S
Litopterna					
Macrauchenia	.	.	x	.	S
Xenorhinotherium	x	.	.	.	S
Proboscidea					
Haplomastodon	.	x	x	x	S
Notimastodon	.	x	.	.	S
Cuvieronius	.	x	x	x	S
Mammuthus	.	.	.	x	S
Mammut	.	.	.	x	F
Perissodactyla					
Tapirus	x	x	x	x	S/F
Equus (Amerhippus)	x	.	x	x	S
Hippidion	x	.	.	.	S
Onohippidion	.	.	x	.	S
Artiodactyla					
Tayassu	x	x	x	.	S/F
Braziliochoerus	.	x	.	.	F
Palaeolama	x	x	x	x	S
Lama	.	x	x	.	S
Camelops	.	.	.	x	S
Ozotoceros	.	x	.	.	S
Blastoceros	x	.	.	.	S
Mazama	x	.	x	x	F
Odocoileus	x	.	x	x	F/S
Morenelaphus	x	.	.	.	S
Charitoceros	.	.	x	.	F
Agalmoceros	.	.	x	.	F
Bison	.	.	.	x	S

finds Late Pleistocene evidence of *Eremotherium* (the giant ground sloth) and *Cuvieronius* (a proboscidean) throughout the tropical American lowlands (Webb and Perrigo 1984).

The mammal fauna of the northern (tropical) Andes and the adjoining altiplano, presented in table 12.2, includes such key localities as the Sabana de Bogotá above 2000 meters in Colombia, the Punin Fauna ranging from 2300 to 3100 meters in Ecuador, Pikimachay Cave and several other localities at about 4000 meters in Peru, and Ulloma at nearly 4000 meters in Bolivia. These have been well reviewed by Hoffstetter (1986). The most distinctive members of the Andean large mammal fauna are four endemic deer genera, including modern *Hippocamelus* and *Pudu*, as well as the extinct *Agalmoceros* and *Charitoceros*. (Cartelle [1989] showed that supposed records of *Hippocamelus* from the Atlantic side of Brazil and elsewhere in temperate South America are incorrectly identified and actually represent the extinct genus *Morenelaphus*.) Another important member of the Andean fauna is the horse *Onohippidion*.

Table 12.2 includes a current list of large land mammals from Late Pleistocene sites in the western Amazon Basin (Rancy 1991). The most productive sites (fig. 12.2) occur along the Napo River in Ecuador and Brazil, the Ucayali in Peru, and the Juruá, Acre, Purus, and Madeira Rivers in Brazil. Of interest are several genera not present in the high Andes nearby; these include *Megalonyx*, the two largest mylodont genera, *Lestodon* and *Mylodon*, and many shelled edentates, of which only two are known in the high Andes. The Amazon uniquely supports all three genera of South American proboscidea.

Large mammals from Late Pleistocene sites in the Atlantic coastal region of South America, essentially eastern Brazil, are exceedingly well known. The Lagoa Santa fauna, collected for 150 years from a large number of karst caves in the state of Minas Gerais, provides by far the richest Late Pleistocene sample of terrestrial vertebrates in tropical America (Cartelle 1994). A fundamental difficulty with these samples, however, is that many Holocene taxa are inextricably mixed with those from the Late Pleistocene. Other key sites are Janauba in Bahia (Cartelle 1989) and Touro Passo in Rio Grande do Sul. The most distinctive features of the Late Pleistocene large mammal fauna from this region are various south temperate taxa which extend their northern limits into this region: these include the sloth, *Nothrotheriops;* the armadillo, *Cabassous;* doedicurine glyptodonts; *Arctodus*, the giant extinct bear; and the modern deer, *Blastoceros* and *Ozotoceros*, as well as the extinct *Morenelaphus*.

DISCUSSION

Phase One: Savanna Corridors

The first step in the evolution of the neotropical mammal fauna was initiated by emergence of the isthmian land bridge in Late Pliocene time. Most land mammal participants in the Great American Interchange were savanna-adapted groups that moved from north temperate to south temperate latitudes and vice versa (Webb 1978; Stehli and Webb 1985). Most of the families that participated in this vast faunal interchange had done so by Middle Pleistocene time (Webb 1978; Tonni et al. 1992). This first phase established maximum continuity from north temperate through south temperate latitudes, so that large mammal faunas in Florida bear remarkable resemblance to those in the pampean region of Argentina (Webb 1976).

The immigrant taxa from North America diversified more in South America than the converse (Webb 1978; Marshall et al. 1982). This may be attributable to the vast area of temperate conditions in North America compared with the relatively narrow "southern cone" of South America. Central America played a particularly important role as a tropical staging area in which the northern groups became adapted to tropical American conditions, especially to tropical savannas.

The interamerican biotic interchange coincided with the onset of cooler conditions in the Late Pliocene. At about 2.5 Ma, not only did glaciers advance at high latitudes in the Northern Hemisphere, but also there was a strong trend to cooler and more open habitats in tropical America. In the high plains of Bogotá, for example, Hooghiemstra and Ran (1994, 68) record a rapid temperature decrease of 4–5°C, followed by a further gradual cooling, and they recognize the first widespread paramo vegetation in the northern Andes. Similarly Van der Hammen and Cleef observe that "if some initial stages of the present paramo flora and vegetation developed in Tertiary hilltop 'savannas,' some floristic relation with present-day savannas may be expected to exist" (1986, 173–74). Thus, the high Andes served, not as a barrier to the interchange, but more probably as a congenial ramp for temperate taxa moving in both directions through tropical latitudes. Several mammal groups such as llamas and nothrothere sloths, narrowly adapted to scrub conditions, must have traversed the tropics early when conditions were at their driest.

By the Middle Pleistocene, the Great American Interchange had evidently slowed down considerably (Stehli and Webb 1985; Pascual and Jaureguizar 1990). Many new genera of northern affinities that appear subsequently in South America evolved in place (Webb 1978) as "autochthonous pseudoimmigrants" (Marshall et al. 1982). Patterson and Pascual correctly inferred "that many of the Recent tropical genera of northern ancestry may have been present in northern South America throughout much of Pleistocene time" (1972, 257). By Middle Pleistocene time, easy passage through savanna corridors in the isthmian region was curtailed by rainforest, much as it is at present.

Phase Two: Rainforest Corridors

The second phase of neotropical mammal fauna evolution was initiated when savanna corridors through the isthmian region were closed by prevailing rainforest, probably during the Middle Pleistocene. Presumably, the dominance of rainforest biota was most pronounced during interglacial times when, as now, warmer, more equable conditions favored its expansion in most of the American lowland tropics (Webb 1991).

This second phase allowed the rich Amazonian fauna, with its ancient South American stocks, to spread northward into Central America, producing for the first time uniformity of rainforest faunas throughout the American tropics. Since the tropical area in South America (across the Brazilian bulge) is about five times that in North America, dominance by Amazonian forms was predictable. These intertropical immigrants consisted mainly of the ancient South American stocks which make up slightly less than half of the modern mammal genera in that continent. One-third (30%) of those genera came from Oligocene waif dispersers (hystricognath rodents and ceboid primates), and one-sixth (17%) from the two surviving orders that were already present at the beginning of the Cenozoic, namely, marsupials (possums) and xenarthrans (armadillos, anteaters, and sloths).

Meanwhile, in South America the northern immigrant mammals continued to diversify throughout the Pleistocene, and they accounted for more than half (53%) of modern land mammal genera (Webb and Marshall 1982; Stehli and Webb 1985). Thus, for the temperate groups that had marched right through the tropics one or two million years earlier, the Late Pleistocene became a time of evolutionary differentiation.

The regional data compiled above provide some preliminary impressions of how faunal differentiation proceeded throughout South and Central America during this second phase of neotropical mammal evolution. The most striking feature of the Late Quaternary large mammal faunas is their overall homogeneity throughout tropical America. Aside from the marginal incursions of north temperate forms such as mammoth, mastodon, bison, and *Camelops* into nuclear Central America, and of south temperate forms such as pampas deer and pampean kinds of glyptodonts and armadillos into the cerrado region of southeastern Brazil, regional endemism at the generic level is remarkably limited. Many large mammal genera ranged right through the tropics from the United States into Argentina or vice versa. Table 12.3 lists these transtropical genera of the Late Pleistocene.

Nevertheless, the Pleistocene record of tropical mammals begins to

Table 12.3 Widespread Genera of American Tropics: Ecological Valences

Glossotherium*	Savanna
Eremotherium*	Forest edge
Propraopus*	Forest edge
Pampatherium*	Savanna
Hydrochoerus	Wet savanna
Protocyon	Mixed
Arctodus*	Savanna
Smilodon*	Mixed
Mixotoxodon*	Savanna
Haplomastodon*	Savanna
Tapirus	Mixed
Equus (Amerhippus)*	Savanna
Tayassu	Mixed
Palaeolama*	Savanna
Mazama	Forest
Odocoileus	Forest edge

*Extinct

indicate some regional differentiation. The New World deer, the only group of ruminants that participated in the interchange, provide the best paleontological example of such Pleistocene differentiation. Three distinct sets of genera appear in tropical and subtropical terrain. In the equatorial lowlands were *Odocoileus* and *Mazama*, probably derived from Central American origins. The Andes cradled two extinct genera, *Agalmoceros* and *Charitoceros*, and two extant genera *Hippocamelus* and *Pudu* (now each subdivided into distinct species between the altiplano to the north and the puna to the south). The southeastern temperate region of South America produced three endemic genera, namely, the extinct *Morenelaphus* and the living *Blastoceros* and *Ozotoceros*, adapted respectively to cerrado, forest edge, and pampas environments.

Other mammalian families of northern origin similarly diversified throughout the American tropics. The Canidae ranged widely and differentiated into some seven genera (Berta 1988). The Cricetidae produced at least fifty neotropical genera. Even granting some pre-interchange staging in Central America, these rodents must have evolved explosively once the interchange began (Webb 1985).

In phase one of the interchange, the Andes Mountains evidently provided the "high road" by which temperate savanna forms dispersed through equatorial South America (Webb 1978). In phase two, however, the Andes formed an important barrier to some taxa, thus imposing increasing provincialism between cisandean and transandean faunas in later Quaternary and Recent times. At the generic level, such pairs as *Megatherium-Eremotherium, Toxodon-Mixotoxodon, Cuvieronius-Haplomastodon* and possibly *Glyptodon-Glypotherium* indicate this separation.

Amazonian paleoecology. In recent decades, the long-term stability of neotropical rainforests has been called into question by a variety of biological and geological evidence. Damuth and Fairbridge (1970) showed that arkosic sediments were moved in very large volumes off the mouth of the Amazon during the last glacial interval, and Ab'Saber (1977) demonstrated the extensive aeolian features of the Amazon that could account for such sediment transport. Climatologists pointed out that weather patterns necessary to promote a shift to predominant savanna conditions involved only subtle strengthening of the southern trade winds during the dry season (austral winter) (Meehl 1992; Iriondo and Latrubesse 1994). The strongest evidence of savanna expansion came from studies of modern faunal and floral disjunctions (Haffer 1969; Vanzolini and Williams 1970; Vuilleumier 1971; Whitmore and Prance 1987), and in part reflected the work of Moreau (1966) and others in equatorial Africa. Episodes of increased aridity and expansive savannas during Pleistocene glacial intervals were proposed as the mechanism promoting the present rich diversity within Amazonia.

The Late Pleistocene mammal fauna from western Amazonia lends support to this savanna hypothesis. In table 12.3 we list the probable ecological significance of the Amazon Basin large mammal genera. The abundance and diversity of pastoral large mammals are striking and give strong evidence of an open landscape, probably a tropical savanna. Many of these same genera range into the Atlantic region of Brazil and into south temperate latitudes in Bonairean and Patagonian parts of Argentina. In temperate latitudes, the same genera are recognized as savanna or steppe-adapted fauna (Pascual and Jaureguizar 1990). Thus, during certain intervals of the Late Pleistocene, savanna habitats extended widely through the Amazon Basin (Haffer 1969; Colinvaux 1987, 1989). Similarly, Paula Couto had previously observed that "the Hylaea seems to be of relatively Recent origin, for that [Late Pleistocene vertebrate] fauna was not adapted to the life in so dense a forest. It was perhaps a fauna of savanna" (1982, 7).

The new dynamic view of Amazon history still had to account for the continued existence of a very rich rainforest biota, far richer than that of Africa (Meggers et al. 1973). For that reason, many students of Amazonian biodiversity (e.g., Haffer 1969; Vanzolini and Williams 1970; Whitmore and Prance 1987) postulated rainforest refugia. These large subcircular areas of high present-day precipitation were mapped and designated as the critical loci in which equatorial multistratal rainforests had survived the dry glacial intervals.

A fundamental weakness of this rainforest refuge theory is that the supporting evidence is derived from Recent biotic patterns. We believe, however, that a theory about Quaternary conditions ought to be tested by Quaternary evidence. Figure 12.2 locates the most important Quaternary mammal sites of the western Amazon Basin (Rancy 1991). Seven of these sites fall directly within the areas designated as Late Pleistocene forest refugia (Haffer 1969; Whitmore and Prance 1987). The mammals from these sites, however, are predominantly savanna mammals, thus directly contravening those Late Pleistocene loci of "pure" rainforest.

We cite three examples from Rancy 1991. First, the Napo River paleofauna consists of *Glyptodon, Eremotherium, Mylodon, Haplomastodon, Cuvieronius,* and *Tayassu,* a set of large mammals well adapted to savanna habitat and largely contradictory to multistratal rainforest. The second example involves the largest sample of Pleistocene mammals from the Amazon Basin. The upper Juruá River in the Brazilian state of Acre produces twenty-three genera, predominantly grazing and mixed-feeding herbivores, including the extinct camelid *Palaeolama.* The upper Juruá fauna occurs in the East Peruvian Refugium of various authors (e.g., Whitmore and Prance 1987). Third, the Pleistocene mammal collection from the middle Ucayali River calls into question yet another part of the East Peruvian Refugium. These mammal faunas suggest that the Amazon Basin did not conserve large blocks of rainforest refugia during the Late Quaternary, but rather that the rainforest meandered in ribbonlike fashion, following river galleries through savanna areas much more extensive than at present.

Even though the large mammals from the western Amazon are predominantly savanna forms and directly contradict rainforest refugia, they must not be overinterpreted in the opposite direction. These faunas do not dictate that the whole landscape was open savanna. It is instructive to consider the five genera from the Late Pleistocene record that continue to dwell in the Amazon, namely, *Dasypus, Hydrochoerus, Eira, Tapirus,* and *Tayassu.* The ecological feature shared by these survivors is eurytopy: today they live in rainforest, deciduous forest, cerrado, chaco, and caatinga. The first known fossil of the avian genus *Opisthocoma* (the hoatzin) also occurs in the western Amazon, and it indicates a riparian forest-edge habitat (Rancy and Olson, unpublished data).

Another feature of the Amazon large mammal record is the absence of equids. This is probably not an artifact, because equids usually have a high probability of entering the fossil record because their cheek teeth are relatively indestructible. The most likely explanation for the absence of equids is that Amazon habitats were seasonally flooded, for this is the one problem that might have been insurmountable for equids. Certainly, several equid genera are well known in adjacent habitats in the lower Andean slopes and also in the diagonal of open habitats across eastern and southern Brazil. Thus, the western Amazon did not offer its mammal fauna a pure savanna setting, but rather a mobile habitat mosaic that included meandering belts of rainforest and perhaps seasonal flooding. Geomorphologists studying these same fluvial systems have proposed similar conclusions (Räsänen et al. 1987).

Phase Three: Late Pleistocene Extinctions

A striking feature of Late Pleistocene large mammal genera of the Americas is the predominance of extinct over extant genera. Of the seventy tropical American large mammal genera listed in table 12.2, only nineteen survive into the Recent. Thus nearly 75% of the fauna became extinct. In the Amazon Basin, only four of twenty-three known genera survived (table 12.3 above). The Late Pleistocene extinctions almost exclusively destroyed large mammals. Therefore, by focusing on large mammals, our study has emphasized the remarkable Late Pleistocene faunal decline that occurred rather abruptly about 11,000 years ago. This pattern is not uniquely tropical but encompassed all of the New World (Martin and Klein 1984; Webb and Barnosky 1985). The greatest losses, in terms of the taxonomic hierarchy, were the endemic neotropical orders Notoungulata and Litopterna. All of the larger Edentata, notably the many kinds of gound sloths and many of the great shelled forms such as glypotodonts, were lost. All of the New World Proboscidea and all but one of the New World Perissodactyla also vanished. By far the most successful order of large herbivores at resisting the wave of extinctions were the Artiodactyla in which camelids, tayassuids, and especially cervids persisted, giving a survivorship rate of about half of the Pleistocene genera.

The explanation of these dramatic large mammal extinctions lies beyond the scope of this chapter. Despite extensive efforts to resolve the debate as to whether these extinctions can be attributed to abrupt climatic change or to the onslaught of humans entering the New World, a clear consensus remains elusive (Martin and Klein 1984). The two primary explanations are not mutually exclusive, although uncritical acceptance of both seems unsatisfactory. If human hunting and human habitat modification provided the *coup de grâce* to so many large-bodied, widespread taxa, then more detailed evidence of these effects ought to be available. On the other hand, as Azzaroli (1992) recently pointed out with regard to monodactyl horses, large herds were exceedingly well adapted to a wide variety of habitats in both the New World and the Old. He reasoned, therefore, that their extinction "can hardly have been caused by climatic factors alone and is believed to be largely the result of prehistoric overkill" (Azzaroli 1992, 151).

Here it is more relevant to consider the consequences, rather than the causes, of these devastating extinctions. Janzen and Martin (1982) pointed out the ecological importance of the missing large herbivores. For example, they noted the present lack of natural dispersal agents for tropical trees with big seeds and large fleshy fruit, such as the Guana-caste tree (*Enterolobium cyclocarpum*) and palms of the genus *Scheelia*. They also remark on the anachronistic appearance of very large spines and other defenses in several tropical tree species. Today introduced horses play a role analogous to those of the extinct herbivores, greatly increasing the distribution of large-fruited trees. Indeed, Janzen and Wilson recognize that "a mixed grassland-forest, populated by range cattle and horses, is probably a more 'natural' habitat for these plants than the pre-Columbian pure forest habitats [now] being protected" (1983, 437). The additional ecological effects of a large herbivore fauna in opening tropical forests by grazing and by encouraging fires can only be guessed.

CONCLUSIONS

The Great American Interchange produced an extensive transtropical mingling of land mammal taxa at the end of the Pliocene and into the Early Pleistocene. In this first phase, savanna-adapted groups of land mammals were predominant. By Middle Pleistocene time, the isthmian savanna corridor was closed by persistent rainforest, much like the present.

This introduced phase two of neotropical mammal evolution. The vast faunal resources of Amazonia extended their ranges into Central America. Regional evolutionary differentiation of the mingled fauna began especially at tropical latitudes. The old northern immigrants continued to diversify in the tropics and in temperate South America. For example, the deer (family Cervidae) document rapid regional differentiation at the generic level. At least six groups of large mammals developed cisandean and transandean geminate pairs of genera, reflecting the importance of the Andes as a barrier to lowland equatorial taxa.

In the Amazon, contrary to earlier stabilist views, geologists and biologists have demonstrated the strong influence of wet/dry climatic cycles, keyed in complex fashion to global patterns of glacial/interglacial cycles and Milankovich curves. The Late Pleistocene expansion of savanna at the expense of rainforest is supported by large mammal data from the western Amazon. Furthermore, seven Late Pleistocene mammal sites there directly contradict the proposed rainforest refugia in the Napo Refugium and the East Peruvian Refugium. This does not imply, however, that these areas were exclusively covered by savanna, for the vertebrate data also indicate an ecotonal forest-edge habitat. Taken together, the data suggest to us that the rainforest biota survived interstitially with the savanna biota; presumably it followed a dendritic pattern, occupying ribbons of gallery forest that meandered with the Amazonian tributaries.

The third phase in the evolution of the neotropical mammal fauna involved the Late Pleistocene extinctions. Approximately 75% of Late Quaternary large mammals of the American tropics vanished. Presumably these cataclysmic results reflect the combined effects of man and climatic change. The consequences of these sudden extinctions are still evident in the absence of critical distributors for large-seeded trees, and in the greatly reduced degree of grazing and fire-suppression of forest succession. Were the megafauna still alive, present tropical forest habitats would be more open, as partly indicated by the effects of horse and cattle grazing in historic time.

ACKNOWLEDGMENTS

We thank our many collaborators in the field, past and present, who helped produce the critical fossil collections that are beginning to elucidate tropical American prehistory. These include our two revered mentors, George Gaylord Simpson and Carlos de Paula Couto. We acknowledge the help and companionship of John Mawby, Howel Williams, and Steve Perrigo in Central America, and John Eisenberg in Amazonia. We are exceedingly grateful to John Eisenberg, Rosendo Pascual, Bruce MacFadden, Kent Redford, Castor Cartelle, and Paul Colinvaux for helpful discussions. We thank the editors for calling together the symposium that resulted in this book and seeing it through to completion. This research was supported by National Science Foundation grant BSR 891806 to David Webb. This chapter is contribution number 465 in Paleobiology from the Florida Museum of Natural History.

REFERENCES

Ab'Saber, A. N. 1977. Espaços ocupados pela expansão dos climas secos na America do Sul, por ocasião dos periodos glaciais quaternários. *Paleoclimas* 3:1–19.

Azzaroli, A. 1992. Ascent and Decline of Monodactyl Equids: A case for prehistoric overkill. *Ann. Zool. Fennici* 28:151–163.

Berta, A. 1988. Quaternary evolution and biogeography of the large South American Canidae. *Univ. Calif. Pub. Geol. Sci.* 132:1–149.

Cartelle, C. 1989. Sobre uma pequena Coleção des restos fosseis de mamíferos do Pleistoceno-Holoceno de Janauba. *Resumo das Comunicaçoes, XI Congresso Brasileiro de Paleontologia,* Curitiba, 120–21.

———. 1994. *Tempo passado: Mamíferos do Pleistoceno em Minas Gerais.* Belo Horizonte: Editora Palco.

Colinvaux, P. A. 1987. Amazon diversity in light of the paleoecological record. *Quatern. Sci. Revs.* 6:93–114.

———. 1989. Ice-age Amazon revisited. *Science* 340:188–89.

Damuth, J. E., and R. W. Fairbridge. 1970. Equatorial Atlantic deep-sea arkosic sand and ice-age aridity in tropical South America. *Geol. Soc. Am. Bull.* 81:189–206.

De Onis, J. 1992. *The green cathedral: Sustainable development of Amazonia.* Oxford: Oxford University Press.

Eisenberg, J. F. 1989. *Mammals of the neotropics.* Vol. 1. Chicago: University of Chicago Press.

Emmons, L. H. 1990. *Neotropical rainforest mammals.* Chicago: University of Chicago Press.

Frailey, C. D. 1986. Late Miocene and Holocene mammals, exclusive of the Notoungulata, of the Rio Acre region, western Amazonia. *Contrib. Sci., L.A. Count. Mus.* 374:1–46.

Galusha, T., N. M. Johnson, E. H. Lindsay, N. D. Opdyke, and R. H. Tedford. 1984. Biostratigraphy and magnetostratigraphy, Late Pliocene rocks, 111 Ranch, Arizona. *Geol. Soc. Am. Bull.* 95:714–22.

Haffer, J. 1969. Speciation in Amazonian forest birds. *Science* 165:131–37.

Heine, K. 1994. The Mera site revisited: Ice-age Amazon in the light of new evidence. *Quatern. Int.* 21:113–19.

Hoffstetter, R. 1986. High Andean mammalian faunas during the Plio-Pleistocene. In *High altitude tropical biogeography,* ed. F. Vuilleumier and M. Monasterio, 218–44. Oxford: Oxford University Press.

Hooghiemstra, H., and E. T. H. Ran. 1994. Late Pliocene–Pleistocene high resolution pollen sequence of Colombia: An overview of climatic change. *Quatern. Int.* 21:63–80.

Iriondo, M., and E. M. Latrubesse. 1994. A probable scenario for a dry climate in central Amazonia during the Late Quaternary. *Quatern. Int.* 21:121–28.

Janzen, D. H., and P. S. Martin. 1982. Neotropical anachronisms: The fruits the gomphotheres ate. *Science* 215:19–27.

Janzen, D. H., and D. E. Wilson. 1983. Mammals: Introduction. *In Costa Rican natural history*, ed. D. H. Janzen, 426–42. Chicago: University of Chicago Press.

MacFadden, B. J., F. Anaya, and J. Argollo. 1993. Magnetic polarity stratigraphy of Inchasi: A Pliocene mammal-bearing locality from the Bolivian Andes deposited just before the Great American Interchange. *Earth Planet. Sci. Lett.* 114:229–41.

Marshall, L. G., A. Berta, R. Hoffstetter, R. Pascual, O. A. Reig, M. Bombin, and A. Mones. 1984. *Mammals and stratigraphy: Geochronology of the continental mammal-bearing Quaternary of South America.* Palaeovertebrata, Mémoire Extraordinaire.

Marshall, L. G., S. D. Webb, J. J. Sepkoski, and D. M. Raup. 1982. Mammalian evolution and the Great American Interchange. *Science* 215:1351–57.

Martin, P. S., and R. G. Klein, eds. 1984. *Quaternary Extinctions: A Prehistoric Revolution.* Tucson: University of Arizona Press.

Matthew, W. D. 1915. Climate and Evolution. *Ann. N.Y. Acad. Sci.* 24:171–318.

Meehl, G. A. 1992. Effect of tropical topography on global climate. *Ann. Rev. Earth Planet. Sci.* 20:85–112.

Meggers, B. J., E. S. Ayensu, and W. D. Duckworth, eds. 1973. *Tropical forest ecosystems in Africa and South America: A comparative review.* Washington, D.C.: Smithsonian Institution Press.

Moreau, R. E. 1952. Africa since the Mesozoic, with particular reference to some biological problems. *Proc. Zool. Soc. Lond.* 121:869–13.

Moreau, R. E. 1966. *The bird faunas of Africa and its islands.* New York: Academic Press.

Orgeira, M. J. 1990. Paleomagnetism of Late Cenozoic fossiliferous sediments from Barranca de Los Lobos (Buenos Aires Province, Argentina): The magnetic age of the South American land mammal ages. *Phys. Earth Planet. Int.* 64:121–32.

Pascual, R., and E. O. Jaurezigar. 1990. Evolving climates and mammal faunas in Cenozoic South America. *J. Hum. Evol.* 19:23–60.

Patterson, B., and R. Pascual. 1972. The fossil mammal fauna of South America. In *Evolution, mammals, and southern continents*, ed. A. Keast, F. C. Erk, and G. Glass, 247–310. Albany: State University of New York Press.

Paula Couto, C. de. 1982. Fossil mammals from the Cenozoic of Acre, Brazil. Part 5, Notoungulata Nesodontinae, Haplodontheriinae and Litopterna, Pyrotheria, and Astrapotheria. *Iheringia*, ser. geol. 7:5–43.

Prance, G. T. 1973. Phytogeographic support for the theory of Pleistocene forest refuges in the Amazon basin, based on evidence from distribution patterns in Caryocaraceae, Chrysobalanaceae, Dichapetalaceae, and Lecythidaceae. *Acta Amazon.* 3:5–28.

Rancy, Alceu. 1991. Pleistocene mammals and paleoecology of the Western Amazon. Ph.D. diss., University of Florida, Gainesville.

Räsänen, E. M., J. S. Salo, and R. J. Kalliola. 1987. Fluvial perturbance in the western Amazon Basin: Regulation by long-term sub-Andean tectonics. *Science* 238:1398–1401.

Raup, D. M. 1988. Diversity crises in the geological past. In *Biodiversity*, ed. E. O. Wilson, 51–57. Washington, D.C.: National Academy Press.

Reig, O. A. 1980. A new fossil genus of South American Cricetid rodents allied to *Wiedomys*, with an assessment of the Sigmodontinae. *J. Zool. Soc. Lond.* 192:257–81.

Scott, W. B. 1937. *A history of land mammals in the Western Hemisphere*. Revised ed. New York: Hafner.

Shaw, C. A., and H. G. McDonald. 1987. First record of giant anteater (Xenarthra: Myrmecophagidae) in North America. *Science* 236:186–88.

Simpson, G. G. 1940. Mammals and land bridges. *J. Wash. Acad. of Sci.* 30: 137–63.

Simpson, G. G., and C. de Paula Couto. 1981. Fossil mammals from the Cenozoic of Acre, Brazil. Part 3, Pleistocene Edentata Pilosa, Proboscidea, Sirenia, Perissodactyla, and Artiodactyla. *Iheringia*, ser. geol. 6:11–73.

Soruco, R. S., ed. 1991. Fósiles y facies de Bolivia. Part 1, Vertebrados. *Rev. Téc. Yac. Petról. Fis. Bol.* 12:357–718.

Stehli, F. G., and S. D. Webb, eds. 1985. *The Great American Biotic Interchange*. New York: Plenum Press.

Tonni, E. P., M. T. Alberdi, J. L. Prado, M. S. Bargo, and A. L. Cione. 1992. Changes of mammal assemblages in the pampean region (Argentina) and their relation with the Plio-Pleistocene boundary. *Paleogeogr., Palaeoclimatol., Palaeoecol.* 95:179–94.

Van der Hammen, T., and A. M. Cleef. 1986. Development of the High Andean paramo flora and vegetation. In *High altitude tropical biogeography*, ed. F. Vuilleumier and M. Monasterio, 153–201. New York: Oxford University Press.

Vanzolini, P. E., and E. E. Williams. 1970. South American anoles: Geographic differentiation and evolution of the *Anolis chrysolepis* species group (Sauria: Iguanidae). *Arq. Zool.* 19:1–289.

Vuilleumier, B. S. 1971. Pleistocene changes in the fauna and flora of South America. *Science* 173:771–80.

Wallace, A. R. 1876. The geographical distribution of animals. 2 vols. London: Macmillan. Reprint, New York: Hafner, 1962.

Webb, S. D. 1976. Mammalian faunal dynamics of the Great American Interchange. *Paleobiology* 2:220–34.

———. 1978. A history of savanna vertebrates in the New World. Part 2, South America and the Great Interchange. *Ann. Rev. Ecol. Syst.* 9:393–426.

———. 1985. Late Cenozoic mammal dispersals between the Americas. In *The Great American Biotic Interchange*, ed. F. G. Stehli and S. D. Webb, 357–86. New York: Plenum Press.

———. 1991. Ecogeography and the Great American Interchange. *Paleobiology* 17:266–80.

Webb, S. D., and A. D. Barnosky. 1985. Faunal dynamics of Quaternary mammals. *Ann. Rev. Earth Planet. Sci.* 17:413–39.

Webb, S. D., and L. G. Marshall. 1982. Historical biogeography of Recent South American land mammals. *Spec. Pub. Pymatuning Lab. Ecol.* 6:39–52.

Webb, S. D., and S. C. Perrigo. 1984. Late Cenozoic vertebrates from Honduras and El Salvador. *J. Vert. Paleontol.* 4:237–54.

———. 1985. New magalonychid sloths from El Salvador. In *The evolution and ecology of armadillos, sloths, and vermilinguas,* ed. G. G. Montgomery, 113–20. Washington, D.C.: Smithsonian Institution Press.

Whitmore, T. C., and G. T. Prance, eds. 1987. *Biogeography and Quaternary history in tropical America.* Oxford Monographs in Biogeography, vol. 3. Oxford: Clarendon Press.

Wilson, E. O. 1992. *The diversity of life.* Cambridge: Harvard University Press.

13

Quaternary Environmental History and Forest Diversity in the Neotropics

Paul A. Colinvaux

INTRODUCTION

Ice-age climates have prevailed over most of the last million or so years, punctuated with short-lived interglacial episodes like the Holocene in which we live. It follows that most species of contemporary tropical forests have passed more of their history in ice-age environments than in the familiar environments of the warm present. Thus, the distribution and abundance of life on the contemporary earth is in some large part set by adaptations that have served in the ice-age climates prevailing through most of the Quaternary. Patterns of diversity in modern forests were set by subsequent reassortment of ice-age communities in the changed conditions of the Holocene.

Within the Neotropical Region, by far the largest accumulation of species is in the Amazon Basin, by repute the most diverse ecosystem on earth. We should not expect to understand the causes of this remarkable diversity without first understanding the ice-age environments in which Amazonian species have lived for most of the last million years. But the climate and environment of the ice-age Amazon is essentially unknown.

From within the Amazon Basin, only four paleoecological sections unequivocally radiocarbon-dated to the last glacial period have been published (Liu and Colinvaux 1985; Bush et al. 1990; Absy et al. 1991; Colinvaux et al. n.d.), though these are supplemented by data from Amazon deltaic deposits recovered in submarine cores from off the mouth of the Amazon River (Damuth and Fairbridge 1970; Haberle n.d.). Even without problems in the interpretation of these data, four sections cannot possibly be regarded as sufficient to reconstruct the environmental history of a diverse region as large as the continental United States.

Long pollen records of neotropical lowlands outside the Amazon system are almost equally few, being confined to isolated sections from southeastern Brazil, Guatemala, and Panama (Deevey et al. 1983;

Leyden 1984; Bush and Colinvaux 1990, Bush et al. 1992; Ledru 1992, 1993; De Oliveira 1992), supplemented with shorter lake or swamp sections from near the Caribbean coast of South America (Salgado-Laboriau 1980; Leyden 1985; Wijmstra and Van der Hammen 1966).

Four Plausible Changes for the Tropics of Glacial Times

Arguments have been made in favor of four different properties for the ice-age climates of the neotropics: reduced carbon dioxide, reduced temperature, lowland aridity, and lowland flooding. Evidence in support of these several hypotheses is not equally strong and is to some extent mutually exclusive.

The environmental property most safely applied to the ice-age Amazon is that the ambient air had a concentration of CO_2 that was significantly less than that of Holocene times. Data from the Vostok ice core from Antarctica establish that a CO_2 concentration of about 0.02% by volume was the global norm for much of the glacial period (Leuenberger et al. 1992). It would be rash to expect that the distribution and abundance of plant species would have been the same in a 0.02% CO_2 atmosphere as in the 0.03% CO_2 atmosphere that has apparently pertained throughout the Holocene.

One likely consequence of reduced CO_2 is that plants with C4 photosynthesis will be favored over C3 plants in marginally arid environments. The minority of plants with C4 photosynthesis are able to extract CO_2 against steeper concentration gradients than can the C3 plants that make up the majority (Barber and Baker 1985). On the contemporary earth, the C4 mechanism appears to yield most advantage through the water conservation made possible by efficient CO_2 uptake, and hence shortened periods with open stomates. C4 photosynthesis therefore is largely confined to hot deserts and tropical savannas or grasslands where water stress is a limiting factor of plant life. When ambient CO_2 concentrations were reduced in ice-age time, however, the C4 ability to extract CO_2 against steeper concentration gradients should have extended the range of C4 plants. Expansion of savanna grasses in regions with marked dry seasons in glacial times therefore might as well have resulted from diminished CO_2 as from increased aridity.

A second property of ice-age climates that might well be expected to influence all parts of the globe, the tropical isthmus and the Amazon included, is the cooling associated with an ice age. Interest in possible cooling of glacial lowlands, however, has been limited because of the CLIMAP (1976, 1981) reconstruction that required the surface temperature of tropical oceans to be only moderately depressed. But accumulating pollen data from tropical rainforests now provide strong evidence that tropical lands were in fact significantly cooler in glacial times, whatever the explanation of the marine data. Cooling associated with northern glaciations appears to have reduced mean annual temperatures in the lowlands of both Panama and the Amazon 4–9°C at different intervals of the last glacial cycle, as discussed below. Cooling in the range of 6°C for much of the 80,000 years or so of the last glacial period would have significant implications for community composition in the rainforests of the neotropics.

Contemporary opinion has concentrated on the possibility that aridity rather than cooling was the predominant environmental forcing in the neotropics as part of a pantropical condition of ice-age aridity (Tricart 1977; Street and Grove 1979; Prance 1982; Whitmore and Prance 1987; Clapperton 1993). In part, this view derives from extrapolating the ice-age aridity of Africa to the New World and the discovery of land forms suggestive of past arid episodes, but the real weight of the aridity hypothesis depends, not on direct evidence, but rather on a biogeographic scenario of the past known as the Pleistocene refuge hypothesis (Haffer 1969, 1974). As I propose to show, this hypothesis is an unsafe vehicle on which to base reconstructions of past environments. Altered local patterns of rainfall over the neotropics in glacial times were more likely than continentwide changes. To the extent that rainfall patterns were different, species ranges would have changed repeatedly with glacial cycles, with profound consequences for the maintenance of diversity.

Other speculations about past Amazonian environments depend on the propensity of the lowlands to flooding. The Amazon Basin is both low-lying and closely dissected by rivers drawing much of their water from the flanks of Andean mountains rising 3000 m to 6000 m above the lowlands at their feet. The most spectacular of flooding hypotheses requires the breaching of the ice dam of an Andean proglacial lake, an event that could have had a catastrophic effect on communities downstream (Campbell and Frailey 1984). One expansion of this hypothesis allows a large freshwater lake to occupy much of the Amazon Basin in late glacial or Early Holocene times (Frailey et al. 1988). This postulated inundation of much of the Amazon Basin is still without such confirming data as evidence of old shorelines or the identification of the necessary dam. Compelling evidence against the hypothesis is that Räsänen and colleagues (Räsänen et al. 1992; Räsänen et al. 1995) have now shown that lacustrine deposits in Amazon lowlands have such varied ages that their deposition in an ancient lake seems to be precluded, and Clapperton (1993) has demonstrated that glaciers were not in position to have provided the required ice dam.

Regardless of such unsubstantiated hypotheses, flooding has been important in maintaining Amazonian diversity. In the western Amazon, shifting rivers disturb forests on their floodplains with intervals comparable to the life spans of individual trees, and more widespread floods are possible (Bates 1863; Salo et al. 1986; Salo 1987; Frost and Miller 1987; Frost 1988; Colinvaux et al. 1985; Colinvaux 1987).

Of the four environmental changes, reduced CO_2 seems the most firmly established, though the consequences for plant communities are still uncertain. Flooding is established with certainty only as a disturbance phenomenon on century and millennial time scales. But both tropical aridity and cooling in an ice age are contentious concepts with profound implications for the history of tropical forests. The next sections assess the evidence suggesting either cooling or aridity in the neotropical environments of an ice age, at first setting aside biogeographic arguments. The Pleistocene refuge hypothesis can then be evaluated by comparing its environmental predictions against the actual record.

On Neotropical Palynology

A principal tool of the studies used in this review is pollen analysis. Experience is showing that the pollen tool is at its most precise in the tropics, despite early doubts bred from experience in the very different floras of the temperate zones where pollen analysis was invented. Temperate trees whose signals enter largely into the pollen record are typically wind pollinated, and their species are few in number. The result is that a temperate pollen diagram is a play with few actors. That the shuffling of these few actors yields strong signals for secular climate change is demonstrated by the way that transfer functions for climate variables can be applied to statistics as crude as the percentage of pollen of a tree genus in the sum of total pollen (Prentice et al. 1991).

In the tropical rainforest, by far the larger number of trees are pollinated by animals—by insects mostly, but also by birds and even by bats. The tree species are immensely more numerous—in the western Amazon up to 300 species per hectare (Gentry 1988), nearly all of them animal pollinated. But a few genera are, or can be, pollinated by wind, even in the tropical rainforest. The anemophilous trees tend to be disturbance plants—many of them in the families Moraceae and Urticaceae or some of the ulmacious (elm family) genera like *Trema* and *Celtis*. Perhaps more important still, all Gramineae (grass family) remain wind pollinated, whether they are tiny annual grasses, small perennials of the forest floor, C4 grasses of a tropical savanna, aquatics of a floating grass mat in a river, or bamboos ten meters high. Small populations of any kind of grass can potentially flood the signals from a major population of a rainforest tree, leading the unwary to false conclusions.

But despite all initial doubts, the neotropical pollen record is proving to be strikingly informative. The first surprise is that pollen production is large, so copious indeed that pollen concentrations in the sediments of Amazonian lakes are larger, sometimes much larger, than the concentrations to which we are accustomed in north temperate lakes (Liu and Colinvaux 1985). Apparently, tropical animal pollination changes delivery system without diverting from production. And lakes receive pollen less from the wind and more from land runoff, sometimes perhaps after the deaths of insect vectors. We identify even *Ficus* pollen in lake sediments, although it is produced inside the closed figs and can reach a lake only after the fig has rotted or on the corpse of its porter-wasp.

As long as one is wary about grass pollen, or such wind-pollinated pioneer plants as *Cecropia*, the presence of some wind-pollinated disturbance trees does more to provide a signal than to hide the record of that great majority of insect-pollinated plants. Bush (1991) used largely the record of the more wind-pollinated taxa, which make up perhaps half the pollen rain in many pollen spectra, to demonstrate that such vegetation types as tierra firme forest, seasonal savannas, and the periodically inundated varzea and igapo forests can be clearly separated by TWIN-SPAN and DCA ordinations of pollen data (fig. 13.1). It was from Bush's study that we learned that conspicuous percentages of Urticaceae/Moraceae, or 40% of distinctive taxa occurring as rare types, are signals of tropical forest.

Pollen designed for animal transport tends to be distinctive, making possible accurate identifications. The difficulty, of course, is learning where to start looking in a flora of 80,000 species. We are tackling this problem by a program of pollen trapping in forest plots where a thorough botanical survey has been done. We set out simple pollen traps, essentially just plastic funnels filled with fiber, in arrays of twenty to one hundred, and replace them at yearly intervals (Bush 1992). We now have three years of data from the 50 ha plot on Barro Colorado Island in Panama, from the Cuyabeno reserve in Ecuadorian Amazonia, and from the Minimum Critical Area plots at Manaus, and are now trapping at as many of our Amazon sites as possible. Pollen influx averages about 7000 grains cm^2/yr. Rainforest trees disperse their pollen through the air to the traps as far as 50 m. On top of providing data on pollen signatures of communities, the trapping program wonderfully reduces the

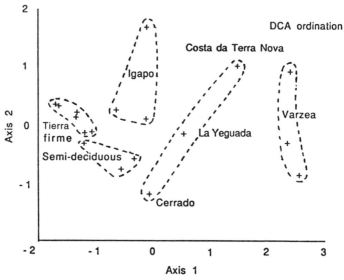

Fig. 13.1   DCA scattergram of pollen percentages in Amazonian surface samples. Samples are mud-water interfaces of lakes (from Bush 1991).

difficulties of pollen identification, since we have short species lists in which to hunt unknowns. We build our reference collection around the species lists of the study plots.

In the long pollen records from the Six Lakes site in central Amazonia (described below, fig. 13.4), these methods let us identify 187 pollen types, most of them animal-pollinated trees.

Thus, we have two sorts of pollen signals in the tropics: the percentage composition of the few anemophilous taxa (or other copious pollen producers like the basically insect-pollinated family Melastomataceae) and many animal-pollinated taxa whose presence we note as index fossils. Both kinds of pollen data can be used in multivariate ordinations, eventually to yield signals more precise than those available from temperate pollen.

It remains true that great care must be taken with interpretations when, as is usual, pollen is identified only to genus or family rank. Difficulties with the meaning of enhanced grass-pollen percentages, as described above, are a notable example. But many tropical genera have taxa adapted to a variety of habitats. Trees of the genus *Caryocar*, for instance, are commonly associated with cerrado in Amazonian Brazil, but *C. brasiliense* is associated with moist *Araucaria* woodland in southeastern Brazil (Prance 1990), so that a *Caryocar* signal must be interpreted with care. *Cuphea* pollen in the glacial interval of marsh deposits on the Serra Carajás of the eastern Amazon Basin has been taken as evidence for dry or open savanna vegetation (Van der Hammen and Absy 1994), but the pollen are just as likely (in our opinion, more likely) to come from *Cuphea* species that are common as understory shrubs of tierra firme rainforest, or from the damp sands of the drained marsh.

Possibly, pollen analysis always gets its results from what a legal scholar would call "a preponderance of the evidence," though the "preponderance" can be rigorously helped by multivariate ordinations. In the neotropics, the present resolution of pollen analysis as a paleoecological tool is more limited by inadequate knowledge of species distributions or tolerances than by the mechanics of the method. Already tropical pollen analysis is a powerful tool for reconstructing floristic histories. With a decade or two of work on modern pollen distributions, the method should be even more useful for paleoclimatology than it has been in temperate latitudes.

COOLING AT THE ICE-AGE EQUATOR

Glaciers on tropical mountains in Africa, New Guinea, and South America descended by 1000–1500 m in rough synchrony with the spread of continental ice sheets of the last (Wisconsinan) glacial advance (Livingstone 1962; Mercer and Palacios 1977; Flenley 1979; Clapperton 1987). This would seem to be prima facie evidence that the atmospheric cooling of glacial times was ubiquitous, being experienced at the equator just as in mid and high latitudes. This straightforward conclusion, however, was generally taken to be falsified by the CLIMAP (1976, 1981) reconstructions of sea surface temperatures, which allowed equatorial oceans to cool only in the range of 2–4°C.

It is well established that Andean tree lines descended in parallel with the descent of mountain glaciers. Pollen from the sediments of an extinct lake under the city of Bogotá allow reconstruction of the relative height of tree line throughout complete glacial cycles, showing that the Andean tree line descended 1500 m (Hooghiemstra 1984; Hooghiemstra et al. 1992). Thus, well-established data demonstrate significant (depression in the order of 6°C) atmospheric cooling over the upper elevations of tropical mountains, whereas widely accepted paleoceanographic reconstructions suggested minimal cooling of the tropical sea surface (Rind and Peteet 1985).

Some attempts to resolve the apparent contradiction between the terrestrial and marine temperature records rest on assumptions about ice-age precipitation. Owing to high insolation and melt rates, the growth of tropical glaciers is known to be critically dependent on precipitation (Thompson et al. 1985). Thus, increased snowfall or cloud cover, or both, could drive glacial fronts down tropical mountains with little or no change in atmospheric temperature. An obvious difficulty with temperature-free explanations for glacial descents, however, is that they scarcely account for the associated descents of tree line. Tree lines are certainly temperature dependent (Tranquillini 1979). One possible resolution of this dilemma is that tree lines are driven down by local cooling induced by the descent of the ice itself. This may be called the "catabatic winds hypothesis," in which cold air flowing downslope from the glaciers drives the tree line down with it (Colinvaux 1987).

If glacial and tree-line descents are accepted as true indicators of atmospheric cooling, and marine reconstructions describe the temperature of tropical lands at sea level as well as of the sea itself, then temperature gradients must have been compressed at the lower elevations on the flanks of tropical mountains. We are then presented with the phenomenon of warm tropical lowlands, above them perhaps 1500 m of mountain slopes across which runs a temperature gradient much steeper than that on the flanks of contemporary mountains, and above this the colder upper regions of the mountains. This scenario is theoretically possible if the lower slopes of the mountains were comparatively arid and the upper slopes were moist. With altitude, temperature lapse rates in dry air are larger than corresponding lapse rates in moist air. The lapse rate (called the adiabatic lapse rate) depends on the water content of the air as a function of the energy used to heat the included water vapor. The actual ratio of adiabatic lapse rates between dry and wet air is roughly in a ratio of 10–6°C/km. Therefore, if ice-age air at low elevations was drier than interglacial air, temperature gradients with altitude should have been steeper, and the apparent contradiction between marine and terrestrial records is resolved (Van der Hammen 1974). This explanation thus requires arid but warm tropical lowlands to have persisted under wet but also relatively warm tropical highlands with a steep temperature gradient between (Rind and Peteet 1985; Bonnefille et al. 1990). In rejecting both this concept of extra steep temperature gradients in glacial times and the postulate that Pleistocene glacial descents could be driven by excess moisture without significant temperature depression, Broecker and Denton (1990) concluded that the apparent temperature disparity between equatorial land and sea in glacial times was an "enigma" for which they had no explanation.

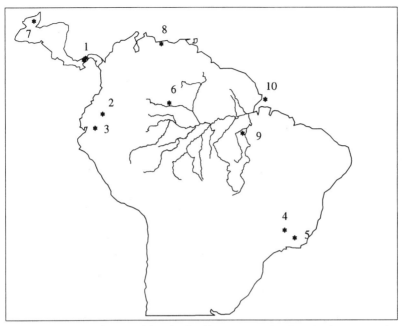

Fig. 13.2   Neotropical sites with long lowland histories: (1) El Valle and Lake Le Yeguada, Panama; (2) Mera, Ecuador; (3) San Juan Bosco, Ecuador; (4) Lagoa Serra Negra, Minas Gerais, Brazil; (5) Lagoa dos Olhos, Minas Gerais, Brazil; (6) Hill of the Six Lakes (Lakes Pata and Dragão), Brazil; (7) Petén; (8) Lake Valencia, Venezuela; (9) Carajás Plateau; (10) Amazon Fan.

The hypothesis of compressed temperature gradients has been tested directly. When applied to equatorial South America, the hypothesis requires that part of the Amazon Basin abutting the Andes to have been arid or semiarid in glacial times. Data from the Mera and San Juan Bosco sites in Ecuador show that this requirement cannot be met (Liu and Colinvaux 1985; Bush et al. 1990). The two sites are about 160 km apart, both in the upper reaches of the modern lowland Amazonian rainforest between 900 and 1100 m elevations (fig. 13.2). Pollen, phytoliths, and wood samples together show that the local vegetation at the time of deposition had much in common with the modern tropical rainforest of the region, although having been invaded by a number of what are now Andean tree taxa. Modern precipitation is in the order of 5 m per annum. The paleobotanical data therefore leave no doubt that the climate at the time of deposition was extremely moist (fig. 13.3). Two radiocarbon dates on wood samples from each section show that

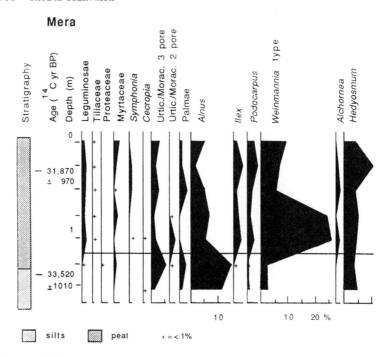

Fig. 13.3 Pollen percentage diagram of glacial-age section from the foot of the Andes at 1100 m in the Amazon rainforest at Mera, Ecuador. Most of the pollen grains in the category "other" are of forest trees. The pollen spectra show that plants of the modern

they date to between 26,000 and 33,000 B.P. The four radiocarbon dates have quoted errors in the 1000-year range and no "infinite" dates (before 40,000 B.P.) are reported (table 13.1). The dates are therefore likely to be as reliable as any radiocarbon determinations in this time interval can be, allowing a strong presumption that the deposits are indeed of glacial age. Use of the pollen spectra of the San Juan Bosco and Mera sections as time-stratigraphic markers correlating with well-dated lake cores from the central Amazon now confirms the correctness of the dating (fig. 13.4; Colinvaux et al. n.d.; Colinvaux et al. 1996). Thus, the Amazon lowlands at the foot of the equatorial Andes are shown to have been wet in the last glacial cycle as in the Holocene, and the application of dry-air lapse rates becomes untenable.

The Mera and San Juan Bosco data also provide direct evidence that all elevations of the equatorial lands were cooled equally. At both sites, alpine taxa such as *Alnus* and festucoid grasses (identified from phytoliths by D. R. Piperno) appeared to have been abundant in the local vegetation, though these taxa are now confined to altitudes above 2500 m

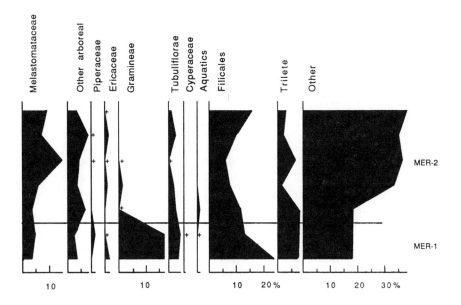

tropical rainforest grew in association with taxa like *Alnus, Podocarpus,* and *Weinmannia* now confined to high elevations in the Andes (Bush et al. 1990).

or even 3000 m (Bush et al. 1990). Temperature-sensitive descents are thus suggested for all elevations on the eastern flank of the equatorial Andes and are recorded from the paramo to the tropical rainforest.

Applying standard moist-air lapse rates, or actual measured modern lapse rates, to the Mera and San Juan Bosco data yields temperature depressions between 7°C and 9°C for the western Amazon between 33,000 and 26,000 B.P. These estimates suggest temperatures even lower than those calculated by CLIMAP (1976) for high-latitude oceans (about 6°C lower than modern). Yet the estimates are consistent with up to 9°C decrease in temperature calculated for the greatest descent of Andean glaciers (Clapperton 1987; Hastenrath and Kutzbach 1985).

Using the Mera and San Juan Bosco data to argue for widespread cooling in the Amazonian lowlands is inherently vulnerable to dismissal by the catabatic winds hypothesis. If descent of tree line can be attributed to cooling in winds blowing from glaciers, so can the further descent of the more sensitive among alpine taxa. It thus might be thought possible to explain away the peculiar ice-age forest communities at the

Table 13.1 Radiocarbon Dates from Wood Samples in the Mera and San Juan Bosco
Sections, Amazonian Ecuador

Mera	
26,530 ± 270 (B-10170)	First exposure
33,520 ± 1,010 (B-9618)	Second exposure
San Juan Bosco	
26,020 ± 300 (B-27144)	Uppermost layer
30,990 ± 350 (B-27145)	Near base of deposit

foot of the eastern flank of the Andes as a purely local phenomenon
dependent on cold air descending from extended alpine glaciers. Now,
however, similar cooling, and the unfamiliar forest communities that
result, can be shown to have existed in the Amazon lowlands far removed from the Andes Mountains.

Cooling and Forests of the Central Amazon Lowlands

The Hill of Six Lakes is a low inselberg, rising to 300 m out of Amazon
bottomlands that are locally only 75 m above modern sea level (fig.
13.2). A number of lakes occupy closed basins of unknown origin on the
hill. The basins have been likened to karst structures, but the water is
pH 4, and underlying carbonatite plugs are at depths in the order of 250
m (Justo and De Souza 1984). Thus, the origin of the basins by solution
and collapse seems unlikely. Cores from Lakes Pata and Dragão of this
complex have yielded long, perhaps transglacial, pollen histories of the
Amazon lowlands (Colinvaux et al. n.d.; fig. 13.4).

Isotope stage 2, the last glacial maximum (LGM), is identified in the
Lagoa Pata core by direct radiocarbon dating to be represented by a
pollen zone defined by cluster analysis (fig. 13.4; zone P2). Like the
pollen spectra from glacial age San Juan Bosco and Mera at the foot of
the Andes (fig. 13.3), the spectra of the LGM at Lagoa Pata show a
mixture of tropical rainforest with pollen signatures of populations now
confined to cooler elevations.

The Pata LGM spectra have up to 10% *Podocarpus* pollen, though
but a single grain was found in all the Holocene samples combined, and
no such high *Podocarpus* percentage is known from any modern neotropical lowland forest. With the *Podocarpus* peak of the LGM are the cool-
adapted taxa *Humiria* and Ericaceae, as well as maxima of the pollen
type *Weinmannia* (a grain that looks like the cloud-forest taxon *Weinmannia* but the identity of which is not certain), copious Melastomataceae, *Hedyosmum*, *Rapanea*, and *Ilex*. These signals for cooling are em-

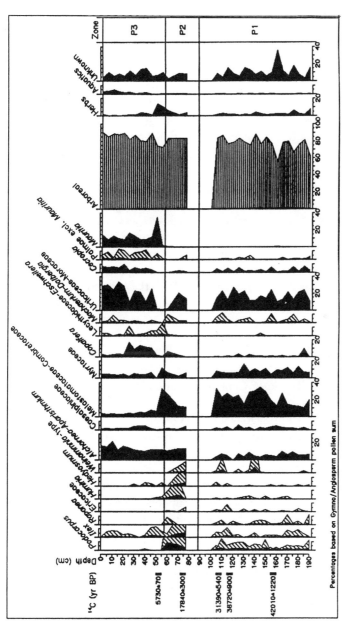

Fig. 13.4 Pollen percentage diagram from Lagoa Pata, Hill of the Six Lakes, central Amazonia. The lake lies below the 300 m contour on an inselberg rising out of the Amazon Plain at 75 m elevation, making this the first glacial-period pollen diagram from the lowland Amazon forest. Rainforest was present throughout the last 40,000 years, but forests of the glacial period included taxa like *Podocarpus* now only found at high elevations on Mt. Pico da Neblina 90 km away. The gap in the record is sediment lost in coring. (Pollen analysis by P. E. De Oliveira.)

bedded in, and added to, the tropical rainforest signal that persists in both glacial and nonglacial parts of the record.

The Pata pollen data thus serve to extend the reconstruction of a cooler, but still moist, tropical rainforest climate from the foot of the Andes at Mera to the central Amazon lowlands. The presence of cold-adapted taxa in lake deposits dated directly to the LGM serves as a time-stratigraphic marker to confirm the glacial age of the Mera deposits. And it also allows direct calculation of the temperature lowering required to account for significant populations of more montane species in the Amazon lowlands.

All reports of *Podocarpus* populations in the region are from Pico da Neblina, a mountain rising to 3014 m, 90 km northeast of the Hill of Six Lakes on the border with Venezuela. Five species of *Podocarpus* are known from Pico da Neblina, but none have been found below 1100 m. Minimal descents of *Podocarpus* in glacial times thus were 800 m to the summit of the Six Lakes hill and 1000 m to the surrounding lowlands. Using the moist-air lapse rates required by the presence of tropical rainforest yields a minimum temperature depression of 6°C (Colinvaux et al. n.d.).

Cooling and Forests of Southeast Brazil North of Capricorn

Beyond the edge of the Amazon Basin proper, in southeastern Brazil yet still north of the Tropic of Capricorn, pollen histories from the sediments of two lakes demonstrate glacial cooling in broad synchrony with the data from Mera and Lagoa Pata (De Oliveira 1992). Basal sediments at Lagoa dos Olhos are directly radiocarbon-dated to the LGM at about 19,000 B.P. and a 6 m core from Lagoa Serra Negra is radiocarbon infinite beneath the second meter. Both lakes are now in semidry landscapes the presettlement vegetation of which was cerrado or cerrado woodland. Pollen data show that both supported coniferous woodland within the last glacial period (figs. 13.5, 13.6). The present limits of the conifers are several hundred kilometers to the south, where many data are available to show that their distribution is temperature limited. The northward advances of these conifers in glacial times therefore allow calculation of temperature depressions without the necessity of assuming past adiabatic lapse rates.

The coldest part of the glacial period in the Serra Negra record has radiocarbon "infinite" ages, terminating in an upper date of 31,000 B.P. These deposits underly an upper sedimentary sequence dated from 14,000 B.P. to the Late Holocene (fig. 13.5). No unconformity is visible in the sediments, but the proximity of the dates allows a strong inference

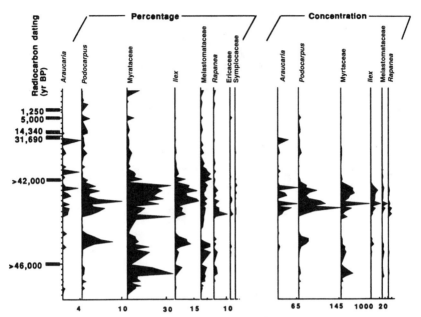

Fig. 13.5 Partial pollen percentages and concentrations from Lagoa Serra Negra, Minas Gerais, southeastern Brazil. The diagrams show only the pollen attributed to *Araucaria-Podocarpus* forest now found 700 km to the south (De Oliveira 1992).

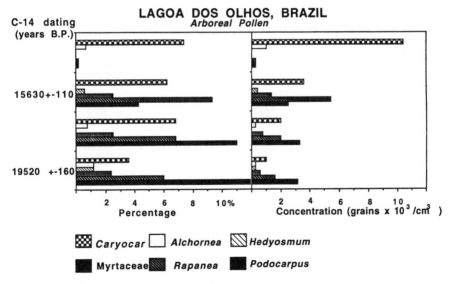

Fig. 13.6 Partial pollen percentages and concentrations from Lagoa dos Olhos, Minas Gerais, southeastern Brazil. The figure shows pollen of the principal trees over the section spanning the last glacial maximum (LGM) (De Oliveira 1992).

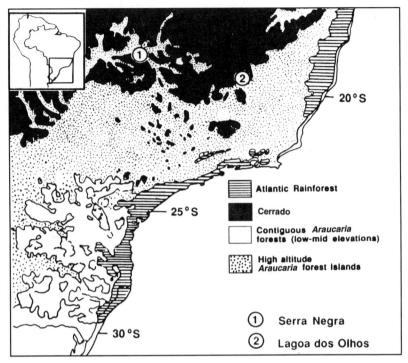

Fig. 13.7 Vegetation of southeastern Brazil at the Tropic of Capricorn. The pollen data of figures 13.5 and 13.6 show that the contiguous *Araucaria* forest advanced to north of Capricorn in glacial times, to the latitude of Lagoa Serra Negra (1).

of a gap in the sedimentary record spanning the LGM itself. Pollen zones SN4 and SN3 (fig. 13.5) thus appear to represent local vegetation in the earlier part of the last glaciation. High percentages of *Araucaria, Podocarpus, Ilex, Rapanea,* Melastomataceae and Myrtaceae, together with the presence of *Symplocus, Drimys, Xylosma, Esenbeckia,* and *Daphnopsis,* indicate closed *Araucaria* forest occupying the immediate landscape. To find similar modern *Araucaria* forest it is necessary to travel 700 km to the south (fig. 13.7). Thus a northward forest advance is suggested for the earlier part of the last glacial cycle. Because sediments from the LGM itself (18,000 B.P.) cannot be unambiguously identified in the Serra Negra cores, nor from the neighboring bog record at Salitre (Ledru 1993), we must rely on the record at Lagoa dos Olhos (fig. 13.6).

The basal pollen zone at Lagoa dos Olhos spans from 19,520 to 13,685 B.P. (fig. 13.6; De Oliveira 1992). Abundant spores of the algae *Cosmarium* and *Mougeotia,* together with lesser amounts of *Zygnema* and

Debarya demonstrate a climate sufficiently moist to support an open-water lake, then as now. Important trees in the LGM landscape were *Caryocar, Alchornea, Hedyosmum, Rapanea, Podocarpus,* and Myrtaceae. This assembly lacks the *Araucaria* pollen characteristic of earlier glacial times at Serra Negra, though it has the cold-sensitive *Podocarpus.* Lagoa dos Olhos is the drier of the two sites today. The pollen suggest it always has been drier, in particular having grass pollen percentages that were as high at the LGM as they were throughout the Holocene. This allows the working hypothesis that, although cool and moist enough for *Podocarpus* at the LGM, the Lagoa dos Olhos site was marginally too dry for the *Araucaria* that was a co-dominant in the coniferous forest further inland.

The cooling required to move contiguous *Araucaria* forest 700 km northward can be calculated by matching isotherms to modern forest boundaries, from the temperature regime of the farthest north of modern high-elevation *Araucaria* woodland, or from the narrowly known physiological tolerances of *Araucaria angustifolia,* yielding depressions of mean winter temperature of 6°C, 10°C, and 9°C respectively. The more modest of these estimates, 6°C cooler in winter, is consistent with the depressions calculated from lapse rates in the central Amazon and at the foot of the Andes.

Adding these two records to that of Ledru (1993) from the Salitre bog does not yet provide a complete record for the region, for of the three, the only record with sediment unambiguously dated to the LGM is Lagoa dos Olhos, more to the east and in a slightly different climatic and vegetation region (fig. 13.7). The cooling signal given by *Podocarpus* at Lagoa dos Olhos, though clear, cannot be calibrated with present data. The *Araucaria* cooling signal from Lagoa Serra Negra is perhaps the best calibrated cooling signal in the neotropics, but is available only for early in the glaciation. Nevertheless, the data are sufficient to allow tentative acceptance of the most modest *Araucaria* calculation of 6°C temperature depression at the LGM, at least as a working hypothesis (Colinvaux et al. 1996).

Cooling and Forests of Lowland Panama

A long pollen record is available from the Pacific lowlands of Panama. Not only does it provide an estimate of the range of environments encountered by land biota passing through the Panama bottleneck of the isthmus at all stages of glacial cycles, but it also provides a test of the catabatic winds hypothesis for cooling at the base of the Andes that is independent of the Six Lakes records from central Brazil. None of the mountains in Panama were high enough to support glaciers at any time in the Pleistocene, and yet alpine taxa descended in Panama just as they did in Ecuador.

The Panamanian site at El Valle (fig. 13.2) lies at 500 m elevation, being the floor of an old caldera that formerly held a lake. Local vegetation is tropical woodland in climate with alternating wet and dry seasons, though tropical moist forest appears to have been the natural vegetation of the surrounding lowlands. A 55 m sediment core apparently spans most of the last glacial cycle, ending about 8000 B.P. with the final draining of the lake (fig. 13.8; Bush and Colinvaux 1990). A pollen record for the last 14,000 years from existing Lake La Yeguada serves to extend the record both in time and space (fig. 13.9; Bush et al. 1992). As at the Ecuadorian and central Amazon sites, the pollen and phytolith histories from the Panamanian deposits reveal large descents of temperature-sensitive taxa in glacial times.

At both El Valle and La Yeguada, oaks (*Quercus*), *Magnolia*, *Ilex*, *Symplocus*, *Gunnera*, Caryophyllaceae, and Ranunculaceae were shown by pollen and phytolith data to have been prominent parts of the local flora, though in modern times significant populations of these taxa occurring simultaneously are found only more than 800 m higher in the Chiriquí Highlands. As at Mera, San Juan Bosco, and the Six Lakes sites, the pollen data also demonstrate that forest cover was uninterrupted throughout the records, requiring that moist-air lapse rates be used to calculate temperature depression in the order of 4–6°C (Bush and Colinvaux 1990).

No other long pollen records provide temperature signals from the lowlands of Central America and the isthmus, but an oxygen isotope record from calcareous sediments in a karst lake in Guatemala suggests an 8°C temperature depression (Leyden et al. 1993).

The Panamanian data follow those from Mera and San Juan Bosco in Ecuador in showing that glacial-age temperature depression on mountainsides is a property of all elevations. It becomes parsimonious, therefore, to postulate low-elevation cooling whenever cooling upslope is demonstrated. This is an operationally useful concept because many more records are available at high elevations. In Costa Rica, for instance, high elevation (1400–2300 m) sites of glacial age yield pollen data suggestive of 8°C temperature depression (Hooghiemstra et al. 1992). Thus, cooling in the order of 6°C is demonstrated for the lowlands of Central America and the isthmus as a property of glacial climates, no less than in the Amazon Basin.

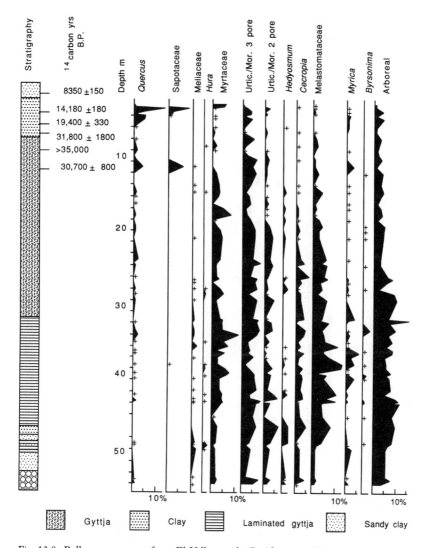

Fig. 13.8 Pollen percentages from El Valle, on the Pacific coast of Panama. The sediments are from an ancient caldera lake at an altitude of 500 m on which the village of El Valle is built. The lake drained about 8000 years ago. Oaks (*Quercus*) grew in the caldera in the glacial period, though they are now confined to the highlands of Chiriquí (Bush and Colinvaux 1990).

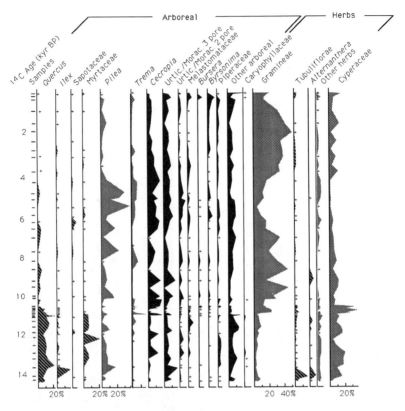

Fig. 13.9 Pollen percentages from Lake La Yeguada, Panama. The vertical axis is plotted as radiocarbon years B.P., instead of depth. In the late glacial period, from 14,000 B.P. to about 11,000 B.P., oaks (*Quercus*) and other taxa from the cooler uplands (heavy shading) were still present in the lowlands. Tropical forest taxa are in black. The diagram should be read in conjunction with figure 13.8 (Bush et al. 1992).

Rapprochement of Land and Sea

That cooling was widespread in the neotropic lands in glacial times almost certainly implies that cooling was ubiquitous. The forcing can, of course, be locally modified. It is possible that cooling was less pronounced in Pacific Panama than, say, in the central Amazon, because of the insulating effect of the oceans that might tend to even out seasonal extremes. And the glacial cooling of tropical southeastern Brazil that sent the coniferous forest northward for 700 km possibly records, in part, northward displacement of the south polar front that should have had only regional significance. Yet general cooling of the neotropics in glacial times becomes increasingly difficult to deny (Colinvaux et al. 1996).

The latest measure of glacial temperature from tropical Brazil uses the ratios of noble gasses dissolved in glacial-age groundwater as a thermometer, finding a temperature reduction of 5°C at the LGM (Stute et al. 1995). Paleoecological evidence of low-elevation cooling is coming from Africa, where the record for cooling in pollen diagrams can now be separated from the record of the widespread African aridity (Livingstone 1993; Maley 1991). Descent of forest taxa in pollen profiles radiocarbon-dated to the last glacial interval in Southeast Asia has also been documented (Stuijts et al. 1988). More generally, the demonstration that cooling on tropical high mountains is associated with parallel cooling in neighboring lowlands extends the records of cooling of tropical lowlands to wherever montane glaciers have been shown to descend at the LGM. Tropical lands therefore cooled in an ice age.

What, then, of the CLIMAP (1976, 1981) consensus that sea surface temperature (SST) cooled only marginally in the tropics? It may be time to ask if this consensus is mistaken, as Emiliani (1992) has argued. Recent use of strontium-calcium ratios as a thermometer for coral reefs conclude that SST over Barbados reefs was reduced by 5°C at the LGM (Guilderson et al. 1994). Simultaneous cooling over land and sea in glacial times is the best resolution of the "enigma" of cold high mountains but relatively warm sea surface of which Broecker and Denton (1990) wrote.

The conclusion of a tropical paleoecologist, based on the presently available data, must be that global cooling of ice ages was experienced by tropical ecosystems of all elevations, certainly on the land, probably in the surface oceans (Colinvaux et al. 1996). Cooling, like reduced CO_2 but unlike reduced precipitation, was a ubiquitous property of the ice-age earth experienced in the tropics no less than in mid-latitudes. Profound reassortments of species populations in response to this cooling are to be expected.

THE ARIDITY HYPOTHESIS

In the quarter century since the first publication of Haffer's (1969) refuge hypothesis, prevailing opinion has been, and remains, that the Amazon Basin was arid in glacial times, the latest strong statement of this conclusion being in a treatise by Clapperton (1993). This makes the Amazon region conform to the concept that the ice-age earth was generally a dry place, partly because cooling of the oceans reduced the rate of evaporation, and hence the transfer of moisture from sea to land, partly from changed patterns of oceanic circulation (Broeker 1995b).

The resulting earth was dusty, as is shown, for instance, by dust in ice cores from both Greenland and Antarctica and by huge deposits of loess in China. Recently, dust has been found in a core of glacial-age ice from the high Andes of Peru, bringing the evidence of dust very close to the Amazon Basin itself (Thompson et al. 1995; Broecker 1995a). More impressively, evidence of reduced lake levels across tropical Africa suggests that the equatorial portions of that continent were so much drier in ice-age times than their present dry state as to warrant the adjective "arid" (Livingstone 1975; Street and Grove 1979). Despite these lines of argument, however, it will be the conclusion of this review that the Amazon lowlands remained forested throughout glacial cycles. Any reductions in precipitation that there might have been were not sufficient to replace the forest with savanna or other open vegetation.

Paleoecological Evidence from Peripheral Areas
Peripheral areas of the neotropics, like the Galápagos Islands and the Caribbean coast undoubtedly had reduced rainfall in glacial times, as attested by well-dated lake level and other paleolimnological data (Colinvaux 1972; Bradbury et al. 1981; Leyden 1984).

Aridity around the Caribbean coasts is likely to have had much in common with the increased aridity of East Africa, which can be accounted for in climate models by a 10–20% reduction in monsoonal rains that effectively lengthened dry seasons (Kutzbach and Guetter 1986). Lengthened dry seasons thus could account for drier or more seasonal climates in the interval between about 20,000 and 9,000 B.P. for the Petén of Guatemala, and for the coasts of Venezuela and the Guianas (Wijmstra and Van der Hammen 1966; Salgado-Laboriau 1980; Bradbury et al. 1981; Leyden 1985). This aridity of Caribbean sites was localized, however, for it did not apply to lowland Panama, where lakes and forests are shown by paleoecological and paleolimnological data to have persisted despite some evidence for lengthened dry seasons (Bush and Colinvaux 1990; Bush et al. 1992; and see below, p.398). The Caribbean data therefore in no way suggest that the neotropical interior, even of the Isthmus of Panama, was arid. In particular, the data give no support to the idea that the distant and much wetter lowlands of the Amazon Basin should have been arid.

Equally peripheral is evidence for equatorial aridity from the Galápagos Islands and the western flank of the Ecuadorian Andes. The Galápagos Islands, some thousand kilometers offshore from equatorial South America, have a peculiar arid climate resulting from patterns of wind and cold, surface ocean currents that produce stable inversions for most of the year. Only with annual descent of the Intertropical Convergence Zone (ITCZ) or with warm surface water associated with El Niño/ Southern Oscillation (ENSO) events is the inversion overridden to cause heavy rain. Sediments of a single lake in the archipelago, El Junco on Isla San Cristóbal, hold a record of environmental events from the present to before radiocarbon infinity. Lake El Junco dried completely sometime before 25,000 B.P., not to become a permanent lake again until 10,000 B.P. (Colinvaux 1972). The cause of this increased ice-age aridity of the Galápagos Islands is still not completely known. I have argued (Colinvaux 1972) that the ITCZ must have failed to reach the islands at any part of the year, and Newell (1973) suggested that this could have happened if the ITCZ was displaced far to the south by increased Hadley cell circulation. Houvenaghel (1974) and Simpson (1975) invoked increased oceanic circulation, which would have extended the present system of cold surface water and bottom-heavy, stable air masses. Whatever the true cause, Galápagos ice-age aridity is clearly related to regional events in the eastern Pacific Ocean.

The Interandean Plateau and the Equatorial Pacific Coast
From the Pacific coast of the equatorial Andes, signals for precipitation are mixed. Yaguarcocha, a lake in the Interandean Plateau of Ecuador, provides a history of this drier part of the modern Andean system (Colinvaux et al. 1988). The Yaguarcocha pollen diagram apparently defines the Pleistocene-Holocene boundary with a replacement of near-desert vegetation, recorded by high percentages of Chenopodiaceae and Amaranthaceae pollen, with the dry woodland and open vegetation characteristic of much of the Interandean Plateau in modern times. Radiocarbon dating of the bottom of the section is unsatisfactory because of a series of inversions. The sediments are superbly banded, showing that no physical inversion or postcollection contamination of sediments has occurred (Colinvaux et al. 1988). Extrapolation from the sequential radiocarbon dates of the upper part of the section yields an age of about 10,000 B.P. for the suspected Pleistocene-Holocene boundary. A recent radiocarbon determination of 30,000 B.P. from organic material in the lava-flow dam of the lake is consistent with a glacial age for the lowermost sediment (M. L. Hall, personal communication).

   The dry glacial-age landscape suggested by the basal pollen zone at Yaguarcocha suggests the climate in which the "cangagua" deposits accumulated in the Interandean Plateau of Ecuador and southern Colombia (Sauer 1965). The cangagua formation is a loess-like mantle that overlies glacial moraines, thicker on older moraines than on younger.

Although derived from airfall volcanic deposits, the cangagua is interbedded with tephra layers and is presumably reworked from older airfall volcanics. Some cangagua exposures have abundant fossils of Pleistocene megafauna, nests of dung beetles usually associated with the dung of grazing animals, and nodules of silica that suggest periodic evaporation of groundwater.

Clapperton (1993) suggests a glacial age for the cangagua, partly because the thickest deposits overlie old moraines. The Yaguarcocha pollen record is consistent with this conclusion in that it suggests an arid steppe-like landscape toward the end of the last glaciation. The simplest explanation for this dry vegetation and climate is that it is a product of glacial cooling. The 6°C cooling of glacial time that drove cold-adapted plants 1000 m down Andean slopes raised the relative height of the Interandean Plateau by 1000 m, and a cold alpine desert was the result. The aridity of the Interandean Plateau recorded by the cangagua formation and the Yaguarcocha sediments, therefore, was a direct consequence of cooling that was important because of the high elevation of the plateau and the presence of shadowing mountains to both east and west. Like the aridity of the glacial-age Galápagos Islands, glacial aridity in the central valley of the Andes was due to local causes and does not provide a signal for regional aridity.

A general history of the Pacific coast of Ecuador is provided by pollen studies of offshore sediments contributed by the Guayas River (Heusser and Shackleton 1994). Pollen in the glacial-age sediments show increased percentages of *Podocarpus* and *Alnus*, reflecting enlarged populations on the lower slopes in response to cooling, but no significant increase in the Chenopodiaceae, Gramineae, or ecologically related taxa of arid habitats. As for the Mera and San Juan Bosco deposits of the eastern flank of the equatorial Andes, therefore, the Guayas River deposits from the western flank signal regional cooling, not regional aridity.

Pollen from Tropical Southeastern Brazil

Three pollen records from Minas Gerais State are radiocarbon dated to the last glaciation, the record from Salitre bog (Ledru 1993) and two lake records, Lagoa Serra Negra and Lagoa dos Olhos described above for their strong and well-calibrated record of cooling in at least part of the last glacial cycle. But the cooling signal, the 700 km northward advance of *Araucaria* forest, is also a precipitation signal. These were coniferous forests that flourished in what are now the habitats of cerrado and cerrado forest. It seems inescapable that the climate was moderately moist, as well as colder, at the times of *Araucaria* and *Podocarpus* advance.

A caveat to this conclusion is that the record is least complete for the LGM itself: neither at Lagoa Serra Negra nor at the Salitre bog are sediments dated directly to the LGM, both sites having yielded Holocene or late-glacial sections over deposits dated to 30,000 B.P. or earlier. At Lagoa Serra Negra, a sample dated at 14,000 B.P. has less *Araucaria* than the samples from 30,000 B.P., though more than the traces (consistent with long-distance transport by wind) in the Holocene deposits (fig. 13.5). Samples dated to the Early Holocene in the Salitre bog retain *Araucaria*, but the upper layers of the bog are penetrated by modern roots, suggesting that the Holocene ages of this part of the deposit might be spurious (Ledru 1992). Only at Lagoa dos Olhos are parts of a continuous sediment column dated directly to the LGM, and there the pollen record is of forest with *Podocarpus* but without *Araucaria* (fig. 13.6).

Minas Gerais State, together with São Paulo State to the south of it, have been the subject of extensive geomorphological studies that have been interpreted as implying past aridity in the whole of the Brazilian Highlands. Surface sediments, known as the "Cobertura dendritica," include unsorted sands, sandy clays, and gravel in an unconsolidated regolith. Where this mantle occurs, it has been interpreted as colluvium deposited at a time when the vegetative cover could not have been complete (Bigarella et al. 1969). It has been impossible to date the material, which is essentially without organic matter, but the arid-land hypothesis of its origin has been widely accepted as evidence that the region was arid in glacial times.

That aridity and bare ground could have characterized Minas Gerais while Lagoa Serra Negra and Lagoa dos Olhos were depositing sediments with pollen of coniferous trees seems most unlikely. It can of course be argued that sediments of LGM age missing from the Salitre bog and Lagoa Serra Negra represent the postulated arid time, which then becomes a function of isotope stage 2 (LGM) only. However, the paleolimnology of Lagoa dos Olhos shows that an open-water lake was present in the forested landscape of the LGM, showing that some parts of the landscape were not arid. Thus, if the Cobertura dendritica is correctly identified as colluvium of glacial age deposited because vegetative cover was incomplete, climates must have been patchy both in time and space. Obviously more well-dated records are required, but meanwhile the aridity hypothesis for tropical southeastern Brazil should be treated with caution.

Geomorphological Evidence for Neotropical Aridity

Fossil sand dunes are visible on satellite and sidescreen radar imagery over wide areas of the Orinoco llanos of Colombia, northwest of the Amazon Basin and in the Pantanal region at the boundaries of Bolivia, Paraguay, and Brazil southwest of the basin (Tricart 1974, 1977). Because dunes are not forming under modern conditions, these dune fields suggest past climates in those regions so arid that vegetation was virtually absent. No independent dating of the fossil dunes is available, but a glacial age is consistent with the general concept of tropical aridity. The orientation of the dunes yields paleowind directions from the northeast, letting Clapperton (1993) suggest that the northeast trade winds penetrated into the continent and were deflected along the Andes.

The most general of the geomorphological arguments for glacial aridity throughout the Amazon Basin (and indeed throughout most of Brazil) is the existence of "stone lines" under an overburden of one to several meters of surface sediment. Ab'Saber (1982) has famously argued that these stone lines are widespread, and that wherever they are found, they represent burial of local stones by colluvium at times of scant vegetation cover, or even that they represent true desert pavements. Stone lines that have been examined by others within the Amazon Basin appear to be in fact concretions formed within latosols under a humid environment (Irion 1982; Sternberg 1982). No independent assessment of the hypothesis of the ubiquity, origin, and antiquity of stone lines appears to have been made, without which it must be premature to accept them as evidence for formerly arid conditions throughout Brazil.

Dissected colluvium on slopes that are now well vegetated in the Brazilian highlands have been interpreted by Bigarella and colleagues (1969) as evidence that the landscape was partially devegetated, as described above. Similar arguments have been offered for dissected deposits in parts of the Amazon basin (De Meis 1971).

These various geomorphological arguments for lowland Amazon aridity were given an apparent chronological context by the discovery of arkosic sands in radiocarbon-dated piston cores from the Amazon Fan (Damuth and Fairbridge 1970). The arkosic sands suggested an origin on a land surface that had not been subjected to humid weathering, and hence an origin in arid glacial times. However, the feldspars of these deposits are parsimoniously understood as being of Andean origin, or as the product of deep erosion of the Amazon River channel following eustatic lowering of sea level (Milliman et al. 1975; Irion 1984). The arkosic sands thus do not require an arid land surface in the Amazon Basin. More recent data from the Amazon Fan, based on long cores obtained by the Ocean Drilling Project (ODP), demonstrate a history of uninterrupted forest in the Amazon lowlands through glacial cycles (see below).

The Refuge Hypothesis

Contemporary belief in a dry rather than a cool ice-age Amazon depends more on a biogeographic scenario, the refuge hypothesis, than it does on extrapolations from the periphery, although the fact that some dry sites of glacial age (Yucatán, Lake Valencia, coastal Guiana, sand dunes in the llanos) can be found in the periphery gives heart to the hypothesis.

The biogeographic data behind the refuge hypothesis are disjunct distributions of related species in different parts of the Amazon Basin. Endemicity in Amazonian forest birds was noted first and used by Haffer (1969, 1974) in setting up the hypothesis, but distributions of helioconid butterflies are known in more detail, allowing maps like that in Brown 1987 (fig. 13.10A).

Each center of endemism appears to lie on higher (therefore wetter) terrain, although the regions come close together or actually abut in the lowlands. Often centers of bird or even plant distributions can be mapped as being concentric with the butterfly centers of dispersal. The refuge hypothesis suggests that this pattern is a relict of past geography when the centers of modern endemism had been physically separated by terrain inhospitable to the forest communities in which they lived. Vicariant speciation then proceeded in the postulated forest isolates. The favored mechanism to produce this vanished archipelago of rainforest refuges is a drying of the basin that fragmented the forest, leaving patches (refuges) on the elevated centers of endemism thought to have escaped the universal drought by reason of their orographic rainfall (Prance 1982; Whitmore and Prance 1987).

So powerful has been the refuge hypothesis that it has led, not so much to attempts to test it with paleoecological data, as to a hunt for confirming evidence of past Amazonian aridity. The geomorphological evidence or arguments summarized in the last section, particularly the stone lines, have been valuable to the hypothesis. The case for tropical aridity had wide appeal; the refugial model provides an elegant explanation for both local endemisms in the Amazon and even perhaps for the fundamental high diversity of the system itself. Therefore, says the hypothesis, the Amazon lowlands must have been dry.

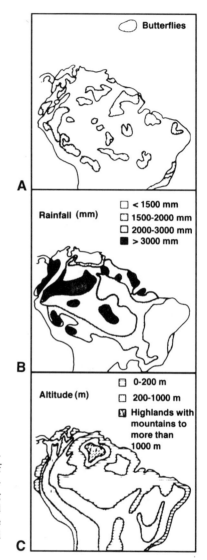

Fig. 13.10 Physiography, rainfall, and endemism in the neotropics. The pattern of endemism in helioconid butterflies (A) reflects rainfall (B) and relief (C), suggesting that environmental factors unconnected with glacial cycles provide vicariance for the separation of species in taxa with minimal powers of dispersal (from Colinvaux 1979).

Evidence from the Amazon Fan: Ice-Age Savannas Denied

The refuge hypothesis is a good hypothesis in that it makes falsifiable predictions: in particular, it makes statements about the glacial geography of the Amazon that can be tested by paleoecological reconstructions. The ideal test shall be based on pollen analysis of a grid of enough ancient lake cores from all parts of the Amazon Basin to demonstrate the true geography of the ice-age Amazon, with the Hill of Six Lakes cores described above being the first part of the test from the Amazon bottomlands (the core from the Serra Carajás, reviewed below, is from a 700–900 m plateau). But a shortcut is possible by using sediments of the Amazon Fan to extract pollen spectra that are integrals of the entire basin.

In glacial maxima with lowered sea level, sediments from the Amazon River system flowed across the continental shelf through a submarine canyon, eventually to be deposited in deep water 700 km seaward of the continental shelf as the Amazon Fan (Stow et al. 1985). The fan has been collecting since the Miocene orogeny emplaced the Andes, diverting the rivers to flow eastward to the Atlantic (Hoorn et al. 1995). During interglacial intervals of high sea level, Amazon sediments were deposited on the continental shelf (Nittrouer et al. 1995).

In 1994, on leg 155 of the Ocean Drilling Program (ODP), a series of long cores were raised from the Amazon Fan. Haberle (n.d.) subsampled sediments of the last glacial cycle from the fan cores. To provide a Holocene voucher, Haberle also sampled sediment suspended in the water of the modern Solemoes-Amazon system on a traverse of the whole basin from west to east by riverboat (fig. 13.11).

Strikingly, at no time in the glacial cycle was grass pollen significantly more abundant than in the Holocene river samples (fig. 13.11). The overwhelming pollen signal throughout is for tropical forest. The Amazon lowlands were not converted from forest to savanna in glacial times as required by the refugial hypothesis.

Almost equally striking is the consistent signal for cooling in the forest communities, particularly the forests of the LGM. *Podocarpus* is routinely present in the fan deposits, reaching to 3–4% at the LGM. With the *Podocarpus* are increases in *Hedyosmum* and Melastomataceae of the kind that appear as cooling signals at the Six Lakes site and at the foot of the Andes at Mera. Even *Alnus*, now a plant of the high Andes, was constantly present at 2–3%. These data allow the strong signal for cooling found in the pollen spectra of Lagoa Pata at the Six Lakes site to be extrapolated to the entire Amazon forest system.

The fan deposits are integrals of the pollen rains of the entire Amazon Basin at the times of deposition. Undoubtedly, they are biased toward plant communities lining the riverbanks, though the Amazon River system is so elaborated as to leave minimal areas as interfluves. In addition, the predominant contribution of the Andes to total system runoff might account, in part, for the importance of such taxa as *Alnus* in the fan pollen. But if widespread grassy savannas had existed in the

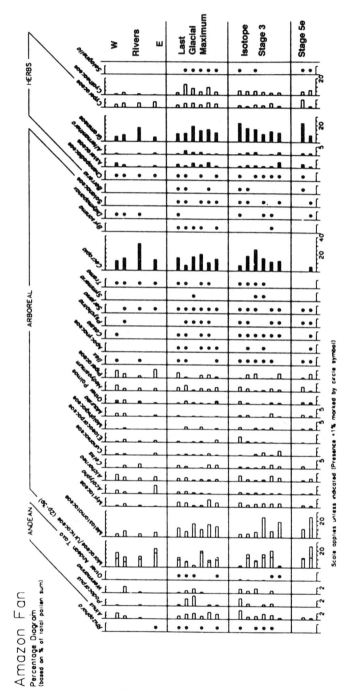

Fig. 13.11 Pollen percentages from the Amazon Fan. Core sections from the last glacial maximum and isotope stages 3 and 5e were identified by oceanographic criteria. The upper section of the diagram records pollen in suspension in the river system on a traverse of the Amazon Basin. Notice that Gramineae pollen remains roughly constant, thus demonstrating that savannas were not more extensive at any stage of a glacial cycle (Haberle n.d).

basin, their wind-dispersed pollen should certainly have been prominent in the pollen entering the rivers, even if entomophilous trees lined the banks. The signature of savannas could not have been drowned out by strips of riverine forest.

The pollen record from the Amazon Fan is the first direct and comprehensive data set describing the relative proportions of forest and open vegetation in glacial times. The data show unequivocally that the Amazon lowlands were forested in glacial times as they are now (Haberle n.d.). This is as expected from the evidence for cooling, which requires the more thermophilic of Amazon species to occupy the lowlands rather than elevated places of orographic rainfall assigned to them by refugial theorists (Liu and Colinvaux 1985). The Amazon lowlands were forested throughout the last glacial cycle and, by inference, throughout the Pleistocene.

On Carajás and Rondônia

On the face of it, a conflict arises between the pollen evidence from the Amazon Fan that the lowlands were without extensive savannas and paleoecological data from Serra Carajás that have been interpreted as requiring widespread aridity. Together with a short pollen section from Rondônia, the Carajás record has been used to reaffirm the argument that the whole of the Amazon lowlands were covered with savanna in glacial times (Van der Hammen and Absy 1994).

Serra Carajás is a large plateau 700–890 m high in east-central Amazonia (fig. 13.2). Below the plateau at about 200 m, the Amazon lowlands are vegetated with tierra firme forest, but drier forest of the cerrado type, or actual savanna, is present on the plateau itself in a climate with a pronounced dry season. Numerous small closed lakes occupy karst basins on the plateau, some of them said to be interconnected and ponded with groundwater. A 6 m core was taken from one of the basins that was occupied not by a lake but by a marsh. The core spans more than 30,000 radiocarbon years. Both mineralogical and pollen studies have been published (Soubies et al. 1989; Absy et al. 1991).

The persuasive evidence for past aridity has been two sandy layers rich in siderite, mixed with quartz and hematite, one of them bounded by radiocarbon dates of 22,870 (+2,540/−1,930) B.P. and 10,460 (+850/−770) B.P., the other being slightly below a date of 31,120 (+3,420/−2,400) B.P. (Soubies et al. 1989). The dates for the younger siderite layer thus suggest the LGM. Thus, we have at the LGM a drained marsh basin collecting slope-washed sand from the watershed.

Van der Hammen and Absy (1994) rely on pollen data, principally maxima of Gramineae (grass), Compositae, *Borreria*, and especially *Cuphea*, to conclude that the siderite layers imply regional climatic aridity. But the pollen data are open to different interpretations. Percentages of Gramineae fluctuate widely throughout the pollen diagram. Although there are grass maxima in the sand/siderite layers, the largest grass maximum is in the Middle Holocene; as elsewhere in the tropics, signals from grass pollen are ambiguous. *Cuphea* could as well represent forest understory as savanna. *C. carthagensis* and *C. mesostemon* are typically found in waterlogged soils adjacent to rivers. *Cuphea* therefore suggests either waterlogged soil or forest understory: it is not a reliable indicator of savanna. The *Borreria* and Compositae are as likely to be weeds growing in a drained lake basin as they are to represent open savanna. Indeed, there is no more reason for using *Borreria* as a savanna indicator than there is of using *Cuphea*. Thus, the parsimonious explanation of the modest herb maxima, including Gramineae, at the siderite layers is that they represent the herb community growing on the drained lake bed. In addition, no evidence is offered that the influx of tree pollen blown up from the forest below the plateau falls with the siderite layers, merely that the percentage of total tree pollen is lower. Such is the nature of percentage statistics that this would follow automatically if grass influx is high from weeds of the basin.

It is noteworthy that the Carajás pollen diagram actually duplicates some of the signals for cooling that are found at the Six Lakes site, the Amazon Fan, and elsewhere, particularly in high Melastomataceae percentages throughout the glacial period, together with episodes within the glacial period when *Ilex* pollen is more than 20%. That no *Podocarpus* was found is probably because Carajás is, and always was, a dry plateau unsuited to conifers at any stage of a glacial cycle.

The Carajás Plateau is between two significant tributaries of the Amazon, the Xingu and the Tocantins, both of which must contribute pollen from surrounding vegetation to the Amazon and thence, in glacial times, to the fan. As already noted, no significant increase of grass pollen (let alone *Borreria* and *Cuphea*) are recorded in the fan sediments. It is necessary to conclude that there was no extensive savanna in the lowlands around Carajás in glacial times. The most that the Carajás paleoecological record suggests is local reduction of precipitation on an already rather dry and ecotonal plateau.

The second datum used by Van der Hammen and Absy (1994) is from boreholes in a creek valley from Rondônia at the southwestern edge of the Amazon Basin. The borings penetrated to sedimentary clays under 25 m of lateritic surface. AMS dates obtained on the buried materials are in the 40–50,000 B.P. range. Included pollen has high grass percentages, although the surrounding modern vegetation is tierra firme forest. As with the Carajás deposits, admixtures of *Borreria* and *Cuphea* are taken to be savanna indicators. Again, as at Carajás these genera are more plausibly assigned to stream-bed vegetation than to regional savanna. In view of the Amazon Fan data, these stream-bed drill cores from Rondônia probably represent an edaphic habitat or a local climate shift too small to be detected when pollen from all over the Amazon are integrated.

Extinct Amazonian Mammals

Webb and Rancy (this volume, chap. 12) report the discovery of an extinct Pleistocene fauna of mammals adapted to grazing or browsing in savanna or other open vegetation in the western Amazon. A prime site for the mammal remains is the region along the Napo River where it drains the Andes in the wettest parts of Amazonia. Modern precipitation in the region is in the 5–7 m range. Paleoclimatic proxy data are available for the region from the Mera and San Juan Bosco sites in Ecuador, both dated to glacial times directly by radiocarbon and also by time-stratigraphic correlations to the Six Lakes, Serra Negra, and Amazon Fan records. These data demonstrate that glacial climates were wet and that closed forest persisted in the region through glacial times (Liu and Colinvaux 1985; Bush et al. 1990). Unfortunately, the mammal remains are not dated by radiometric or stratigraphic means. In the absence of other evidence that the region was arid at any time in a glacial cycle, the fossil deposits would be more easily understood if their true age were to be Late Pliocene or Early Pleistocene, at an earlier stage in the Andean orogeny and before the full establishment of Pleistocene glacial climates.

A Demonstration that the Glacial Amazon Lowlands Were Not Arid

Paleoecological data are now sufficient to repudiate the hypothesis of widespread tropical aridity in glacial times within the Amazon lowlands now covered by tropical rainforest. The definitive data are pollen histories of sediments from the Amazon Fan that integrate pollen signals from all the watersheds of the Amazon Basin. Local climate shifts must certainly have occurred, as fronts and air masses were adjusted to the atmospheric circulation of the ice-age earth, with some localities being moister and some drier. But the overall pattern was one of little change. The Amazon forest was not fragmented in glacial times as proposed by refugialist theory.

This interpretation is supported by the long pollen section from Lagoa Pata, the first from a lowland site in the Amazon rainforest. The Lagoa Pata pollen diagram shows that lowland tropical rainforest was present at all the levels sampled back to radiocarbon infinity (fig. 13.4). Analyses by P. E. De Oliveira (personal communication) from the even older underlying 4 m section show the same rainforest signature as all the pollen spectra of figure 13.4. At no level in the sediments is there the slightest evidence of expansion of savanna or cerrado vegetation. This includes sediment dated directly to the LGM. The one uncertainty in the record arises from the gap below the LGM sample in figure 13.4 due to sediment lost while coring. Because of the extreme slowness of the sedimentation rate, this gap spans several thousand years. It seems, however, implausible to suggest that a dry savanna episode has been fortuitously missed when pollen spectra bracketing the gap are so similar. In view of the fact that no such savanna episode can be detected in pollen from the fan, the possibility can be discounted.

Recent reconstructions of the glacial climate of South America persist in mapping the Amazon lowlands as being savanna covered. One such reconstruction is that of Clapperton (1993) shown in figure 13.12A. A map by Van der Hammen and Absy (1994) is similar to Clapperton's map. Both maps are heavily influenced by the geomorphological arguments summarized earlier, and by the interpretation of the Carajás pollen record as requiring a regional savanna. An alternative reconstruction by Bush (1994) is shown in figure 13.12B and assumes precipitation reduced by 20% as a direct consequence of cooling. The Bush map shows the western and central parts of the Amazon Basin to have been occupied by rainforest, thus being consistent with the evidence from Lagoa Pata, but it allows savanna in the lower Amazon in direct conflict to the pollen evidence from the Amazon Fan (fig. 13.11). In figure 13.12C I offer a reconstruction suggested by the new data in which all the main drainage of the Amazon lowlands is forested.

On the other hand, paleoecological data do provide strong inference that temperature was depressed in the order of 6°C all across the Amazon system. The more thermophilic of rainforest taxa probably maintained their largest populations at the lowest elevations. In the bottomlands, these thermophilic rainforest taxa were joined by more cold-adapted taxa, which now maintain their largest populations at higher elevations. The cooling was sufficient to depress Andean tree lines by

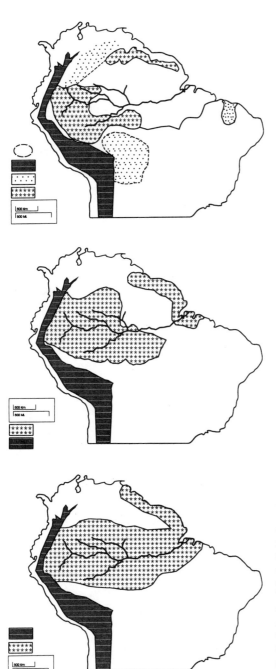

Fig. 13.12 Changing re-
constructions of the extent
of forest in the Amazon
lowlands during glacial
times. (A) The reconstruction by Clapperton (1993)
relies on undated geomor-
phological structures such
as stone lines to infer past
aridity. (B) The recon-
struction by Bush (1994)
attempts to plot the extent
of forest if rainfall was
reduced by 20% as a con-
sequence of 6°C cooling
of the lowlands. (C) The
reconstruction in (C) is
based on the direct evi-
dence of pollen from the
Amazon Fan and from
the Hill of Six Lakes that
the Amazon lowlands re-
mained forested through-
out the last glacial cycle.

1500 m, and to let many Andean trees enter what is now the habitat of tropical rainforest 1500 m below their present range. Floristically, these ice-age Amazon forests were subtly different from their modern replacements, with the tropical conifer *Podocarpus* achieving populations unknown to modern botanists.

An Amazon even 5°C colder than present would be most limiting to thermophilic species during periodic incursions of Antarctic air, the phenomenon known to Brazilian meteorology as "friagems" when near-freezing temperatures are encountered in the Amazon lowlands (Nimer 1989). Friagem incursions of Antarctic air are likely to have been at least as frequent in glacial times, when their arrival into already cooler forests would have limited the elevations to which thermophilic species could aspire, thus reinforcing bottomland habits on rainforest trees.

The paleoecological reconstruction of the ice-age Amazon as having reduced temperature, forest in the bottomlands, and any aridity probably local is not sympathetic to the refugial hypothesis (fig. 13.12C). The refuge hypothesis is intellectually compelling despite conflicts with data, however, because it explains endemism by vicariance through geographical isolation. Denial of the refugial hypothesis on paleoecological grounds therefore would be more acceptable if an alternative hypothesis providing adequate vicariance could be constructed.

THE PLEISTOCENE AMAZON WITHOUT REFUGIA

The Amazon Basin represents in an extreme form the general condition of tropic lands of having huge areas with similar ranges in annual temperatures. A few degrees of latitude at the equator makes very little difference to insolation received or to temperature extremes experienced, whereas the same few degrees at high latitudes involve a move from one temperature regime to another (Osman and Whitlatch 1978; Terborgh 1973). The practical consequence of this is that all equatorial life has relatively large areas of potentially exploitable habitat. In temperate latitudes, adjacent regions to the north or south might be too hot or too cold. Within the tropics, migration is always to a place of suitable temperature.

Large areas with year-round habitable temperature regimes are a property of all tropical regions, but the Amazon Basin compounds this general property by its huge size. A first necessity to understanding the biogeography of the Amazon Basin, therefore, is to envisage a land comparable in size to the continental United States with essentially one temperature range. The only Amazonian localities that might be isolated from the rest by temperature differences are the more elevated portions subject to adiabatic cooling. It is not without significance that these elevated portions turn out to be centers of endemism.

In only the crudest sense, however, can the vast Amazon Basin be thought of as a single habitat. Maps of relief, soils, precipitation, or seasonality show great subregional and local variety, all amply reflected in vegetation maps of less than basinwide scale. Temperature extremes almost everywhere tolerable to all Amazonian organisms in the lowlands do not free the biota from isolation by distance and suboptimal habitat. And to this inherent vicariance is added the dissections of the great river system, providing effective barriers to dispersal to many forest animals.

The implications of the size and variability of the Amazon system were reviewed by Endler (1977, 1982), who showed that parapatric speciation, or separation of clines along the most obvious geographical discontinuities, or the propensity of peripheral areas to receive gene flows unequally on different sides, or all three combined, should account for the observed biogeographic patterns of the Amazon Basin. The power of Endler's argument is readily apparent by inspection of any map that sets the observed areas of endemism against Amazon physiography or climate (fig. 13.10 above).

Endler's analysis allows the ecological commonsense conclusion that the broad patterns of Amazon diversity reflect current processes in conditions of modern climate. Patterns of diversity and geography merge. Furthermore, at least some patterns generated in the different environments of ice-age time would have been overwhelmed by dispersal and selection throughout the 10 ky of Holocene time through which the earth has just passed, depending on the generation times and dispersal ranges of the species involved.

Despite showing that the refuge hypothesis was not the simplest explanation of the observed distributional data, Endler (1982) could not discriminate between predictions of the refuge hypothesis and those of the three complementary hypotheses of parapatric speciation, separation of clines at geographic barriers, and peripheral effects because the resolution of available data was inadequate. By using the principle of parsimony it was possible to discard the refuge hypothesis as the least satisfactory of the four because it required cycles of forest fragmentation that were not required by the other three. But it was not possible to reject the refuge hypothesis on biogeographic evidence alone, as once its assumptions were met, it explained the observed distributions as well as the other three. Endler's response to this state of affairs was to call for more precise biogeographic data. A closer look at the paleoecological data, however, might serve this purpose better.

The environmental characteristics of the ice-age Amazon suggested by the paleoecological data so far available were lowered temperature, lowered CO_2, local changes in patterns of rain including more even precipitation in the southeast, and the likelihood of continual climatic changes on millennial time scales. In this history, the role of temperature is likely to be decisive for the persistence of organisms in many habitats, particularly the lowest temperatures experienced in a run of years.

In the shorter intervals of extreme cold that come once or twice in each glacial cycle, extremes of temperature could have been as much as 7–9°C below modern levels. At those times, many rainforest taxa would probably be excluded from the more elevated parts of their present range altogether.

A critical consequence of lowered temperature was that elevated regions, particularly those adjacent to the high mountains of the Amazon periphery, were sites of disturbance on virtually all time scales. Yet these are the very areas described as refuges from climatic change in the Haffer-Prance hypothesis. The observation that "refuges" actually occupy the sites of prime disturbance obviously impoverishes the argument that they were the habitats of ice-age refugees. More important, the observation also allows a new hypothesis to account for their high species richness, and hence high endemism. Species collected in the elevated regions of disturbance because these were sites of repeated invasions and retreats with the vicissitudes of climatic change.

Where the wet elevated regions abut true mountains, as in the Andean forelands, the dynamics are at their strongest as montane and lowland floras are squeezed together, repeatedly invading each other. This is the pattern recorded directly by the pollen diagrams from Mera and San Juan Bosco in Ecuador (Liu and Colinvaux 1985; Bush et al. 1990). It is thus understandable that the Andean foothills in the Napo region of Ecuador and Peru appear to have the highest diversity of any part of the Amazon Basin, perhaps indeed of the world (Gentry 1988). Tree species persist because of the frequent disturbances that allow as frequent invasions or recoveries. For taxa that can speciate more rapidly than forest trees, like butterflies and forest birds, these are the sites of repetitive disturbance through climatic change, where competitive exclusion and character displacements (called biological accommodations by some ecologists) are likely consequences of invasions (Connell 1978; May 1986; Colinvaux 1993).

Paleoecological data thus show that the areas taken by Haffer (1969) to be refuges from ice-age events are in fact the areas where ice-age events had most impact. Critical among the disturbances of the glacial cycles are the imposition of those short-lived times of warmth (like the present) that we call an interglacial. And the process of invasion and reassortment of species proceeds in elevated or peripheral centers of endemism in modern times, just as it did throughout Pleistocene time. We merely see one of the extreme conditions.

In one way, the paleoecological data amplify Endler's (1982) analysis because they enlarge the possibilities of vicariance beyond just the effects of distance and geographical discontinuities that he remarked. Local regions of higher precipitation and relief are indeed separated as the "refugial" maps show. But their isolating properties are their exposures to changes in temperature. Invasions, retreats, and reinvasions in the elevated isolates proceed with the rhythms of glacial advances as the air cools and warms, probably with accompanying local changes in precipitation or seasonality. Many of the interactions between the invaders are necessarily competitive, thus favoring species isolations.

Thus, a paleoecological explanation of the observed patches of endemism in the Amazon is that they are centered, not on places of refuge where nothing happened, but rather on the dynamic centers of disturbance where oscillations of temperature on millennial time scales forced multiple invasions, and hence the biological accommodations that could lead to speciation. The model is as much a model of vicariance as the refugial model itself, because it works on the same scattered patches of high relief. But it is vicariance with disturbance, what may be called the "vicariance-disturbance" hypothesis (Colinvaux 1993).

What, then, of Endler's (1982) prediction that the biogeographic patterns of the ice-age Amazon would have tended to be erased in the 10,000 years of Holocene time? The answer may lie in the fact that the centers of endemism are still differentiated by temperature, particularly on the scale of tree lifetimes, when the intermittent cold days of the friagem result from cold Antarctic air flowing over the basin. The pattern of cool moist uplands dotted over and around warmer moist lowlands is the perennial condition of the Amazon, though most strongly expressed in glacial periods. This permanent condition of the Amazon system is mirrored in its modern biogeography.

THE ISTHMUS OF PANAMA AS A PLEISTOCENE FILTER

This general analysis of Pleistocene vegetation and climate of the neotropics is relevant to the passage of land animals through the isthmus. As elsewhere in the neotropics, glacial climates of the isthmus were colder and had reduced CO_2 concentrations. But the isthmus has also yielded evidence for glacial aridity that is far more persuasive than any data that have been offered from the Amazon Basin, thus allowing the postulate that passage of the isthmus might sometimes have been possible along a savanna corridor unimpeded by closed forest.

Strong evidence for ice-age aridity comes from lakes in eastern Guatemala and the Yucatán, where pollen and limnological data from lake sediments show unequivocally that land now covered with moist tropical forest supported only arid savannas in glacial times (Deevey et al. 1983; Leyden 1984). Thus, there can be no doubt that open landscapes were available in the broad northwestern parts of the land bridge, say, from Honduras west to Mexico. But our data from Panama do not easily allow an open corridor in the narrow eastern land bridge of Panama itself.

Central Panama is now forested on both its northern (Caribbean) and southern (Pacific) slopes, with the climate on the Pacific side being the more seasonal. Furthermore, the low pass through the central cordillera through which the Panama Canal runs supports moist tropical forest from coast to coast. As long as this lowland forest is intact, there can be no savanna corridor through the isthmus. An important question, therefore, is whether the aridity demonstrated in Yucatán and Guatemala was sufficiently regional to disrupt lowland Panamanian forests, either on one of the coasts or in the region of the Panama Canal. Contemporary argument for regional aridity depends on extrapolating from the demonstrated glacial aridity of eastern Guatemala to data for the reduced level of Lake Valencia in Venezuela in late glacial times (Salgado-Labouriau 1980; Bradbury et al. 1981). However, this extrapolation seems incompatible with pollen and paleolimnological data from the two long Panamanian sections so far available.

The combined record from Lake La Yeguada and El Valle shows that tropical forest occupied the Pacific coast of Panama between 500 and 700 m at all stages of the last glacial cycle and Holocene (Bush and Colinvaux 1990; Bush et al. 1992; see figs. 13.8–9 and discussion above). Forest composition changed continually, particularly as oaks descended to become important elements in glacial-age forests or were restricted to higher elevations in the Holocene. Nevertheless, the vegetation on the Pacific slope was maintained in woodland throughout the history.

The paleolimnological record at these two Panama sites also is not consistent with past aridity. An open-water lake was maintained in the El Valle crater from early in the last glacial cycle until the Early Holocene (Bush and Colinvaux 1990). First ponding at Lake La Yeguada following emplacement of the dam was about 14,000 B.P. and resulted in a late-glacial lake deeper than the later Holocene lake (Bush et al. 1992). Thus, the combined history shows sufficient precipitation on the Pacific slope of Panama to maintain open-water lakes and forest throughout the glacial cycle. The sole hint of changed precipitation is from fine pink banding over part of the El Valle long-core, which has been interpreted (Bush and Colinvaux 1990) to represent inputs of clay in rainstorms following dry seasons more pronounced than those of the present day. With those observations, we reach the limits of our present data set. We can add to this only arguments based on the evidence for cooling.

In Panama, as in the Amazon Basin, ice-age cooling in the order of 5–6°C tended to lower the altitudinal ranges of plant populations. Some of what are now montane taxa occupied lower slopes at least 800 m below their present ranges. Similar descents would have brought the more thermophilic of rainforest taxa below sea level. Short of this impossibility, it is reasonable to expect to find the tropical forest taxa in the lowest elevations available, in particular, across the isthmus in the lowlands now the route of the Panama Canal.

The few paleoecological data so far available from Panama thus give little encouragement to the idea of an open savanna causeway running the whole way up the isthmus. This need not mean, however, that the way through the isthmus was beset for megafauna with impassable forest barriers. Some tropical forest would have to be penetrated in the lowlands at all stages in a glacial cycle, but most of the isthmus was elevated, with oak woodland on midslopes. There was never a causeway of open savanna-land to connect the continents; rather, there was an ever-changing patchwork of plant communities that included tropical forest in the lowlands. This patchwork, constantly changing through succeeding glacial cycles, offered a filter bridge, possibly constraining the migrations of some animals more than others.

REFERENCES

Ab'Saber, A. N. 1982. The paleoclimate and paleoecology of Brazilian Amazonia. In *Biological diversification in the tropics*, ed. G. T. Prance, 41–59. New York: Columbia University Press.

Absy, M. L., A. Cleef, M. Fornier, M. Servant, A. Siffedine, M. F. da Silva, F. Soubies, K. Suguio, B. Turcq, and T. Van der Hammen. 1991. Mise en évidence de quatre phases d'ouverture de la forêt dense dans le sud-est de l'Amazonie au cours des 60,000 dernières années: Première comparaison avec d'autres regions tropicales. *C.R. Acad. Sci. Paris* 313:673–78.

Absy, M. L., and Van der Hammen, T. 1976. Some paleoecological data from Rondônia, southern part of the Amazon Basin. *Acta Amazon.* 6:293–99.

Bates, H. W. 1863. *The naturalist on the river Amazons.* Vol. 2. London: John Murray.

Bigarella, J. J., M. R. Mousinho, and J. X. de Silva. 1969. Processes and environments of the Brazilian Quaternary. In *The periglacial environment.* ed. T. L. Péwé, 417–87. Montreal: Arctic Institute of America, McGill-Queens University Press.

Bonnefille, R., J. C. Roeland, and J. Guiot. 1990. Temperature and rainfall estimates for the past 40,000 years in equatorial Africa. *Nature* 346:347–49.

Bradbury, J. P., B. Leyden, M. Salgado-Laboriau, W. Lewis, C. Schubert, M. Binford, D. Frey, D. Whitehead, and F. Weibezahn. 1981. Late Quaternary environmental history of Lake Valencia, Venezuela. *Science* 214:1299–1305.

Broecker, W. S. 1995a. Cooling the tropics. *Nature* 376:212–13.

———. 1995b. *The glacial world according to Wally.* New York: Lamont-Doherty Earth Observatory of Columbia University.

Broecker, W. S., and G. H. Denton. 1990. The role of ocean-atmosphere reorganizations in glacial cycles. *Quatern. Sci. Revs.* 9:305–42.

Brown, K. S. 1987. Soils and vegetation. In *Biogeography and Quaternary history in tropical America,* ed. T. C. Whitmore and G. T. Prance, 19–45. Oxford: Oxford University Press.

Bush, M. B. 1991. Modern pollen rain data from South and Central America: A test of the feasibility of fine-resolution lowland tropical palynology. *Holocene* 1:162–67.

———. 1992. A simple yet efficient pollen trap for use in vegetation studies. *J. Veg. Sci.* 3:275–76.

Bush, M. B., and P. A. Colinvaux. 1990. A pollen record of a complete glacial cycle from lowland Panama. *J. Veg. Sci.* 1:105–18.

Bush, M. B., P. A. Colinvaux, M. C. Weimann, D. R. Piperno, and K-b. Liu. 1990. Late Pleistocene temperature depression and vegetation change in Ecuadorian Amazonia. *Quatern. Res.* 34:330–45.

Bush, M. B., D. R. Piperno, P. A. Colinvaux, L. Krissek, P. E. De Oliveira, M. C. Miller, and W. E. Rowe. 1992. A 14,300-yr paleoecological profile of a lowland tropical lake in Panama. *Ecol. Monog.* 62:251–75.

Campbell, K. E., and D. Frailey. 1984. Holocene flooding and species diversity in southwestern Amazonia. *Quatern. Res.* 21:369–75.

Clapperton, C. M. 1987. Maximum extent of the Late Wisconsin glaciation in the Ecuadorian Andes. In *Quaternary of South America and Antarctic Peninsula,* vol. 5, ed. J. Rabassa, 165–80. Rotterdam: Balkema.

———. 1993. *Quaternary geology and geomorphology of South America.* Amsterdam: Elsevier.

CLIMAP Project Members. 1976. The surface of the ice-age Earth. *Science* 191:1131–37.

———. 1981. *Seasonal reconstruction of the Earth's surface at the last glacial maximum.* Geological Society of America Map and Chart series, MC-36. Boulder, Colo.

Colinvaux, P. A. 1972. Climate and the Galápagos Islands. *Nature* 240:17–20.

———. 1979. Ice age Amazon. *Nature* 278:399–400.

———. 1987. Amazon diversity in light of the paleoecological record. *Quatern. Sci. Revs.* 6:93–114.

———. 1993. Pleistocene biogeography and diversity in tropical forests of South America. In *Biological relationships between Africa and South America*, ed. P. Goldblatt, 473–99. New Haven: Yale University Press.

Colinvaux, P. A., J. E. Moreno, M. C. Miller, P. E. De Oliveira, and M. B. Bush. N.d. A long pollen record from lowland Amazonia: Cooling and forest in glacial times. *Science.* In press.

Colinvaux, P. A., K-b. Liu, P. E. De Oliveira, M. B. Bush, M. C. Miller, and M. Steinitz-Kannan. 1996. Temperature depression in the lowland tropics in glacial times. *Climatic Change* 32:19–33.

Colinvaux, P. A., M. C. Miller, K-b. Liu, M. Steinits-Kannan, and I. Frost. 1985. Discovery of permanent Amazon lakes and hydraulic disturbance in the upper Amazon Basin. *Nature* 313:42–45.

Colinvaux, P. A., K. Olson, and K-b. Liu. 1988. Late glacial and Holocene pollen diagrams from two endorheic lakes of the Inter-Andean plateau of Ecuador. *Rev. Paleo. Palyn.* 55:83–100.

Connell, J. H. 1978. Diversity in tropical rain forests and coral reefs. *Science* 199:1302–10.

Damuth, J. E., and R. W. Fairbridge. 1970. Equatorial Atlantic deep-sea arkosic sands and ice-age aridity in tropical South America. *Bull. Geol. Soc. Am.* 81:189–206.

De Meis, R. M. 1971. Upper Quaternary process changes of the middle Amazon area. *Geol. Soc. Am. Bull.* 82:1079–84.

De Oliveira, P. E. 1992. A palynological record of Late Quaternary vegetation and climatic change in Southeastern Brazil. Ph.D. diss., Ohio State University, Columbus, Ohio.

Deevey, E. S., M. Brenner, and M. W. Binford. 1983. Paleolimnology of the Petén Lake district, Guatemala. Part 3, Late Pleistocene and Gamblian environments of the Maya area. *Hydrobiologia* 103:211–16.

Emiliani, C. 1992. Pleistocene paleotemperatures. *Science* 154:851–56.

Endler, J. A. 1977. *Geographic variation, speciation, and clines.* Princeton: Princeton University Press.

———. 1982. Pleistocene forest refuges: Fact or fancy. In *Biological diversification in the tropics*, 641–57. New York: Columbia University Press.

Flenley, J. 1979. *The equatorial rain forest: A geological history.* London: Butterworths.

Frailey, C. D., E. L. Lavina, A. Rancy, and J. P. de Souza Filho. 1988. A proposed Pleistocene/Holocene lake in the Amazon Basin and its significance to Amazonian geology and biogeography. *Acta Amazon.* 18:119–43.

Frost, I. 1988. A Holocene sedimentary record from Anañgucocha in the Ecuadorian Amazon. *Ecology* 69:66–73.

Frost, I., and M. C. Miller. 1987. Late Holocene flooding in the Ecuadorian rain forest. *Freshwater Biol.* 18:443–53.

Gentry, A. H. 1988. Tree species richness of upper Amazonian forests. *Proc. Nat. Acad. Sci.* 85:156–59.

Guilderson, T. P., R. G. Fairbanks, and J. L. Rubenstone. 1994. Tropical temperature variations since 20,000 years ago: Modulating interhemispheric climate change. *Science* 263:663–665.

Haberle, S. N.d. Late Quaternary vegetation and climate history of the Amazon Basin: Correlating marine and terrestrial pollen. *Ocean Drilling Prog. Repts.* In press.

Haffer, J. 1969. Speciation in Amazonian forest birds. *Science* 165:131–37.

———. 1974. *Avian speciation in tropical South America*. Cambridge: Nuttall Ornithological Club.

Hastenrath, S., and J. Kutzbach. 1985. Late Pleistocene climate and water budget of the South American altiplano. *Quatern. Res.* 24:249–56.

Heusser, L. E., and N. J. Shackleton. 1994. Tropical climatic variation on the Pacific slopes of the Ecuadorian Andes based on a 25,000-year pollen record from deep-sea sediment core Tri 163–31B. *Quat. Res.* 42(2):222–25.

Hooghiemstra, H. 1984. *Vegetational and climatic history of the high plain of Bogotá, Colombia: A continuous record of the last 3.4 million years*. Vaduz: Strauss and Cramer.

Hooghiemstra, H., A. M. Cleef, G. W. Noldus, and M. Kappelle. 1992. Upper Quaternary vegetation dynamics and paleoclimatology of the La Chonta bog area (Cordillera de Talamanca, Costa Rica). *J. Quatern. Sci.* 3:205–25.

Hoorn, C., J. Guerrero, G. A. Sarmiento, and M. A. Lorente. 1995. Andean tectonics as a cause for changing drainage patterns in Miocene northern South America. *Geology* 23:237–40.

Houvenaghel, G. T. 1974. Equatorial undercurrent and climate in the Galápagos Islands. *Nature* 250:565–66.

Irion, G. 1982. Mineralogical and geochemical contribution to climatic history in central Amazonia during Quaternary time. *Trop. Ecol.* 23:76–85.

———. 1984. Sedimentation and sediments of Amazonian rivers and evolution of the Amazonian landscape since Pliocene times. In *The Amazon limnology and landscape of a mighty tropical river and its basin*, ed. H. Sioli, 201–14. Dordecht. Dr. W. Junk.

Justo, L. C., and M. M. De Souza. 1984. Jazida de Niobio do Morro dos Seis Lagos, Amazonas. In *Capitulo 27: Principias depositos minerais do Brasil*, vol. 2. Rio de Janeiro: DNPM.

Kutzbach, J. E., and P. J. Guetter. 1986. The influence of changing orbital parameters and surface boundary conditions on climate simulations for the past 18,000 yrs. *J. Atmospher. Sci.* 43:1726–59.

Ledru, M. 1992. Modifications de la végétation du Brésil Central entre la dernière époque glaciaire et l'interglaciaire actuelle. *C.R. Acad. Sci. Paris* 314: 117–23.

———. 1993. Late Quaternary environmental and climatic changes in central Brazil. *Quat. Res.* 39(1):90–98.

Leuenberger, M., U. Siegenthaler, and C. C. Langway. 1992. Carbon isotope composition of atmospheric CO_2 during the last ice age from an Antarctic ice core. *Nature* 357:488–90.

Leyden, B. 1984. Guatemalan forest synthesis after Pleistocene aridity. *Proc. Nat. Acad. Sci.* 81:4856–59.

———. 1985. Late Quaternary aridity and Holocene moisture fluctuations in the Lake Valencia basin, Venezuela. *Ecology* 66:1279–95.

Leyden, B. W., M. Brenner, D. A. Hodell, and J. H. Curtis. 1993. Late Pleistocene climate in the Central American lowlands. *Geophys. Monogr.* 78:165–78.

Liu, K-b., and P. A. Colinvaux. 1985. Forest changes in the Amazon Basin during the last glacial maximum. *Nature* 318:556–57.

Livingstone, D. A. 1962. Age of deglaciation in the Ruwenzori Range, Uganda. *Nature* 194:859–60.

———. 1975. Late Quaternary climatic change in Africa. *Ann. Rev. Ecol. Syst.* 6:249–80.

———. 1993. Evolution of the African climate. In *Biological relationships between Africa and South America*, ed. P. Goldblatt, 455–72. New Haven: Yale University Press.

Maley, J. 1991. The African rainforest vegetation and paleoenvironments during Late Quaternary. *Clim. Change* 19:79–98.

May, R. M. 1986. The search for patterns in the balance of nature: Advances and retreats. *Ecology* 67:1115–26.

Mercer, J. H., and O. Palacios. 1977. Radiocarbon dating of the last glaciation in Peru. *Geology* 5:600–604.

Milliman, J. D., C. P. Summerhayes, and H. T. Barreto. 1975. Quaternary sedimentation on the Amazon continental margin: A model. *Geol. Soc. Am. Bull.* 86:610–14.

Newell, R. E. 1973. Climate and the Galápagos Islands. *Nature* 245:91–92.

Nimer, E. 1989. *Climatologia do Brasil*. Rio de Janeiro. IBGE.

Nittrouer, C. A., S. A. Kuehl, R. W. Sternberg, A. G. Figueiredo, Jr., and L. E. C. Faria. 1995. An introduction to the geological significance of sediment transport and accumulation on the Amazon continental shelf. *Mar. Geol.* 125:177–92.

Osman, R. W., and R. B. Whitlach. 1978. Patterns of species diversity: Fact or artefact? *Paleobiology* 4:41–54.

Piperno, D. R., M. B. Bush, and P. A. Colinvaux. 1990. *Quatern. Res.* 33:108–16.

Prance, G. T. 1982, ed. *Biological diversification in the tropics*. New York: Columbia University Press.

Prentice, I. C., P. J. Bartlein, and T. Webb. 1991. Vegetation and climate change in eastern North America since the last glacial maximum. *Ecology* 72:2038–56.

Räsänen, M. E., A. M. Linna, J. C. R. Santos, and F. R. Negri. 1995. Late Miocene tidal deposits in the Amazonian foreland basin. *Science* 269:386–90.

Räsänen, M., R. Neller, J. Salo, and H. Junger. 1992. Recent and ancient fluvial deposition systems in the Amazonian foreland basin, Peru. *Geol. Mag.* 129:293–306.

Rind, D., and D. Peteet. 1985. Terrestrial conditions at the last glacial maximum and CLIMAP sea-surface temperature estimates: Are they consistent? *Quatern. Res.* 24:1–22.

Salgado-Laboriau, M. L. 1980. A pollen diagram of the Pleistocene-Holocene boundary of Lake Valencia, Venezuela. *Rev. Paleobot. Palyn.* 30:297–312.

Salo, J. 1987. Pleistocene forest refuges in the Amazon: Evaluation of the biostratigraphical, lithstratographical, and geomorphological data. *Ann. Zool. Fennici* 24:203–11.

Salo, J., R. Kalliola, Y. Hakkinen, P. Niemela, M. Puhakka, and P. D. Coley. 1986. River dynamics and the diversity of Amazon lowland forest. *Nature* 322:254–58.

Sauer, W. 1965. *Geologia del Ecuador.* Quito: Ministerio de Educacion.

Simpson, B. B. 1975. Glacial climates in the eastern tropical South Pacific. *Nature* 253:34–36.

Soubies, F., K. Suguio, L. Martin, et al. 1989. The Quaternary lacustrine deposits of the Serra dos Carajas (State of Para, Brazil)—ages and other preliminary results. *Boletin IG-USP,* Publ. Esp. 8:233–43.

Sternberg, H. 1982. Refugial theory and Amazonian environment. In *Evolution and environment,* ed. V. A. Morak and J. M. Likovsky, 997–98. Prague.

Stow, D. A. W., D. G. Howell, and H. C. Nelson. 1985. Sedimentary, tectonic, and sea level controls. In *Submarine fans and related turbidite systems,* ed. A. H. Bouma, W. R. Normark, and N. E. Barnes, 15–22. New York: Springer-Verlag.

Street, F. A., and A. T. Grove. 1979. Global maps of lake level fluctuations since 30,000 yr B.P. *Quatern. Res.* 12:83–118.

Stuijts, I., J. C. Newsome, and J. R. Flenley. 1988. Evidence for Late Quaternary vegetational change in the Sumatran and Javan highlands. *Rev. Paleobot. Palyn.* 55:207–16.

Stute, M., M. Forster, H. Frischkorn, A. Serejo, J. F. Clark, P. Schlosser, W. S. Broecker, and G. Bonani. 1995. Cooling of tropical Brazil (5°C) during the last glacial maximum. *Science* 269:379–83.

Terborgh, J. 1973. Chance, habitat, and dispersal in the distribution of birds in the West Indies. *Evolution* 27:338–49.

Thompson, L. G., E. Mosley-Thompson, J. F. Bolzan, and B. R. Koci. 1985. A 1500-year record of tropical precipitation in ice cores from the Quelccaya ice cap, Peru. *Science* 229:971–73.

Tranquillini, W. 1979. *Physiological ecology of the alpine timberline.* New York: Springer-Verlag.

Tricart, J. 1974. Existencia de medanos cuaternarios en los Llanos del Orinoco. *Rev. Inst. Geogr. Agustin Codazzi* 5(1):69–79.

———. 1977. Aperçus sur le Quaternaire Amazonien: Recherches françaises sur le Quaternaire. *INQUA. Supp. Bull. AFEQ,* 265–71.

Van der Hammen, T. 1974. The Pleistocene changes of vegetation and climate in tropical South America. *J. Biogeogr.* 1:3–26.

Van der Hammen, T. and M. L. Absy. 1994. Amazonia during the last glacial. *Palaeogeogr., Palaeoclimatol., Palaeoecol.* 109:247–61.

Whitmore, T. C., and G. T. Prance, eds. 1987. *Biogeography and Quaternary history in tropical America.* Oxford Monographs in Biogeography, vol. 3. Oxford: Oxford University Press.

Wijmstra, T. A., and T. Van der Hammen. 1966. Palynological data on the history of tropical savannas in northern South America. *Leid. Geol. Meded.* 38: 71–90.

Contributors

Warren D. Allmon
Paleontological Research Institution
1259 Trumansburg Road
Ithaca, New York 14850

Bryan E. Bemis
Department of Geology
University of California
Davis, California 95616

Ann F. Budd
Department of Geology
The University of Iowa
Iowa City, Iowa 52242

Alan H. Cheetham
Department of Paleobiology
National Museum of Natural History
Smithsonian Institution
Washington, D. C. 20560

Anthony G. Coates
Center for Tropical Paleoecology
  and Archeology
Smithsonian Tropical Research
  Institute
Box 2072, Balboa
Republic of Panama

Paul A. Colinvaux
Center for Tropical Paleoecology
  and Archeology
Smithsonian Tropical Research
  Institute
Box 2072, Balboa
Republic of Panama

Laurel S. Collins
Department of Geology
Florida International University
University Park
Miami, Florida 33199

Timothy Collins
Department of Biological Sciences
Florida International University
University Park
Miami, Florida 33199

Mathew A. Cotton
Field Museum of Natural History
Lake Shore Drive at Roosevelt Road
Chicago, Illinois 60605

Thomas M. Cronin
U.S. Geological Survey
Reston, Virginia 22092

Harry J. Dowsett
U.S. Geological Survey
Reston, Virginia 22092

Dana H. Geary
Department of Geology and
  Geophysics
University of Wisconsin
Madison, Wisconsin 53706

Helena Fortunato
Center for Tropical Paleoecology
  and Archeology
Box 2072, Balboa
Republic of Panama

407

Jeremy B. C. Jackson
Center for Tropical Paleoecology
  and Archeology
Smithsonian Tropical Research
  Institute
Box 2072, Balboa
Republic of Panama

Kenneth G. Johnson
Department of Geology and
  Applied Geology
University of Glasgow
Glasgow G12 8QQ
United Kingdom

Peter Jung
Naturhistorisches Museum
Augustintergasse 2
CH-4001 Basel
Switzerland

Jorge A. Obando
c/o Clinicia Obando
Calle 12–16, Avenida 12
San Jose
Costa Rica

Roger W. Portell
Florida Museum of Natural History
University of Florida
Gainesville, Florida 32611

Alceu Rancy
Departamento de
Geociências Universidade
Federal de Santa Catarina
88040-900
Florianópolis, Brazil

Gary Rosenberg
Academy of Natural Sciences of
  Philadelphia
1900 Benjamin Franklin Parkway
Philadelphia, Pennsylvania 19103

Kevin Schindler
Florida Museum of Natural History
University of Florida
Gainesville, Florida 32611

Thomas A. Stemann
Department of Geology
The University of Iowa
Iowa City, Iowa 52242

Jane L. Teranes
Geological Institute
ETH-Zentrum
CH-8092 Zurich
Switzerland

S. David Webb
Florida Museum of Natural History
University of Florida
P.O. Box 117800
Gainesville, Florida 32611

Index